LEARNING TO READ THE EARTH AND *Sky*

EXPLORATIONS SUPPORTING THE *NGSS*

GRADES 6–12

Learning to Read the EARTH and *Sky*

Explorations Supporting the *NGSS*

GRADES 6–12

Russ Colson
Mary Colson

Arlington, Virginia

Claire Reinburg, Director
Wendy Rubin, Managing Editor
Rachel Ledbetter, Associate Editor
Amanda Van Beuren, Associate Editor
Donna Yudkin, Book Acquisitions Coordinator

ART AND DESIGN
Will Thomas Jr., Director
Himabindu Bichali, Graphic Designer, cover and interior design

PRINTING AND PRODUCTION
Catherine Lorrain, Director

NATIONAL SCIENCE TEACHERS ASSOCIATION
David L. Evans, Executive Director
David Beacom, Publisher

1840 Wilson Blvd., Arlington, VA 22201
www.nsta.org/store
For customer service inquiries, please call 800-277-5300.

20 19 18 17   4 3 2 1

Cataloging-in-Publication Data for this book and the e-book are available from the Library of Congress.

ISBN: 978-1-941316-23-8
e-ISBN: 978-1-941316-68-9

CONTENTS

PART II: THE LANGUAGE OF THE EARTH

PART III: YOU CAN DO IT!

ACKNOWLEDGMENTS

We are indebted to more people than we can list here, but we would like to mention several groups of people whose contributions to this book were profound. We thank the many, many colleagues, friends, and mentors who have helped shape our thoughts on science and science teaching, especially those who read and commented on earlier drafts of this book. Among this group we include our students, who in learning with us have also often taught us. We thank our parents, who first inspired us to notice the world around us and to wonder and explore. We thank our children, who explored the world with us on many family adventures and expanded our understanding of it. We thank the editorial team and staff at NSTA Press, whose support and work have made this book possible. Thanks also go to Patricia Freedman for her meticulous editorial work and to Ken Roy for his contribution of the safety tips and text throughout the book.

ABOUT THE AUTHORS

Russ Colson and Mary Colson have spent much of their careers at the intersection between scientific research and science teaching. They have coauthored papers in peer-reviewed journals on topics ranging from the mysterious death of hadrosaurs in South Dakota to the nature of chromium dimers in silicate melts. They have applied those research insights to classroom investigation, publishing activities for teachers and giving multiple workshops on authentic science in the classroom. Many of the insights and activities in this book come from conversations and arguments on the nature of science, research, and teaching shared during field trips and lunchtime discussions. Through this experience, they have come to truly believe that science in the classroom and science in the research lab need not differ in their core approach.

Russ Colson has worked for more than 20 years as a professor of geology, planetary science, and meteorology at Minnesota State University Moorhead (MSUM). He has engaged hundreds of future teachers in field trips and laboratory science, including opportunities for many undergraduates to present research at national conferences. He founded two successful new programs at MSUM—Earth Science Teaching and Geosciences—and served as director for two education grant programs—Transforming Teacher Education at MSUM and Research Experiences for Teachers. He was a member of the team for the Minnesota earth science teacher licensure standards. In 2010, he was selected by the Carnegie Foundation and the Council for Advancement and Support of Education as one of four national winners of the U.S. Professors of the Year award.

Russ's research experience includes work as an experimental geochemist at the Johnson Space Center in Houston, Texas, and at Washington University in St. Louis, Missouri, where, among other things, he studied how a lunar colony might mine oxygen from the local rock. He put together an experimental petrology laboratory at MSUM with funding from the NASA and the donation of an electron microprobe from Corning. He twice received the university's top award for research involving undergraduates.

Russ has published 18 science fiction stories and articles and enjoys landscaping and gardening at his rural Minnesota home.

Mary Colson teaches eighth-grade earth science at Horizon Middle School in Moorhead, Minnesota, having taught previously in Texas and Tennessee. Her teaching is characterized

ABOUT THE AUTHORS

by investigation-based curriculum, which she creates herself and changes regularly. She has developed field-based projects, including grant-funded work at a local city park and a water quality research project on local wetland. She was named the middle school recipient of the 2008 Medtronic Foundation's Science Teaching Award presented by the Minnesota Science Teachers Association.

She served on the Minnesota Science Teachers Association Board of Directors for eight years, including a term as president, and then served as the elected District IX director for the National Science Teachers Association Council. She worked as a member of the *Next Generation Science Standards (NGSS)* writing team and was subsequently an educational consultant with Achieve Inc. to develop sample evidence statements and science/math tasks for the middle school and high school *NGSS* for Earth and Space Sciences performance expectations.

Mary holds a master's degree in geological sciences and earned her teaching licensure through the University of Tennessee Lyndhurst Fellowship Program, a competitive paid graduate program designed to recruit teachers from the ranks of trained scientists.

Mary is an avid outdoors person, explaining her wide-ranging field experiences, but also enjoys making a good quilt on a winter evening.

ABOUT THIS BOOK

Learning to Read the Earth and Sky is filled with informative visuals that enhance the book's content. We have provided an online Extras page to host full-color versions of many of the book's images. Feel free to print those images, as needed, for classroom instruction. You can access the Extras page at *www.nsta.org/learningtoread*. Throughout the book, images that are available on the Extras page are marked with the following icon: ✪.

You will also notice that this book differs from other NSTA Press books in its use of *earth* and *Earth*. Whereas other NSTA Press publications use *earth* to refer only to soil and *Earth* in all other instances, we have chosen to strictly reserve *Earth* for references to the planet—including not capitalizing *earth science* as a discipline. This style convention is very important to us and at the heart of what we perceive as a long-standing misconception of what earth science is about. You can read more about our usage decision in "The Language of the Earth" section of the introduction (p. xx).

Bringing the universe into the classroom on a scale that students can investigate and discover

Inspiring teachers to reach beyond prepared curricula and explore science with their students

INTRODUCTION

In 1997, a group of college students, a pickax, and I (Russ) were scrambling along a rural gravel road in western North Dakota. The brisk wind cut through our thin jackets as the Sun fell behind a bank of clouds on the horizon. After six weeks of lectures in Geology in the National Parks, students at last had a chance to discover geology for themselves. They gathered around the young woman with the pickax, eyes drifting from the soft yellows and browns in the nearby buttes to the deepening hole in the grey rock. About a foot below the surface, the pickax began to dredge up crisp, black imprints of willow leaves. Eyes widened and interest quickened. "How could it be wet enough for willow trees to live here on the dry plains?" "How did they get into the rock?" "How long ago was it?" Suddenly, science became something to figure out, not just something to know.[C1]

And that's exactly what science should be, something to figure out, not just something to know. Telling stories to children does not teach them how to read a book, and telling facts, laws, or principles does not teach students to read the stories written in the earth and sky. *Learning to Read the Earth and Sky* explores the *doing* of earth science—how we read the stories written in the earth by applying the practices of science.

Appropriately, the *Next Generation Science Standards* (*NGSS;* NGSS Lead States 2013) emphasize science as a practice, not as a body of knowledge. Science is not about what we know, or think we know, so much as it is about how we know it. It is the person who *knows how we know* that participates in science. Only that person can reasonably discuss whether our understanding of the world is true or false. Anyone else must either accept or reject an idea based on their faith in the person who told them.

Along with science as practice, the *NGSS* emphasize the Earth as a complex, interacting system. In the natural world, everything is connected. John Muir recognized this interacting connectivity when he said "When we try to pick out anything by itself, we find it hitched to everything else in the Universe" (Muir 1911, p. 110).

So, the *NGSS* encourage both doing science in the classroom—the science and engineering practices—and learning the complex interplay of systems over the whole Earth and space beyond—the disciplinary core ideas (DCIs). The problem is that *you can't bring an all-encompassing supersystem into the classroom even if middle or high school students were able to understand it.* In fact, trying to capture the whole sweep of everything at once is contrary to the historical practice of scientific research—scientists break complex problems into bite-size, solvable chunks. In discussing the solution to a complex, interconnected problem in his book *A Brief History of Time,* Stephen Hawking (1996) notes that "it might be impossible to get close to a full solution by investigating parts of the problem in isolation. Nevertheless, it is certainly the way that we have made progress in the past" (p. 12). *Learning to Read the Earth and Sky* offers ways to break the immensity into small chunks that we can bring into the classroom.

Teachers might be concerned that the all-encompassing DCIs of the *NGSS* cut the link to more familiar big ideas of earth science. For example, the *NGSS* do not specifically identify classic ideas such as telling stories from rocks and strata (the wellspring of nearly everything we know about Earth's past), the movement of cyclones and fronts (the traditional foundation for understanding weather), or the processes that shape planetary surfaces (one of the primary new discoveries of the last half century). Instead, the *NGSS* DCIs emphasize interactions and cycles within Earth and space systems. Thus, for example, one of the components of the *NGSS* DCIs becomes this ESS2.A grade band endpoint for grade 8:

> *The planet's systems interact over scales that range from microscopic to global in size, and they operate over fractions of a second to billions of years. These interactions have shaped Earth's history and will determine its future. (NRC 2012, p. 181)*

Another component becomes this ESS2.C grade band endpoint for grade 8:

> *Water continually cycles among land, oceans, and atmosphere via transpiration, evaporation, condensation and crystallization, and precipitation, as well as downhill flows on land. (NRC 2012, p. 185)*

Does this mean that the traditional big ideas are no longer a part of the standards put forth for teaching earth science? No, of course not. They are all in there (along with, no doubt, the kitchen sink). The *NGSS* DCIs are less a limitation on what factual information all students should learn than they are a philosophical proposal that whatever students learn about earth and space processes, they should learn within the context of how that component fits into a bigger picture of a system of interacting subsystems.

FOUR PREMISES OF THIS BOOK

Our goal in writing this book is to provide concrete examples of classroom exploration that meet the ambitious goals of the *NGSS* to both teach science as a practice and reach toward an understanding of how all the small parts fit into a greater whole. We offer some of our own experience in bringing the entire universe into the classroom on a scale that students can test and discover, and we break down the sweeping DCIs into specific examples that students can see, touch, and experience.

We start with the four premises that are described in the following sections:

1. Earth science should engage students with the world they know.
2. Teacher and student are colleagues and fellow scholars.
3. Doing earth science requires breaking big concepts into smaller chunks.
4. The purpose of experimental and observational activities in the classroom is to practice doing science, and not to convey factual information in an active and "hands-on" way.

Engaging Students With the World They Know

Nicolas Desmarest was an influential figure in a heated 18th-century controversy over how rocks form. Did they form by cooling of volcanic lava or by crystallization and settling from seawater? Through careful fieldwork in which he mapped the connection between volcanic rocks and volcanoes, he showed an undeniable link between basaltic rocks and the volcanoes of central France, a contribution that swung the verdict to the belief that some rocks form from volcanic lava. When asked in his old age about the "truth" of the matter, rather than reassert his own views, he responded simply "Go and see!"

Thus, Desmarest reminds us that science doesn't begin with theoretical ideas or facts but rather with the belief that we can understand our universe through observation. Like science research, effective science teaching begins with what students can see, feel, and explore, not with theoretical ideas. Earth science in particular deals with phenomena that people can see and experience all around them—rocks, rivers, clouds, and wind. To improve teaching in the classroom, the DCIs of the *NGSS* must be reduced to a scale that students can "go and see."

Seeing alone is not enough. To do science, students and teachers must understand what they see and what it tells us about how the world works. Earth science is not about knowing the laws of nature, or even knowing the stories of Earth's past. Earth science is about *reading* the stories written in the earth and sky, the *practice* of science.

Addressing aspects of our universe that students see and experience, and teaching students to read those stories on their own, gives them ownership in the process of discovery. They realize that science is not something they are told, coming from high oracles of the mysterious science world. Science becomes something that people *do,* something that *they* can do.

This book is not just a "rule book" of science. It is a book of practice, showing how to dribble, how to pass, and how to shoot in the game of earth science. It provides ways for teachers and students to practice the game together, remembering that science is what we do, not just what we know.[C1]

Teacher and Student as Colleagues and Fellow Scholars

In a brand-new science room in 2003, I (Mary) engaged my eighth graders with an old geology activity—crystallizing thymol in a petri dish. Like magma, thymol produces large crystals when cooled slowly and small crystals when cooled fast, illustrating the foundation for one of the key stories told by igneous rocks. But this time, something wasn't working for one of my groups. No crystals formed in their slow-cooled sample, and when crystals finally did grow, they were small. One student in the group looked at me with disappointment. "What did we do wrong?" he asked.

"I don't know," I said. "We'll have to figure it out."

The teacher doesn't know? She has to figure it out?

Postures shifted. Eyes brightened. We started asking questions. "It's cold, so why isn't it solid?" "What did the other groups do different?" "What can we try new?" The students hunched over the lab bench with renewed interest. Suddenly, the lab no longer dealt with getting the "right" answer from the teacher's key. Now the lab dealt with how they could figure out something that the teacher didn't know. In the accident of the lab "not working," it had become real science.

Some comparisons and experimentation led them to the conclusion that, if they melted the thymol entirely, crystallization was delayed because of an absence of "seed" crystals. When it finally did crystallize from a supercooled state, it did so rapidly, producing small crystals. But the real discovery of the activity was that science is about figuring things out, not waiting for the teacher to hand out the answers.

We believe that the authentic teacher engages in discovery with her students, asking her own questions, developing her own exploratory activities, analyzing and interpreting results that don't always seem to make sense. Sometimes labs developed in this way may not be completely polished, and the results not completely certain, but the challenges that arise are not a problem to be avoided. The challenges and uncertainties are the whole point of the activity. In taking up those challenges, the teacher gives students the valuable learning experience of seeing her doing science, not just assigning activities and following recipes that someone else developed. Not only do the students realize that the teacher values true exploration, but they see and learn from the way the teacher asks questions and tests ideas.

Early in my (Russ's) career at Minnesota State University Moorhead, an older faculty member in the education department characterized a teacher as a pipeline through which knowledge flows to the student. This image didn't work for me. Doing science is no more about passive knowledge than playing basketball is about knowing the rule book. The teacher is better characterized as coach, illustrating good science reasoning skills—by sometimes allowing himself to get stumped and having to figure out a puzzle in front of the students—and then giving students the chance to practice solving their own puzzles.

Teachers often look for polished and well-tested activities to do with their students. This is fine to do on occasion—the teacher only has so much time, and the next class period is already pressing. But the point of teaching is not to make it easy on the teacher or the student. *Easy* is giving students a recipe lab where they follow the simple instructions to the inevitable outcome. *Good* is crafting situations where students struggle to figure out what to do, grapple with concepts, and have to ask lots of questions. Some labs should be of this latter type. And some should be of the teacher's own making.

This book provides some activities that we have tested and find useful for cultivating student reasoning and discourse in the classroom. More important, it provides insight into the process of doing science that can help us all be more authentic teachers, developing our own activities and providing the foundation so we can truly say, "My students and I do science together."

Breaking Big Concepts Into Smaller Chunks

Back in the 1980s, a humorous list of test questions circulated among PhD candidates preparing for their preliminary exams. Each of the questions captured the expectation that candidates should provide comprehensive, detailed answers for vaguely worded, abstract, and far-reaching questions. One of the questions was something like "Explain the universe. Give three examples."

The *NGSS* propose that a key outcome of education should be that every student understands Earth's complex systems and how they interact— that students understand Earth's place in time and space and how human actions impact broad planetary processes in complex ways. Although this is an important goal, it may be seen as vague, abstract, and far-reaching. In our view, it's difficult to get to these big-scale understandings without first engaging in much smaller-scale science exploration. The good news is that the "big-picture" goals of the *NGSS* do not limit the small-scale science that the teacher can bring into the classroom. All of earth science, its core discoveries, its methods of investigation, and its stories of past and present, fit comfortably within the broad learning outcomes of the *NGSS*. That's not to say that all of earth science should be brought into the classroom. Trying to cover "all the material" causes a class to devolve into a listing of facts and concepts without time for the actual practice of science exploration. However, the big-picture *NGSS* goals can be reached through a doable subset of classroom-size science explorations.

In real research, scientists might be studying the chemical evolution of the Moon, but their work will focus on a small subcomponent of how that evolution works. Likewise, students might examine how "water continually cycles among land, ocean, and atmosphere," but in the classroom they will look at how water condenses out of air. The job of the teacher is to make choices that limit the scope of the topic, allowing time for students to truly explore some subset of a larger system, while tying what the students are doing into an understanding of that system. The small-scale classroom work helps students understand *how things work and how we know,* while the bigger picture gives them a conceptual understanding of the elegant workings of our world and universe.

New teachers fresh out of college often feel like they have lots of material to cover. For example, they might have ideas about atmospheric circulation and climate, seasons, movement of energy, ocean currents, and how they all work together. But how do you pare that down to something that middle and high school students can do in the classroom? Maintaining a sense of the big picture without getting lost in the sea of details, while still giving students a real experience in science exploration, depends, like real research, on breaking the big picture down into small, solvable components.

Although teachers need to limit what they bring into the classroom, we should not limit the scope of the discipline by predefining a subset of material that every class must encompass. Rather, we should limit the number of examples we use to illustrate the bigger ideas while maintaining the full scope of the discipline as fair game for learning. This book offers examples of specific, small-scale activities that you can do in the classroom, along with connections to the big-picture ideas of the *NGSS* to which the activity applies.

Using Experimental and Observational Activities in the Classroom to Practice Doing Science

Fads are common in teaching. Some of these fads are of lasting importance, while others fade away, yet each one is portrayed as revolutionary and the "final word" at the time. Some fads introduce important new ideas that may not be fully understood until later. A couple of decades ago, "inquiry-based science" was the big thing, raised to importance by *Benchmarks for Science Literacy* (AAAS 1993) and the *National Science Education Standards* (NRC 1996). Its intention was not unlike the science and engineering practices proposed by the *NGSS,* and thus of lasting importance, but in application the pursuit of inquiry often became confused with the use of activities to convey factual information. For example, instead of a teacher telling students about the thicknesses and character of Earth's core, mantle, and crust, the students might color, cut out, and assemble a premade model of Earth's interior—a pedagogically sound activity for learning a concept but not one that engages students in the scientific process.

Thus, inquiry-based science, intended to prompt teachers to do science with their students, sometimes became an alternative avenue for conveying science knowledge. Why? Because doing real science is a lot harder than conveying information. It's hard to create activities. It's hard to interpret the results. It's hard to interact one-on-one instead of as a whole class. And it's especially hard to re-create in the classroom the sense of science discovery that in real life took thousands of scientists hundreds of years.

Teachers don't have hundreds of years in the classroom, and yet they want to give students a sense of the exploration and discovery inherent in science. The secret is to limit the options that students need to consider. Without sufficient limitations, the classroom lab devolves into random experiments that provide little or no insight into the science. With limitations set too tight, students have no real creative or analytical input and the lab becomes a "lecture by activity" in which the goal is to reinforce the content knowledge or derive the "correct answer."

It is helpful to have specific examples of how to apply limits to classroom investigations. Those limits depend on the ability level of the students and on the materials and time available. It is also helpful to have examples of obstacles that students are likely to encounter and misunderstandings they are likely to entertain. This book offers example activities and stories from the classroom that can help guide the teacher in setting those limitations while still providing a meaningful experience in science discovery.

ORGANIZATION OF THIS BOOK

The main part of this book is organized into three sections: "The Practices of Science," "The Language of the Earth," and "YOU Can Do It!" These sections are described in the pages that follow. After these sections are the afterword and three appendixes. The afterword includes a brief discussion of some aspects of teaching that we did not cover in depth in the main sections. Appendix A lists chapter activities and anecdotes related to *NGSS* performance expectations. Appendix B lists chapter activities and anecdotes related to *NGSS* science and engineering practices, DCIs, crosscutting concepts, and performance expectations. Appendix C provides illustrative quotes for selected ideas presented in the introduction and individual chapters. The quotes highlight a few of the significant ideas in science education that have been explored by teachers and researchers. Throughout the book, discussions corresponding to selected ideas in Appendix C are noted by a superscript C, followed by the note number.

The Practices of Science

Science is something we do, not something we know. The *NGSS* emphasize teaching science as a practice. *Learning to Read the Earth and Sky* devotes nine chapters to examining the practices of science, offering sample earth science activities for the classroom and anecdotes that illustrate student challenges and misunderstandings. Our goal is to help students and teachers understand the basic tools and language of science: how to propose and investigate a question that can be answered through science, how to analyze and interpret data, how to create and use a scientific model, and how to explain and communicate a scientific theory. We explore how an experiment differs from a learning activity, how to use graphs and maps, what scientific modeling means, and how to reason from evidence to theory.

The Language of the Earth

More than any other science, earth science is about stories: stories of Earth's past, stories of how things work, stories of how we know. Most of the great discoveries of geology are hinged on learning how to read those stories. The key "content" is not the conclusions of scientific studies—models, theories, and natural "laws." Rather, the key content is an understanding of how we read the stories. Many of these story-reading skills are unique to earth science—the place where it is set apart from chemistry, physics, and biology.

Learning to Read the Earth and Sky devotes five chapters to methods we use to read the stories written in the earth. Although the *NGSS* embeds these story-reading skills (the grammar of the earth, if you will) within the DCIs, we think there is a need, when applying these story-reading ideas in the classroom, to break them out. Without an understanding of these earth-reading concepts, any effort to examine Earth systems becomes an exercise in accepting the "facts" that someone gives without any real understanding of the underlying science. We consider in particular (1) how we read the story of Earth's past as written in earth—the soil, sediments, and layers of rock that make up our planet; (2) how we figure out the nature of places we can never visit, such as distant stars and the Earth's core; and (3) how we track down the movement of elements through the complex cycles and systems of the Earth.

Earth science emerged as people learned to read the stories of Earth's past and present written in its lithosphere, hydrosphere, atmosphere, and biosphere. Today, we apply these same language skills to reading the stories of other worlds written in their own "earthy" materials. In this book, we leave the first "e" in earth science lowercase to remind ourselves that the skills and practices of earth science now apply to more than planet Earth alone.

YOU Can Do It!

We believe that teachers should do science with their students. *Learning to Read the Earth and Sky* devotes three chapters to examining the teacher's role in this collaboration: teacher as curriculum narrator; teacher as guide in starting where you and your students are; and teacher as mentor, practitioner, and scholar.

The teacher, as *curriculum narrator,* can tie spontaneous and small-scale classroom exploration to the big ideas of science. The DCIs of the *NGSS*—and John Muir—tie everything in the universe to everything else, focusing our attention on the elegant way that the universe works in great systems and cycles.

Teachers can engage students *where they are.* We propose that students are most engaged with discovery when they investigate events and places that they know.

Teachers can be *mentors* and *practitioners of science.* We argue that students should be neither rigidly directed by the curriculum nor allowed to flounder with too little guidance. Instead, students should be provided a middle road where they make real choices in what investigations to pursue and how to pursue them, under the guidance of an expert mentor and practitioner of science—the teacher.

Scientists are sailors on the ship, not passengers, and understanding science is about understanding how to sail the ship. If we as teachers don't do a bit of the sailing with our students, then neither we nor they can ever really understand what science is all about.

Safety Practices in the Science Laboratory and Field

Both inquiry-based classroom and laboratory/field activities that immerse students in the practices of science can be effective and exciting. To ensure the success of these activities, teachers must address potential safety issues relative to engineering controls (ventilation, eye wash, fire extinguishers, showers, etc.), administrative procedures and safety operating procedures, and use of appropriate personal protective equipment (indirectly vented chemical-splash goggles meeting ANSI/ISEA Z87.1 standard, chemical-resistant aprons and nonlatex gloves, etc.). When personal protective equipment is indicated for use in an activity's safety notes, it is required for all phases of the activity, including setup, hands-on investigation, and takedown. Teachers can make it safer for students and themselves by adopting, implementing, and enforcing legal safety standards and better professional safety practices in the science classroom and laboratory/field. Throughout this book, safety notes are provided for activities and need to be adopted and enforced in efforts to provide for a safer learning/teaching experience.

Teachers should also review and follow local policies and protocols used within their school district and/or school, such as a chemical hygiene plan and Board of Education safety policies. Additional applicable standard operating procedures can be found in the National Science Teachers Association's (NSTA) *Safety in the Science Classroom, Laboratory, or Field Sites (www.nsta.org/docs/SafetyInTheScienceClassroomLabAndField.pdf).* Students should be required to review this document or one similar to it under the direction of the teacher. Each student and parent/guardian should then sign the document to acknowledge that they understand the procedures that must be followed for a safer working/learning experience in the laboratory. An additional reference is available for teachers to further explore field trip safety considerations: *Field Trip Safety* by the NSTA Safety Advisory Board *(www.nsta.org/docs/FieldTripSafety.pdf).*

Disclaimer: The safety precautions for each activity are based in part on use of the recommended materials and instructions, legal safety standards, and better professional practices. Selection of alternative materials or procedures for these activities may jeopardize the level of safety and therefore is at the user's own risk.

REFERENCES

American Association for the Advancement of Science (AAAS). 1993. *Benchmarks for science literacy.* New York: Oxford University Press.

Hawking, S. 1996. *A brief history of time.* New York: Bantam Books.

Muir, J. 1911. *My first summer in the Sierra.* Cambridge, MA: Riverside Press.

National Research Council (NRC). 1996. *National science education standards.* Washington, DC: National Academies Press.

National Research Council (NRC). 2012. *A framework for K–12 science education: Practices, crosscutting concepts, and core ideas.* Washington, DC: National Academies Press.

NGSS Lead States. 2013. *Next Generation Science Standards: For states, by states.* Washington, DC: National Academies Press. *www.nextgenscience.org/next-generation-science-standards.*

PART I
THE PRACTICES OF SCIENCE

Buoyancy, air pressure, and condensation are laboratory- and classroom-sized concepts that can be applied to modeling of large systems such as thunderstorms.

PART I

Why is it that, in spite of the fact that teaching by pouring in, learning by a passive absorption, are universally condemned, that they are still so intrenched in practice? That education is not an affair of "telling" and being told, but an active and constructive process, is a principle almost as generally violated in practice as conceded in theory. Is not this deplorable situation due to the fact that the doctrine is itself merely told?

—John Dewey, *Democracy and Education*, 1916

In this first section of *Learning to Read the Earth and Sky,* we offer examples of the practices of science from our own experience, illustrations of the practices of science from history, and encouragement to pursue the practices of science in your own classroom.The *NGSS* science and engineering practices are as follows:

- Asking Questions and Defining Problems
- Developing and Using Models
- Planning and Carrying Out Investigations
- Analyzing and Interpreting Data
- Using Mathematics and Computational Thinking
- Constructing Explanations and Designing Solutions
- Engaging in Argument From Evidence
- Obtaining, Evaluating, and Communicating Information

These science and engineering practices are not "steps" in the pursuit of scientific understanding; rather they are overlapping and nested concepts that collectively make up the scientific enterprise. For example, Asking Questions and Defining Problems and Analyzing and Interpreting Data nest within Planning and Carrying Out Investigations; Analyzing and Interpreting Data merges seamlessly with Constructing Explanations and Designing Solutions; and using Mathematics and Computational Thinking is a subset of Analyzing and Interpreting Data as well as Developing and Using Models. All of the previously listed practices underpin Engaging in Argument From Evidence, which shares some of its space with Obtaining, Evaluating, and Communicating Information.

We have chosen a few aspects of these comprehensive practices to examine in this section of the book:

- Chapters 1–3 address experimental and field observations, which include aspects of Asking Questions and Defining Problems, Planning and Carrying Out Investigations, Constructing Explanations and Designing Solutions, and Engaging in Argument From Evidence.
- Chapters 4 and 5 address the interpretation of data through graphs, maps, and cross-sections, which includes aspects of Developing and Using Models; Analyzing and Interpreting Data; Using Mathematics and Computational Thinking; Engaging in Argument From Evidence; and Obtaining, Evaluating, and Communicating Information.
- Chapters 6 and 7 address the creation and use of conceptual and mathematical models, which include aspects of Developing and Using Models, Using Mathematics and Computational Thinking, and Constructing Explanations and Designing Solutions.
- Chapters 8 and 9 address identification of evidence and arguing from evidence, which include aspects of Developing and Using Models; Analyzing and Interpreting Data; Constructing Explanations and Designing Solutions; Engaging in Argument From Evidence; and Obtaining, Evaluating, and Communicating Information.

GO AND SEE

As geologist Nicolas Desmarest implied in the early 1800s with his "Go and see!" challenge, science is based on observation, not theory. When I (Russ) was in grade school, Larry Taylor, my advisor who was involved in the study of the Moon from the very first Lunar Science Conference in 1970, once told me a story that illustrated how the uncertainties in theoretical thinking can often be resolved only by direct observation.

Back in the mid-1960s, in the early days of the Apollo program, an astronomer from Cornell, Tommy Gold, believed that the impact-pulverized dust on the Moon would be so fine-grained and so bone-dry that it would pile up in deadly fluffy layers thousands of feet thick that would quickly swallow up any landing spacecraft like quicksand. Although other scientists were skeptical, the image of the astronauts vanishing into the "Gold dust" was too horrible to dismiss, so NASA launched a series of missions to test the solidity of the surface. The first, Surveyor 1, did not sink into the Moon, but Gold thought that might be because it landed on a rock. The second mission crashed, but the third mission (Surveyor 3) made a soft landing and did not sink into the dust. Subsequently, manned landings were approved, albeit after considerable additional cost resulting from the Gold dust theory, including redesign of the lander feet to help prevent sinking. Even so, the matter wasn't truly resolved until Neil Armstrong stepped onto the surface of the Moon and declared, "I only go in a small fraction of an inch, maybe an eighth of an inch, but I can see the footprints of my boots and the treads in the fine, sandy particles" (NASA transcript of Apollo 11 spacelog, *http://apollo11.spacelog.org/page/04:13:20:58)*. To truly know, we had to go and see.

It's true that theoretical analyses and predictions based on them are an important part of science. But in science, what we know is based on what we can observe and measure. This is the place that our science explorations with students should begin. How can we help students start to ask questions, not of the teacher, or the textbook, or their peers, or even scientists or theories—but of the observations?[C2]

Asking Questions of the Observations

Asking questions of the observations means not only asking questions that can be answered through observation—a key premise of science—but also asking questions that arise from observations. Are there patterns in the data? What might those patterns mean? Can we identify a cause and effect? What new observations might test our interpretation of cause and effect?

The *NGSS* science and engineering practices—such as Asking Questions and Defining Problems and Analyzing and Interpreting Data—and the *NGSS* crosscutting concepts—such as Patterns and Cause and Effect: Mechanism and Explanation—are not steps in a linear process. Rather, making observations, asking questions, and looking for patterns is an iterative process integrated throughout an investigation. It's necessary to keep this integrated nature of science in mind when implementing three-dimensional learning as envisioned by the *NGSS.*

Appendix F of the *Next Generation Science Standards* (*NGSS;* NGSS Lead States 2013) describes the science and engineering practices, and the first entry in practice 1 (Asking Questions and Defining Problems) matrix for the youngest age group (K–2) says that students should be able to "ask questions based on observations to find more information about the natural and/or designed world(s)." Notice that observation comes first. Then questions. Then more observations to answer the questions. We are asking questions of the observations. This is the foundation for all science investigations from the K–2 classroom up through advanced research by professional scientists.

ASKING PROMPT QUESTIONS AND LIMITING OPTIONS

The problem with saying "science is based on observation" is that just making observations isn't enough. Even before students reach the point of analyzing any data, they need to make useful observations and put those observations in context. To start with, beginning researchers, like our students, have no way to evaluate the relative significance of observations. They may not have the experience to evaluate which observations are common and which are unusual, or which lead to a deeper understanding and which do not.

Consequently, at the start of a teaching unit, students probably won't know enough to make relevant observations, much less ask good questions about their observations. They are only starting to imagine the possibility that they can participate in discovery. Students need to explore a bit and be allowed to make somewhat random observations and to ask questions that may not bring them toward the content the teacher has in mind. They are *getting ready* to ask questions of the observations.

In the early 2000s, I (Mary) tried to get students to ask good questions about Moon craters in preparation for designing some cratering experiments. They'd studied topographic maps earlier in the year, so I started with topo maps and photos of the Moon, hoping students would notice the circular features and their different sizes and depths. None of the students noticed those. They did notice other things and asked questions such as "Why is the sky black?" This would have been a great question to launch a study of optical scattering in a high school or college physics class, but it wasn't the direction I planned to take my eighth-grade earth science class.

Learning to make good observations and ask good questions takes practice and guidance from the teacher. Practicing outside is a particularly good way to reinforce the idea that science is about observing our world and that science is something we all can do.

What do you do with good questions that distract from your end goal? In truth, there are an infinite number of possible questions arising from observations. It took hundreds of years for scientists to ask good questions, make good observations, and come to our present understanding of the universe. Therefore, it's important to limit the possible questions and observations in the classroom so that students can make progress within the limitations of their experience, their ability, and the time available. It's equally important to not limit options so severely that students can't participate in discovery. Prompting students with example questions, or arming them with a set of starting ideas, helps them focus on a narrower subgroup of questions while still giving them opportunities to observe, question, and build understanding from where they start.

Most important, offering prompts gives students examples of good observations and relevant questions that guide them to better questions and observations of their own. The authentic teacher doesn't simply instruct students to make observations and come up with questions. Rather, the teacher acts as a mentor and guide, modeling how one goes about asking questions and making observations. Students see how it's done and become better at doing it themselves.

Sometimes it takes quite a bit of prompting to get a class properly focused, and there is a risk of the teacher's guidance overpowering student initiative. Prompting with questions helps keep student ownership in the activity. In the classroom example above, I prompted my students to notice the landscape of the Moon by asking, "How is the surface different from the surface here in Moorhead, Minnesota?" Some of them responded, "Well, there are no trees or grass on the Moon." True, but that fact came from their existing knowledge base, not from observations of the topo maps. I asked, "How about those circular features, what do you suppose those are?" This brought students around to considering the data at hand. Asking "How deep are the craters—are they all the same?"

can prompt students to ask other quantitative questions such as "What are the crater widths—are they the same or different?" Eventually they might get to harder questions, such as "Are wider craters always deeper?" (The crater depth-to-width ratio is the key to understanding the transition from simple craters to complex craters on the Moon.)

Another key to offering prompts is to start out not with a singular ending in mind, but with *endings*, only some of which are likely to be reached. If you have only one ending, one goal, you have no room for creativity or true exploration. However, if you let the questions and observations go "just wherever," you get aimless efforts and unhelpful results that do not showcase the scientific process in any meaningful way. So endings provide boundaries on the scope of what you plan to pursue while allowing variation and creativity within those boundaries.

The challenge created by multiple endings is that you, the teacher, have to have enough understanding, and be flexible enough, to follow a lead prompted by your students. It requires that you be willing and able to depart from an inflexible lesson plan and fly out into the unknown.

Choosing the type of prompting to offer, and how much or how little, becomes the teacher's main job. Because you are trying to encourage student observations and questions and not force those observations and questions to go in one predetermined direction, there is no way that this process can be "canned" by a well-designed curriculum. Allowing student initiative in choosing where the lesson goes requires that you have a firm understanding of the discipline and the process of scientific discovery. It requires that you understand the specific abilities and limitations of a particular class of students and be able turn their spontaneous curiosities and insights into investigations and participation in the scientific process.

THE RIGHT ANSWER IS A TERRIBLE THING TO OFFER A DEVELOPING YOUNG MIND

It can be a showstopper when students ask questions and you, the teacher, tell them the right answer. The better response depends on where you are in a unit and what set of possible endings you have set as the goal. If you are at a later stage in a unit of study, the right response might be a prompting question: "What can we do to figure that out?" At the beginning of a unit, when you are trying to stir good questions and observations, the right response might be to make the question a springboard for engagement. For example, you might offer new evidence that challenges their existing concept or idea, or use the opportunity to brainstorm more questions that address the first question.

From the Horse's Mouth

At the dawn of the age of science, especially in the 17th, 18th, and early 19th centuries, many of the great philosophers (Robert Hooke, John Locke, David Hume, and others) increasingly came to the view that true knowledge must be based on observations rather than on revelation or philosophical speculation. Among earth scientists, this view was expressed by James Hutton, one of the founders of modern geology, who wrote in his "Theory of the Earth" (a 1788 paper read before the Royal Society of Edinburgh) that "we must not allow ourselves ever to reason without proper data, or to fabricate a system of apparent wisdom in the folly of a hypothetical delusion."

There's a well-known story (of uncertain origin) about a group of scholars arguing over how many teeth a horse has (much like the fabled arguments about the number of angels that can dance on the head of a pin). The scholars considered all manner of evidences: What would be the most perfect number for a horse to have? What did Aristotle (the authority) have to say about it? Is the number of teeth revealed by history books or in scripture? What *ought* it to be?

In the end, a young friar asked why they didn't simply open the horse's mouth and count the teeth.

This funny story has been widely used to illustrate the difference between science, which is founded on observation, and other areas of philosophy that pursue other ways of knowing. The basic idea of science is that our understanding of the universe comes from observations that any of us can choose to make. Our job as teachers is to make sure that this basic idea is not lost in the blizzard of facts and theories that we have derived from those observations.

In 2014, my students and I (Mary) had been studying the effect of the "warm" ocean in moderating climate along a coastline. A student commented, "But when I was in California, the water was cold." I prompted with "What time of year were you there?" which encouraged students to consider how the moderating influence of the ocean reverses in the summer, keeping the coastline cooler. Another student asked, "Was the wind blowing? That can make it feel colder."

A student's personal experience is particularly effective at engaging everyone. But you have to take that experience and ask, where do we go with this?

The real, implied question in the story above is "Why would the water feel cold if it is keeping the climate warm?" The student's observation, that the water felt cold, is valid. It needs to fit into whatever worldview that the science is trying to impart. This is the very type of exploration that can help students turn classroom learning into something that is their own. It stirs their questions about wind, ocean currents, changes in relative temperature of land and water with season or time of day, and so on. It offers the opportunity to propose and challenge ideas. It encourages students to think about the question in multiple ways, turning it over and around every which way until they see what it really means. From this question could spring an entire spontaneous experimental activity—real student-initiated exploration.[C3]

> **"The Fiftieth Birthday of Agassiz," by Henry Wadsworth Longfellow**
>
> And Nature, the old nurse, took
> The child upon her knee,
> Saying: "Here is a story-book
> Thy Father has written for thee."
>
> "Come, wander with me," she said,
> "Into regions yet untrod;
> And read what is still unread
> In the manuscripts of God."

MAKING SCIENCE YOUR OWN

Science sneaked up on us one day in Hawaii. We had been exploring one of the last pristine native forests nestled high on the flanks of Mauna Kea. The wet, musty smell delighted us, and we were fascinated with the variety of life established on a spot that just a few generations before was molten lava. We spent our time learning the Hawaiian names of as many tree species as we could and trying to identify them: for example, the 'Ōhi'a Lehua, the Kōpiko, the Hōpu', and the great Koa. It was fun, and we thought it was science. But when we descended the winding highway back toward the sea, and the ocean horizon filled the sky in front of us, real science nearly swept us off the road. Swerving to a halt on the narrow shoulder, we leaped out of the car and held the straight edge of a piece of paper up to the horizon and saw the curve of the Earth.

We already knew the Earth was round. Of course the Earth is round! We learned it in an elementary school history lesson about Christopher Columbus not falling off the edge of the Earth. But for the first three-plus decades of life, we believed that the Earth is round based on faith. When we went to Hawaii and from high on the mountain slope saw that the ocean horizon was curved and not flat, our belief that the Earth is round became science and not merely faith in the observations of others.

It is in making our own discoveries, and in testing the ideas that scientists and teachers give us, that we transform information into science. Science is found in exploration and discovery, not in simply knowing things.

People often imagine that information is the goal of teaching science. A friend once told me (Russ) that "science writing and science teaching are very different from other writing and teaching because the purpose is only to convey information." No. Even in science, *especially* in science, we teach people to do things, not just to know things. We teach them how to read the language of the Earth. We teach them how to learn—not from the scientists, or from the teacher, or from a book … but from observations.

Louis Agassiz, a 19th-century scientist after whom a now-vanished glacial lake in our region between Minnesota and North Dakota was named, once said, "The book of nature is always open!" He was right, and reading the book of nature begins with noticing things and asking questions.

REFERENCE

NGSS Lead States. 2013. *Next Generation Science Standards: For states, by states.* Washington, DC: National Academies Press. *www.nextgenscience.org/next-generation-science-standards.*

EXAMPLE ACTIVITY DESIGN

GO AND SEE: GEOLOGY IN THE FIELD

An Exploratory Introduction to Observation and Questions

The goal of this example activity design (EAD) is to learn to make observations, ask questions based on observations, and then answer the questions with new observations. You might have students articulate their best observation, what question that observation prompted, and what new observations addressed that question. Have them explain *how* the new observation addressed the question. If students are just starting to learn to ask questions of observations, this EAD will require substantial mentoring from the teacher.

Although prepared activities often tightly constrain what the student is supposed to do, our EAD, in contrast, offers the teacher a possible topic, some example questions, and a few prompts to help focus student observation and facilitate the teacher-student interaction in doing science together. As teacher, it's your choice which observations to prompt and which to let students discover and pursue on their own.

TEACHER PREPARATION AND PLANNING

Do some research on the regional geology of your area. What is it known for? What's unique about it? What do *you* wonder about? Knowing your area can help you know how to prompt students to make more science-relevant observations. Possible resources include your state's geological survey, a local university geology department, or the *Roadside Geology* series published for many states by Mountain Press.

For example, most people in our region between Minnesota and North Dakota know that there was once a large glacial lake that filled the valley and that the scarps rising to the east and west of the valley were once the shorelines of the lake. However, many don't know that one key observation supporting this claim is the difference in sediment size between the valley floor and the shoreline scarps. The former beaches are sandy, whereas the former lake bottom is not. Prompting students to think about the size of particles in the sediment (which they typically confuse with soil) can help them notice the different particle sizes. Variation in particle size is also a key idea in understanding sedimentary rocks in general.

EXAMPLE PROMPTS AND LIMITING OPTIONS

Basic questions about any observation of rocks include the following:

- What are some features seen in the rock?
- How did it form?
- What does it tell us?

Here are some other possible questions, depending on the type of rocks available in your area:

- Are the rocks in place (*bedrock*), or are they lying loose on the ground (*float*)? In-place rocks tell us something about that place in the past, whereas loose rocks may have been brought in from someplace else. In our region, bedrock is hundreds of feet below the sediment left by the glaciers. Our students have to look at rocks that either were brought in thousands of years ago or are being currently moved about by rivers.
- What can you look for to be sure that the rock is in place and not just a really big piece of float? Does it match other rocks in the area? Does it "attach" to the earth? If the rock has layers, do the layers go the same direction as other layers in the area?
- Is the float natural or was it brought in for construction or to prevent erosion on a riverbank? How do you know? Are the rocks different from the underlying rock or sediment? Do they only occur in parts of the river you might expect people to want to protect, such as around the footers of bridges or embankments close to a road?
- Is the float made of rounded or angular boulders? This might give you clues as to whether the boulders were transported by water, rounding off the rough edges, or broken out of the local bedrock by weathering.
- Are loose boulders all the same kind of rock or are they different? In our region, glaciers brought in a large variety of boulders from many different areas so that we have a mix of different kinds of rocks that students can compare.
- Are the rocks made only of one type of particle or of many particles stuck together? Are there crystals? Sand grains? Bands? Layers? Fossils? Any other features that you can identify? In general, students first notice things of lesser importance to the geologist—such as lichen stuck to the rock, the outer

yellowish rind of weathered material, or the "shape" of the rock. You might prompt them to look at a freshly broken surface, which gives a better view of the rock itself, and encourage them to pay attention to the particles inside the rock, not the shape of the rock itself.

- How did the rock form? This took centuries for people to answer, so don't expect students to figure it all out in this exploratory activity. However, you don't want idle speculation either. If there are bubbles in the rock, that might suggest it was once liquid (igneous). If there are fossils, that might suggest the rock formed from sediment with that creature in it. If there are tiny grains of sand in the rock, it might have formed from a beach, a dune, or a stream. Some well-placed prompts can allow students to come up with these ideas themselves.
- How did it get here? This common question might address transport (for float) or formation history (for bedrock).
- Are there big layers of different kinds of rock? Which layer is on top? How is it different? How might layers of different kinds of rock have formed, and what do they tell us?
- Are there interesting weathering or erosional features (meanders in rivers with point bars, cutbanks, rounded boulders, potholes, arches, columns, etc.)? How might they have formed?
- Hope for student questions that you don't know the answers to! See if you can figure it out with your students. What evidence can you identify? How can you constrain the possibilities based on *observations*?

EXAMPLE INTERACTION

In keeping with our philosophy of teacher as practitioner of science, we don't want to define too narrowly how you should set up this EAD, but remember that it is an introductory exploratory activity and so students will likely need quite a bit of prompting to get started on a useful track. Consequently, you might want to interact with small groups or even with the class as a whole to get them started thinking about how to ask questions of the observations. An example interactive dialogue with your students that could get them started thinking might go something like this:

Student question: How did the rocks get here?

Teacher prompt: Well, what can we observe about them—are the rocks loose or attached to the earth? (This gets students to thinking about the possibility that the rocks were brought from someplace else by some unknown process, or, if they are attached to the earth, they must have formed in this very spot.)

Student observation: They're just lying around loose.

Teacher prompt: What does that tell us?

Student interpretation: That they were brought in by glaciers (this is a common expectation in our area).

Teacher prompt: If that's so, what other characteristics should the rocks have?

Student observations and ideas: [Various; observations might include rounding of boulders or that the boulders are not all the same kind of rock.]

Continue to focus student attention on asking questions that observations can answer and thinking about what the observations tell us, letting students take greater initiative as they get the idea. Also give them time to have fun with making observations, asking questions, and looking for answers in new observations.

SUMMARY CHECKLIST FOR TEACHER AS PRACTITIONER OF SCIENCE

- Provide the students with opportunities to make observations.
- Provide the time for students to question their observations.
- Offer example questions and observations to show them how it can be done.
- Provide appropriate prompts that guide their thinking while allowing them room to explore.
- Offer new observations or thinking that challenges their initial interpretations.
- Balance guidance with freedom to be creative to provide an open, yet limited, exploration.

TEACHER REFLECTION

You've had a good field trip: The students made good observations and asked good questions, and no one got hurt or lost. Now what? Reflect on the kinds of things the students noticed. Which observations surprised you? Which ones engaged them? Plan to explore some of those questions in future activities to build an integrated instructional sequence based on your student-initiated questions.

THE CONTROLLED EXPERIMENT

In Chapter 1, "Go and See," we examined the importance of asking questions of observations rather than of the teacher or the textbook. But asking questions is only the beginning of investigation. Entering into the practice of science requires us to push through the first blush of curiosity and find more controlled ways to test our ideas.

Back in the early 1800s, a hunter and mountaineer, Jean-Pierre Perraudin, noticed boulders scattered in the Swiss Alps in places where they didn't seem to belong. He imagined that glaciers must have brought them there (Wright 1898). A preposterous idea! For ice to move rock from one place to another, the solid ice had to flow like a river.

Geologist Louis Agassiz was skeptical of this interpretation of the boulders and was not content to simply observe and speculate. He set out to test the possibility of flowing ice through experiment (Agassiz 1840). He drove stakes into the ice in a line across the Aar glacier. Over the course of the following months and years, he documented the movement of the stakes, watching them creep downslope and form a bow shape as the stakes in the center of the glacier got ahead of the ones on the margins (Figure 2.1). Not only was the ice moving, but the center was moving faster than the edges, just like a river. He drove long tubes into the ice to find the rate of movement deeper in the ice. The poles slowly tilted and deformed, showing that the glacier moved at different speeds at different depths, again like a river.

Figure 2.1
Illustration of Agassiz's Experiment

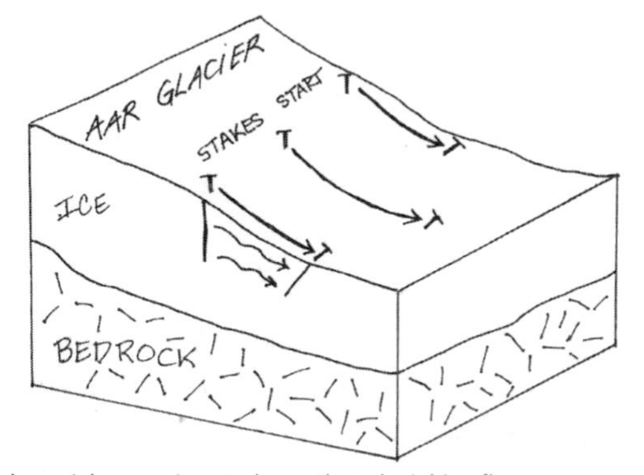

Agassiz's experiment shows that glacial ice flows downslope.

Perraudin and Agassiz imagined the astonishing possibility that ice might move and carry boulders from the mountains down into the plains. That idea left the realm of speculation and entered the realm of science when Agassiz tested it experimentally.

Science as a Way of Knowing

Scientists do experiments and make observations to come to an understanding, which is an idea that sets science apart from other ways of knowing. This idea was a topic of hot debate among scholars back in the 1600s!

"No man's knowledge can go beyond his experience."

—John Locke, *An Essay Concerning Humane Understanding,* 1690

"The Truth is, the Source of Nature has been already too long made only a Work of the Brain and the Fancy: it is now high time that it should return to the plainness and soundness of observations on material and obvious things."

—Robert Hooke, *Micrographia,* 1665

ENCOURAGING EXPERIMENTATION

An Unexpected Experiment

We can encourage our students to explore deeper by showing them our own passion for exploration. When a student comes up with a great question in class, don't give them the answer. Instead, show them how to follow up on it experimentally.

One year, I (Mary) was doing a unit on differential heating between water and land with my students. We left one of the trays of water out overnight so that the temperature of the tap water would equilibrate with room temperature. When we came back the next day, one of my students noticed that the temperature of water in the tray was less than the temperature in the room. "Why didn't it come to the same temperature?" she asked. "Shouldn't it be the same as the room?"

Good question. Is water naturally "cooler" than air? Had the room temperature fallen overnight and the water took longer than the air to heat back up?

We abandoned the planned lesson for a while to explore what might have caused the temperature mismatch. To get us started, and to give my students an example of how questions about the natural world can be answered through experiment, I suggested a test to try.

We left the tray of water out overnight again, but beside it we put a jar of water with the lid tightly closed. When we came back the next day, the tray of water was again cooler than the air, but the water in the jar matched room temperature. This experiment eliminated the possibilities of the mismatch resulting from room temperature falling overnight or water being naturally cooler. We hypothesized that evaporation cooled the water, just like evaporation makes you cold if you stand out in the wind when you're wet. When the jar was closed and water couldn't evaporate into the room, the temperature mismatch disappeared.

Our experiment didn't completely eliminate all other possibilities. For example, our jar was a different shape and material than our water trays. Might that have some effect? We could have continued our experiments by covering one of the trays with plastic wrap. If the temperature mismatch resulted from water evaporation, covering it should eliminate the mismatch, providing a test of our hypothesis. Other experiments might have examined the effect of room humidity on the temperature of the water in the open tray.

One thing to remember is that no one experiment is ever the final word on a question, not in the classroom and not in real research. Teachers and students need to choose how far to pursue an activity, depending on their objectives for the lesson and time available.

The student-initiated question in this example was unplanned. I affirmed the student's question and provided an example of how we can answer questions with experiments instead of from an answer key. However, we want our students to invent experiments, not only do experiments at the prompting of the teacher.[C4] How can we be more deliberate in encouraging questions from students and create situations where students come up with experimental tests of those questions?

Start With Play

In teaching about air pressure at the start of a lesson on weather, I often have my eighth graders play with cups, water, and air, hoping that they discover that air is *something*, exerting force on the world around. In play, students make observations, which become the wellspring for questions. Play is also a great place to start constructing mental models.

Students try to push cups full of air into a bucket of water (upside down, right-side up). They fill cups or graduated cylinders with water and turn them upside down in the water, seeing that the water can "stand up" above the level of the water in the bucket. They put index cards over the open ends of the cups and see that when they turn the cups upside down, the water stays inside the cup. Like magic!

What's holding the water in the cup?

One student, Dan, in figuring out that it had to be the pressure of the air holding it in, asked, "Do you have any straws?" I gave him some straws, and he took duct tape and taped a bunch of straws together until he had a long, flexible tube. Then, with a graduated cylinder full of water and its open end just submerged so that the water in the cylinder stood a foot above the water in the bucket, he inserted one end of his tube to the top of the water in the cylinder, leaving the other end outside the cylinder in air. The water drained as air rushed in. Dan blew air in, noticing that he could drive the level of water in the cylinder below the water level in the bucket. Then, he sucked on the end of the straw still outside the cylinder, drawing the water back up. He was able to feel the pressure needed to pull the water up and experimentally reproduce the forces that were holding up the water.

Play stirred a question about what caused water to rise up in the cylinder. My student constructed a mental model for the cause and invented a way to test that model. His experiment boosted the energy and interest in the whole class as they realized that science was something they could not only do, but initiate. In fact, one envious student asked, "Why did Dan get to do that and I didn't?" Well, he thought it up. You can do that, too!

Safety Notes

1. Have students wear indirectly vented chemical-splash goggles and aprons during the whole activity, including setup, hands-on piece, and takedown.
2. Immediately wipe up any spilled water to avoid a slip and fall hazard.

Even Play Needs Constraints

Lessons in meteorology are particularly conducive to exploratory investigation, particularly lessons involving air pressure, volume, and temperature. However, when encouraging student experimentation, teachers need to keep a close eye on the experiments.

One year I (Russ) had a class of college elementary education majors doing an open inquiry project. One group, knowing that warm air rises because it's less dense than cool air, decided to measure the relationship between air temperature and volume. They had their hot plate, thermometer, and glass Mason jar—its lid tightly sealed—all lined up for their experiment when I circulated past and asked what they were planning to do to measure how the air expands when heated. I tapped the lid of the jar. "How's the air going to expand when it's inside that jar? How will you measure any change in volume?"

"Well ... " They looked at the jar, processing the question with growing dismay. Finally, one student said, "I guess it can't expand." I laughed. "Actually, the air will expand just fine when your jar blows apart from the pressure. What you're making is a bomb!"

After some chagrined collaboration, they adapted their experiment by attaching a balloon to the jar. This allowed air to expand gradually (no bomb) and provided a means to determine the extent of expansion by measuring the diameter of the balloon.

Safety Notes

1. Have students wear indirectly vented chemical-splash goggles and aprons during the whole activity, including setup, hands-on piece, and takedown.
2. Immediately wipe up any spilled water to avoid a slip and fall hazard.

EXPERIMENTAL PRACTICE: MORE THAN HANDS-ON LEARNING

Good experimental practice in your classroom requires learning to control and measure variables *quantitatively* for at least some of your experimental activities. Controlling and manipulating variables is a key element of science and engineering practice 3 of the *NGSS*, Planning and Carrying Out Investigations, as shown in this excerpt from the practice 3 matrix for grades 6–8 in Appendix F of the *NGSS* (NGSS Lead States 2013):

> *Plan an investigation individually and collaboratively, and in the design identify independent and dependent variables and controls, what tools are needed to do the gathering, how measurements will be recorded, and how many data are needed to support a claim.*

Understanding the manipulation of variables also plays a key part in making predictions, as shown in this excerpt from the practice 3 matrix for grades 9–12 in Appendix F: "Make directional hypotheses that specify what happens to a dependent variable when an independent variable is manipulated."

In the 1990s, "hands-on learning" became a popular catchphrase in science education—learning anchors itself deeper in our psyche when we engage not only our minds but our bodies in learning. People soon realized that hands-on activities weren't enough to learn science, and the phrase became "hands-on, brains-on."

The scientific and engineering practices envisioned by *A Framework for K–12 Science Education: Practices, Crosscutting Concepts, and Core Ideas* (NRC 2012) involve even more than hands-on, brains-on learning. The practices of science involve a specific kind of "hands-on, brains-on" activity that parallels the explorations of scientists.

Activity Versus Experiment

An excellent example of a high school hands-on, brains-on activity that teaches humidity and condensation was published in 1993 by the American Meteorological Society as part of Project Atmosphere. The activity involves putting beans in cups of different sizes, with the size of each cup representing the capacity of air to hold water vapor and the beans representing the water vapor (when in the cup) or liquid water (when it spills out of the cup). For each increase of 10°C, the size of the cup doubles, representing a rough doubling of the solubility of water vapor. When saturated air cools, the solubility of water vapor decreases (the size of the cup gets smaller) and some of the water condenses from the air as liquid. Using this imagery, and their cups and beans, students answer a series of questions on relative humidity, dew point, and condensation.

I (Russ) still use a modified version of this activity for my college elementary education majors. But I always emphasize that this is a learning activity, not an experiment. What's the difference? Well, no amount of manipulation of beans and cups would ever convince scientists that water vapor is more soluble in warm air than in cold air. That requires real experiments in the real world. Beans and cups help students to understand a concept that has been figured out by other means, but it does not engage students in the experimental part of the practices of science. It is both hands-on and brains-on, and it engages students in constructing a model for humidity, but it is not an experimental investigation.

Going From Learning Activity to Experiment

The dependence of the solubility of water vapor on air temperature is certainly important in earth science. Solubility of water vapor in air is the key to understanding the formation of fog, dew, frost, clouds, rain, and snow. It is also a key factor in understanding climate change because small changes in atmospheric carbon dioxide (CO_2) will increase temperature, which increases solubility of water vapor—also a greenhouse gas—and thus cause much more warming than accounted for by the CO_2 alone.

Setting up a classroom experimental activity that measures the temperature dependence of water solubility in air is significantly harder than the conceptual activity with beans described above. It's also harder than an experiential demonstration of concept—such as taking a cold can of pop out of the freezer on a humid day and watching the condensation form on the outside of the can. The hard part of measuring the solubility of water in air is figuring out how to control the temperature of air in equilibrium with liquid water (the independent variable) and how to measure the solubility (the dependent variable).

One year during an open inquiry project, I (Russ) had three groups of students interested in measuring the solubility of water vapor in air as a function of temperature. I'd had some experience in previous years with how difficult it is for students to come up with an experimental approach that yields repeatable results. For one reason or another, most of the published high school activities didn't work. So, based on the results of previous classes of students who had tackled the problem with limited success, I suggested a couple of different approaches to try—curious myself which of them might work the best.

The two basic approaches, which students could adapt and modify, were as follows:

- Method 1: Produce saturated air in a 2-liter bottle by equilibrating it with water at a known temperature in a separate bottle attached through a "cyclone" bottle connector. Close off the 2-liter bottle and cool it in an ice-water bath to condense out most of the water vapor. Swab out the water with a paper towel and measure the increase in weight of the paper towel to determine the amount of water that condensed. Repeat for a variety of equilibration temperatures.

- Method 2: Produce air with very low absolute humidity in a 2-liter bottle (my students did this by equilibrating the bottle air with freezer air, but it's not too hard to do with outside air in winter in our region). Add a drop of water to the bottle, close it up, and warm it until the drop completely evaporates. Repeat using additional numbers of drops of water.

Having different groups tackle the same problem using different methods created a sense of collaboration and competition between the groups, something that truly reflects the character of real research. The two methods also approached equilibrium from opposite directions (called an experimental reversal), one from too much water vapor and the other not enough.

Students reported results of their project to the class through posters, and the posters from these three groups are shown in Figures 2.2, 2.3 (p. 24), and 2.4 (p. 24). Although our classroom experiments stood on their own merits, and we didn't talk about how their results matched with "real" science (our goal was to learn to do experiments, not try to reproduce a "right" answer),[C5] the results of the two groups who measured solubility were very close to each other and close to the "real" result.

Figure 2.2
Poster showing experimental results of group 1

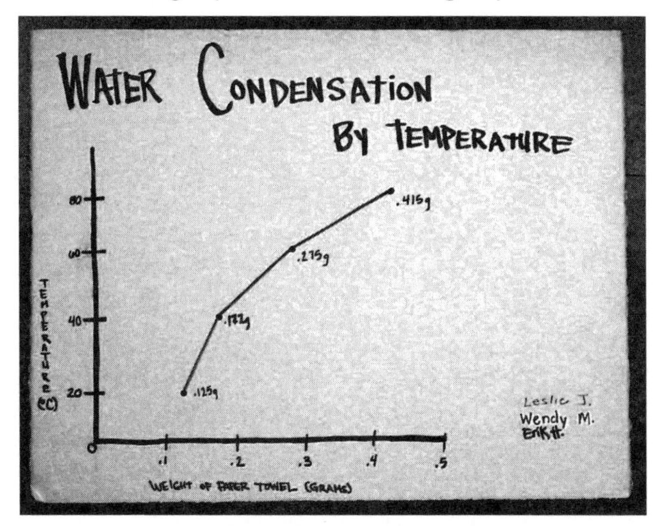

Group 1 used the method of condensing water (H_2O) from air saturated at a known temperature, swabbing out water with a paper towel, and weighing. The group didn't report the volume of air that they used, but the change in temperature is exponential, as expected from published research. They incorrectly labeled the *x*-axis on the poster. The label should be the net weight—wet towel minus dry towel. This group gets about 0.22 g soluble water in air at 50°C. Assuming they used a 2 L bottle (which they did), this corresponds to 0.11 g H_2O/L, compared with the accepted value of about 0.12 g H_2O/L.

STANDING ON THE SHOULDERS OF GIANTS

Isaac Newton famously said that we see farther when standing on the shoulders of giants. This can be true in the classroom as well as the research lab. The experiments with water solubility that I (Russ) did with my students followed several terms of student investigations that produced significantly less-clear results. I had seen a number of trials with previous groups of students fail before I was able to provide the guidance to suggest these experiments. I explained to these new groups that they were building on the results of previous students who had tried other approaches that didn't quite work.

Figure 2.3

Poster showing experimental results of group 2

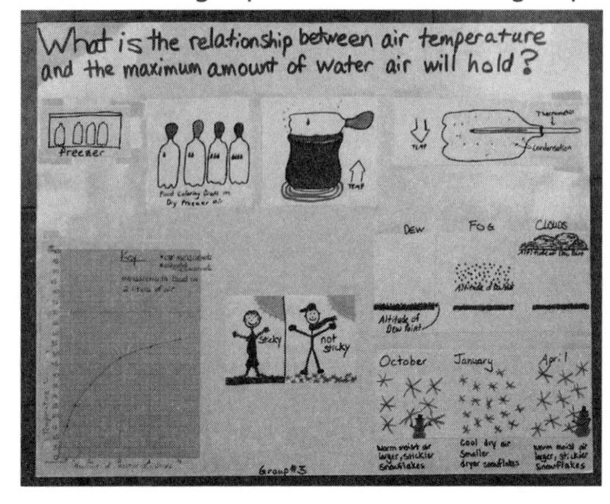

Group 2 used the method of adding a known amount of water (H_2O) to dry air in a bottle and then increasing the temperature of air in the bottle until all of the water evaporated (or was no longer noticeable). The group's graph is not easy to read—it plots temperature increasing upward on the *y*-axis and number of drops of water on the *x*-axis. This group also got an exponential curve, as expected from published research. They reported their air volume but not the mass of a drop of water. If we assume that their water drops are about 0.1 g, a typical value, then they report about 0.22 g H_2O in 2 L at 50°C, or 0.11 g/L, compared with the accepted value of about 0.12 g H_2O/L.

Figure 2.4

Poster showing experimental results of group 3

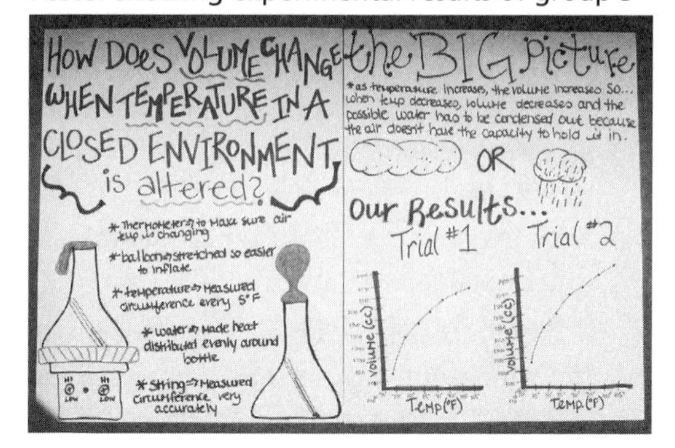

Group 3 chose a different approach than the ones used by groups 1 and 2; this group assumed that the cause of increasing solubility of water in air is the expansion of the air with temperature (making more "room" for water vapor). Although this assumption doesn't match accepted solubility theory, the experiments themselves are valid and showcase the different kinds of approaches that students can take when allowed to run with their ideas. I (Russ) let the group carry out their exploration and didn't correct their misconception. I did offer some questions afterward that prompted them to consider whether their observations really meant what they thought about solubility of water vapor.

Real scientists don't start out with a foolproof experimental plan. They often take years, or even generations, to solve tough experimental challenges, each new effort building on the work that came before. Likewise, it sometimes takes several class periods, or even several different classes of students, to solve a classroom experimental problem. Engaging with this ongoing process of discovery can not only provide a sense of ownership among students, but also engage them in an authentic practice of science—building on work that came before.

This leads us to believe that it isn't necessary for teachers to always start out with a pretested activity that someone else has designed. In some ways, for the best student engagement in science practice, it's better not to.[C4] Instead, choose a question in collaboration with your students, identify the independent and dependent variables that address that question, think about how to control and measure those variables, and set up an experiment that builds on the shoulders of giants.

QUANTITATIVE EXPERIMENTS

Exploratory play is a great place to start experimental investigation. But the practices of science involve more than the initial exploratory play or even the first experimental follow-up of that play. It's good for some of your classroom experimental work to mature toward quantitative measurements.

What is a quantitative experiment? One way to test if your experimental design is quantitative is to ask if the results can be plotted on a graph—one variable that you change in the experiment versus another variable that responds to that change in some measurable way. For example, Louis Agassiz could chart either time or location of a stake on the glacier against its position. In the water vapor solubility experiments, we can chart temperature versus solubility of water vapor. We will consider graphing in Chapter 4, "Analyzing and Interpreting Data, Part 1: Graphing."

FINAL THOUGHTS

This chapter has focused on one portion of a scientific investigation—the controlled experiment. Controlled experiments involve making measurements under conditions in which cause and effect can be established. This is harder to do than merely stating the nature of that relationship or making qualitative observations under less controlled conditions.

Controlled Experiments in Earth Science

Investigations whose main goal is to tell stories—rather than infer fundamental natural laws—set earth science apart from physics and chemistry. Knowing a key theory is rarely the main objective in earth science. Instead, the objective is more often to apply experimental results and theoretical understanding to reading stories of a planet's past or present. For example, the experimental activity measuring solubility of water vapor in air in this chapter does not have the primary goal of exploring the chemistry of solubility, but rather explaining—telling the story of—the water cycle, a part of how the Earth works.

This focus on telling stories rather than learning theories makes earth science particularly well suited to teaching the practices of science, because the methods and approaches for reading the stories are the primary "content" of the discipline. This is what the title of the book, *Learning to Read the Earth and Sky,* means.

For example, measuring the relationships among temperature, pressure, and volume, or measuring the effect of evaporation on water temperature, is harder than simply stating that air expands when warmed or pointing out how cold we feel when a stiff breeze blows across our wet skin. Measuring the relationship between temperature and solubility of water vapor in air is harder than talking about condensation on a cold can of pop and then leaping to the formation of clouds. Despite being harder for both teacher and student, making careful measurements engages students in doing science—and brings them face to face with challenges to science exploration that they will never encounter if they only look for the right answers from conceptual learning activities.

When students realize that they can test the universe around them and not simply rely on scientists to tell them the answers, they gain an appreciation for the process of science that can be reached in no other way. *Learning activities, computer simulations, and science art projects, while teaching science concepts, do not allow students to test science concepts and so do not offer the same learning experience as science experiments.*

One thing to remember in doing experiments with students is that your goal is not to have a foolproof activity that produces the right answer on the first try. In fact, reinforcing the right content knowledge is not even your goal. Your goal is to engage students in good practices of science.

Running into problems engages students in doing science, while performing foolproof cookbook labs does not. One time at the college where I (Russ) work, my department brought in some community geographic information systems experts to talk about how we might prepare students for work in that field. They commented that classroom activities might better prepare students for the workplace if the activities didn't provide all the information needed for a neat and tidy solution. These experts spent most of their workday trying to solve problems that involved missing information and had solutions that weren't neat and tidy at all.

Sometimes it isn't possible to figure out an experimental problem within the time limits of a particular class. Sometimes experiments may continue over years, with different classes building on the results of previous classes, just like scientists build on each other's work in real research. If your efforts to solve experimental problems in the classroom extend over multiple classes or years, you should tell your students this to give them a sense of continuity and purpose—their results might have an impact on students in years to come.

REFERENCES

Agassiz, L. 1840. *Études sur les glaciers.* Edited and translated by A. V. Carozzi. New York: Hafner, 1967.

American Meteorological Society (AMS). 1993. *Water vapor and the water cycle teacher's guide.* Project Atmosphere. Boston: AMS. A modified version of this activity is available at *www.cmosarchives.ca/ProjectAtmosphere/module12_water_vapour_and_the_water_cycle_e.html.*

Hooke, R. 1665. *Micrographia or some physiological descriptions of minute bodies made by magnifying glasses with observations and inquiries thereupon.* London: Council of the Royal Society of London for Improving of Natural Knowledge.

Locke, J. 1690. *An essay concerning humane understanding.* Public domain.

National Research Council (NRC). 2012. *A framework for K–12 science education: Practices, crosscutting concepts, and core ideas.* Washington, DC: National Academies Press.

NGSS Lead States. 2013. *Next Generation Science Standards: For states, by states.* Washington, DC: National Academies Press. *www.nextgenscience.org/next-generation-science-standards.*

Wright, G. F. 1898. Agassiz and the Ice Age. *The American Naturalist* 32 (375): 165–171.

EXAMPLE ACTIVITY DESIGN

THE CONTROLLED EXPERIMENT: IMPACT CRATERING

AN INTRODUCTION TO STUDENT-DRIVEN QUANTITATIVE EXPERIMENTS

We have adapted the following example activity design (EAD) from a well-tested lab on impact cratering.* However, we leave some of the "well-tested" part in the cupboard and instead set the stage for students to formulate experimental questions, design experiments, and figure out for themselves the relationships between variables. In this way, teachers can become mentors as their students muddle through the messiness of a laboratory investigation.

Experimental science should address real-world questions, and it's often helpful to organize lessons around driving questions. This EAD explores the following driving questions:

- Why do the surfaces of rocky planets and moons look so different from one another?
- What processes wear down, reshape, and build up the surface of Earth and other planets?

It's also helpful to fit driving questions into the larger context of your instructional sequence. These two driving questions fit into the big idea that planets change, which in turn is a part of the *NGSS* disciplinary core ideas Earth and the Solar System and The History of Planet Earth.

Note: The main goal of this EAD is not to learn "content knowledge" about real impact craters. In fact, most real impact craters form in hypersonic events, making them quite different in character from the subsonic experiments described here.[C5]

TEACHER PREPARATION AND PLANNING

Review the EAD and do a few experiments ahead of time. Pay attention to your own thought processes and think about how much input you expect from students and how much guidance you want to provide. Student-driven research does not mean that you turn students loose to do whatever they want; doing "whatever" quickly becomes an

*Adapted from G. J. Taylor [project coordinator], *Exploring the Moon: A Teacher's Guide With Activities for Earth and Space Sciences,* EP-306, National Aeronautics and Space Administration, Office of Human Resources and Education, Education Division, and Office of Space Science, Solar Exploration Division, 1994; a more recent version is available at *www.spacegrant.hawaii.edu/class_acts/EP-306.html.*

unproductive class period. Rather, your job is to provide opportunities for science exploration while you act as the coach and mentor, nudging and guiding the students toward the practice of good science.

Your initial preparation should help you consider the boundaries of the exercise so that during the hubbub of experimentation you can provide just the right amount of guidance. Some of your on-the-fly decisions will include whether to nudge students toward more fruitful experimental questions, whether to point out experimental design flaws or let students discover them, and how to facilitate peer-to-peer discussions.

EXAMPLE PROMPTS AND LIMITING OPTIONS

Some basic experimental questions to think about in preparing for student investigation include the following:

- In what ways can an impact crater be characterized? By its shape, or width, or depth?
- What factors can influence the character of an impact crater?
- In what ways does the size of a crater depend on parameters such as velocity, mass, volume, or density of the impactor or on factors such as the character of substrate or angle of impact?
- Does the mass of an impactor have an effect different from the size of an impactor?
- Does mass have a different effect at different velocities?
- Does the nature of the substrate (fine sand, mix of sand sizes, gravel, wet sand, dirt) make a difference?

BEGINNING STUDENT EXPLORATION

- Set out a supply of kitty litter trays with fine sand of uniform grain size, triple beam balances, and impactors of various sizes: steel balls, wood balls, marbles, golf balls, Styrofoam golf balls, trailer hitches, and brooms and dustpans.
- I (Mary) prompt my students with the question "What do you have to do to create craters of different sizes, given these impactors and a basin of sand?" I (Russ) prompt my older students with the question "What variables influence crater size and how do they influence it, quantitatively?" The prompt questions are open-ended, but the scope of the investigation is limited by the

kinds of materials provided and by the focus given in the prompt question. The prompt excludes certain variables such as different sand mixes and other crater characteristics like radius of ejecta blanket. This limitation gives students a sense of "I'm the scientist in charge" without overwhelming them with options.

- Students often begin by throwing the balls at the sand, but they quickly start to articulate scientific questions such as "What will happen if we drop these balls from different heights?" or "What do the craters look like with different sizes of balls?" You might need to remind groups of the prompt question, but this time of exploration infuses a sense of play into the classroom.
- After a sufficient time for unstructured exploration, direct lab groups to consider what might be affecting the size of the craters, which variables they might like to pursue further, and how they might set up their experiments.
- Gather the class for a sharing and brainstorming time. As you facilitate the discussion, you might provide some nudging and mentoring. Are there improvements to the experimental designs that you can offer students from your experience or that students can offer each other? Do you want everyone to do the same experiment? Sometimes there is a class consensus to pursue one particularly exciting experiment, but another good approach is to let groups follow their separate interests. Try to make sure that at least two groups look at the effects of the same variable so each group has another group they can compare against.

Safety Note

Review safety procedures with students by doing a hazards analysis and risks assessment and outlining the safety actions to be taken.

BEGINNING STUDENT EXPERIMENTATION

- Have students *try out* their experimental approach. Have them discover the aspects of their design that don't work and give them time and space to adjust and refine their approach. Be ready with helpful guidance, but not too quick to offer it. Students often ask, "How do we measure the size of the crater?" A reasonable first response is "You figure it out." With trial and error, students realize that measuring the depth of the crater is much less accurate than measuring the diameter, because sticking something into the crater to measure the depth often destroys the crater.

- Here are some other questions or hints that you might provide, depending on your students' questions or actions: Which of the impactors will be the best to use if you want to only look at the effect of mass? How can you design an experiment that separates the effect of mass from the effect of the size of the impactor? How are you going to control your velocity so you can reproduce your results? Sometimes students need the hint that height of fall and velocity of impactor are correlated with each other.
- This might be a good time to tell your students that they need to be careful in making measurements. "Careful" to eighth graders—and college students—often means "we didn't spill the sand … too much!" They might not imagine that measurements should be reproducible to within a half millimeter.
- Take time to debrief as a class, with individuals and groups responding to the question "What have we learned so far?"
- After students have tried out their experimental approach and taken some initial measurements, charge them with organizing their measurements into a data table. Post results for all groups to see. Compare the different approaches and numbers as a class. Come up with good data table formats that everyone can understand. We don't usually have students calculate velocity but rather use the drop height as a proxy for velocity. This avoids some complexity that doesn't really add to the experimental methodology. However, if you are doing the activity in collaboration with a physics or math lesson, you might want to do the conversion, as we suggest for the impact-cratering experiment in the "Example Activity Design" section of Chapter 7 (pp. 136–142).
- If you haven't done so already, you might talk about independent and dependent variables.

THE BIG DAY: FINAL EXPERIMENTS

- Have each group do a series of experiments to test the relationship between the variables they are responsible for. It's fun for a group to be making the same measurements as another group, as this provides for some friendly competition and collaboration as they check out what the other is getting. Encourage repetition of measurements.

- Remind them that the confirmation of scientific results is often found in the ability to make predictions about future events. Tell them that you are going to ask them to predict the size of a new crater based on their results, given the velocity, mass, size, or density of the impactor. This can help give a sense of purpose and focus to their experiments, and maybe encourage careful measurements.

Safety Note

Remind students of safety precautions to be taken, including the use of personal protective equipment (PPE), safety procedures, and engineering controls. PPE is to be worn during the whole activity, including setup, hands-on piece, and takedown.

Note: The graphing, interpreting, and predicting parts of this experimental activity will be discussed in Chapter 4,"Analyzing and Interpreting Data, Part 1: Graphing."

EXAMPLE INTERACTION

Student question: How are we supposed to measure what affects the size of the crater?

Teacher prompt: Well, what are you interested in? What might affect the size?

Student hypothesis: How big the meteorite is?

Teacher prompt: So are you interested in size or mass?

Student question: Are those different?

Teacher prompt: Well, look at the different balls and marbles. What do you think?

Student question (not following teacher's prompt): So do we just do different sizes?

Teacher prompt (trying a different tactic): One thing to think about is how you can change one thing, like the size, or mass, and keep all the others constant.

Student observation: But we found out that the crater is a different size if we drop the ball from different heights.

Teacher prompt: Why do you think that is?

Student explanation: It's moving faster if we drop it from higher up.

Teacher prompt: So, if you want to measure the effect of size of impactor on the crater size, do you want to drop the ball from different heights or from the same height?

Student explanation: Oh, I get it. If we change the height at the same time we can't know which caused the crater to change, the size of the ball or the height.

SUMMARY CHECKLIST FOR TEACHER AS PRACTITIONER OF SCIENCE

- Create an exploratory activity in which students can make mistakes and ask questions.
- Use ideas from these explorations as fodder to develop interesting and testable questions.
- Use discussion sessions to help refine experiments and think about the data.
- Plan to have your students do "dress-rehearsal" experiments to work out design kinks, to practice controlling variables and to practice making measurements.
- Plan time for the "real" experiments in which students measure the relationship between their chosen variables.
- Allow time for repeating and tabulating experimental results.
- Be ready to offer prompts, to encourage deep thinking, to be puzzled yourself, and to value the interests and contributions of the students.

TEACHER REFLECTION

You've had a hectic week with a busy classroom. Take time to reflect on those aspects of the project that worked—or didn't work—for you and your students. Also, take time to think about how this activity moves your students through your instructional sequence. You might want to talk with students about how their experiments fit into the bigger picture. How do the driving questions address the *NGSS* disciplinary core ideas and performance expectations?

If your driving question is related to the processes that wear down, reshape, and build up the surface of Earth and other planetary surfaces, how will your students apply what they've learned to further understanding of that question? For example, during the cratering experiments, students had to smooth the surface of the sand periodically. This action is a visceral metaphor for the effects of erosion on Earth, with its water cycle erasing any remnants of meteorite impacts. You can remind students of this experience when asking "Why doesn't the Earth have many craters compared with other planetary bodies?"

If the driving question compares processes that shape planetary surfaces, you might pursue what agents shape the surfaces of planetary bodies without a water cycle. What is the evidence for those other processes? On Venus and the Moon, lava flows cover older, heavily cratered surfaces.

Whatever your purpose for doing a controlled experiment, watch for ways to integrate student observations and findings into subsequent lessons. Classroom-size experiments whose primary purpose is to engage students in the practices of science are potent *NGSS*-consistent learning experiences, but they need to be fitted into the overall curriculum and into the context of core ideas of the discipline. We talk more about this important *curriculum narration* in Chapter 15, "Teacher as Curriculum Narrator."

FIELD OBSERVATIONS

In the late 1700s, geologist James Hutton claimed that the Earth was really old. Other geologists, such as Charles Lyell, affirmed this claim in the early 1800s. Hutton, Lyell, and the other geologists meant *really* old. Not a thousand years old. Not a million years old. Not even a billion years old. Older. So old that Hutton wrote in 1788 that "we find no vestige of a beginning, no prospect of an end."

The problem was that the geologists couldn't actually measure Earth's age. They based their claim on field observations—such as an unconformity at Siccar Point in Scotland where horizontal layers of red sandstone overlie vertical gray shale layers. They interpreted these and similar observations to tell the story of an area once under the sea whose sediments were uplifted into mountains. And then, grain by grain, the mountains washed into the sea. Eventually a sea marched back over where the mountains had been. Which in turn gave way to mountains again. That washed away to the sea. Over and over. They didn't know how long it took for all that to happen, but all that erosion, deposition, and mountain building had to take longer than a mere billion years.

The 19th century physicists shuddered. No measurement. No experiment. No numbers! Just qualitative field observations.

Lord Kelvin, one of the giants of physics—one of the giants of all of science—took up the problem. He knew that the interior of the Earth was hot. He calculated how hot based on measurements of heat being released from inside the Earth—just as you can estimate the temperature of a skillet on a stove by how much heat you feel radiating from it. With a wealth of experimental data about the movement of heat in spherical bodies, and an impeccable understanding of the mathematics of that movement, he calculated the very oldest that the Earth could possibly be if it had been cooling off since the beginning from an initial molten state. He placed the age of the Earth at about 100 million years, and certainly between 20 million and 400 million years (Young 1986), much too young for the geologists.

Measurements and lab experiments seemed to trump field observation. At the time, Mark Twain wrote this about Lord Kelvin:

> *Some of the great scientists, carefully ciphering the evidences furnished by geology, have arrived at the conviction that our world is prodigiously old, and they may be right but Lord Kelvin is not of their opinion. He takes the cautious, conservative view, in order to be on the safe side, and feels sure it is not so old as they think. As Lord*

Kelvin is the highest authority in science now living, I think we must yield to him and accept his views. (Twain 1903/1962, p. 212)

Didn't this mean that the geologists with their observation of field relationships were wrong? Ultimately, the answer proved to be no, it didn't.

The discovery of radioactivity at the end of the 1800s demolished Kelvin's calculation—he did not know about the heat added by radioactive decay and had not accounted for it. Ironically, radioactive decay later provided the means to measure the age of the Earth, which was billions of years old (Patterson, Tilton, and Inghram 1955). Much, much older than 100 million years.

The lesson from Hutton, Lyell, and Kelvin is that learning how to make valid scientific observations and create repeatable tests of ideas from field data are skills as important in earth science as the controlled experiment is in physics.

Walking across the land and mapping the underlying geology, as Hutton did at Siccar Point, is where much of geological investigation begins and is the foundation for understanding geological processes. For the scientist, mapping has led to understanding where to build canals, where to find coal, how rocks form, and the immense age of the Earth.

For the student, observing the geology and organizing those observations on a map provide perhaps the strongest insight into *how* we have discovered what we know about the earth beneath our feet.[C6] In learning to do fieldwork, the geologist develops the ability to recognize different kinds of rock, their associations and orientations in the ground, and how the clues in the rocks fit together into a coherent story. He or she must construct a three-dimensional mental image of the rocks in the earth and begin to see possibilities of how they came to be the way they are.

In learning to do fieldwork, science students have to hone in on important characteristics and discard extraneous details. Extraneous details are almost unlimited in the field, making this skill difficult to master. Mastery requires experience, including repeated opportunities to gather observations and make interpretations in the field.

FIELDWORK: A VISCERAL ENCOUNTER WITH NATURE

Kim Kastens, a marine geologist at Lamont-Doherty Earth Observatory, has done substantial research on the importance of fieldwork in education and emphasizes the fundamental educational differences between activities in the lab or classroom and fieldwork. Fieldwork engages students in interaction with raw observation, while classroom work involves interaction with various codified ideas (which she calls *inscriptions*):[C7] features of rock become symbols on maps, measurements become data on a graph, and conceptual ideas become drawings or explanations or mathematical expressions.

Those Are *My* Hills

The perky voice of Google's narrator on the highlights video lures the viewer in: "Let's begin by visiting the Lamont seamounts." The screen view plunges dramatically down through the sea surface, and brings us to a line of three seamounts off the coast of Mexico.

To my eye, these weren't just any old three seamounts. I know these seamounts well. Actually, I discovered them. ...

I had just that one night alone with my seamounts, before other human eyes got to see them, and before proliferating multibeam sonars had mapped thousands of other lovely seamounts worldwide.

And yet, twenty-eight years later, I still feel a tingle of proprietary ownership of these three seafloor lumps when I see their distinctive morphology. ...

Navigation is a very deep and old part of the human cognitive system, reaching way back into our evolutionary heritage as vertebrates. I'm toying with the idea that when discoveries are indexed into this deep-seated navigation system in real time, they get arrayed in mental space and attached to you in a way that is different than when you merely see the same information in a finished representation such as a map.

Source: Kastens 2011.

Only fieldwork, the very first step in the long process of codifying understanding, brings students into direct contact with the open book of nature. The importance of fieldwork is affirmed by *A Framework for K–12 Science Education: Practices, Crosscutting Concepts, and Core Ideas* (*Framework;* NRC 2012, p. 60):

> *[Students] need experiences that help them recognize that the laboratory is not the sole domain for legitimate scientific inquiry and that, for many scientists (e.g., earth scientists, ethologists, ecologists), the "laboratory" is the natural world where experiments are conducted and data are collected in the field.*

Diane Allen, a friend and colleague, came to geology later in her career, having begun as a biologist and life science teacher. Fieldwork proved to be the key to her struggle to pick up earth science on the fly. Later, she wrote to us about her experience:

> *As a teacher, I tried to have as many hands-on labs as I could to help students understand Earth Science topics. Although these types of learning activities are good, they are not as meaningful as a trip to observe natural events that demonstrate the principle ... [But] field work can be frustrating. As a student, I walked the state park*

area trying to figure out where the ancient beach areas were. It isn't as "clean or easy" as activities in the classroom, but the learning was very meaningful.

Field observations are also key to learning from the student perspective. Sabrina Cervantez, a ninth grader who participated in a GeoFORCE project in 2005, commented years later as a graduate student that "actually being able to go out there and get hands-on experience really made it stick in my mind." (GeoFORCE is a four-year geoscience program for high school students run by the Jackson School of Geosciences at the The University of Texas at Austin.)

STEPS TO ENGAGING STUDENTS IN FIELDWORK

Reinforce Classroom Learning With Field Experiences

Seeing geology in the field can reinforce classroom learning by providing both a visceral experience with the earth around us and a library of observations to call on when trying to make sense of classroom explanations and interpretations. Herbert Harold Read, a well-known geologist in the mid-20th century, once said that the best geologist is the one who has seen the most rocks (Read 1957).

We believe that field trips are valid learning activities even when the goal is simply to see lots of geology. Over the years, I (Russ) have taken hundreds of college students in introductory classes on two-day geology field trips. Rather than spend those two days taking measurements or mapping, we try to see as much geology as possible. We look at rocks formed in ancient rivers, or along the shore of an ancient advancing ocean, or in the hearts of ancient mountains. We examine the features left behind by glaciers or water erosion.

Every place on Earth has geology you can share with your students, even places without rocks. Back in the 1990s, I (Mary) got a service-learning grant for work in a local city park along the river. My students helped mark and clear trails, and in the process they got to see features along the river that we'd talked about in class: channel, floodplain, point bars. There had been a flood, and students could see its effects in the flattened grass and debris caught high in the trees. Some places had been scoured by the flooding river and others had been buried in new sediments that covered up the grass and cocklebures that grew along the banks.

A couple of students found a deposit of new sediment where the floodwaters had slowed enough for sediment to settle. The top of the mud had dried and looked solid. The first of them—an intrepid experimentalist—stepped onto the new sediment and immediately sank hip deep in soft, water-saturated mud. He managed to pull free, but his boot remained behind in the mud. The two came running up, incredibly excited at their visceral discovery of the processes they had studied in class. "Mrs. Colson! Mrs. Colson! We found some mud deposits, and I lost my boot in the mud!" A lost possession had never been more exciting or deposits of sediment more real. They never got the boot back—perhaps it will be found fossilized in some far-future layer of shale.

Simply getting out into the field to see some of the features and processes you've talked about in class engages your students in the field and gives them an understanding of how our conclusions in geology are tied to field observational data. Learn some of the most significant geological stories of your region, and take your students to see the rocks and features that tell those stories.

Test a Hypothesis With Field Observations

The practice of field science involves more than simply looking at features and connecting those features to classroom learning (the "stop, gawk, and talk on the rock" approach). With a bit more effort, the field observations can become data for your students. Data collection is a key practice of the *Next Generation Science Standards:* "Collect data … to serve as the basis for evidence to answer scientific questions" (NGSS Lead States 2013, Appendix F, science and engineering practice 3 [Planning and Carrying Out Investigations] matrix for grades 6–8).

Not all data come in the form of numbers from a laboratory. The whole Earth is the geologist's laboratory, and data are provided in the features produced by the experiments that nature has already carried out for us.

To be science, an idea or theory must be testable; that is, it must predict some new observation that could be shown to be false. This concept of falsifiability was brought to popular attention by philosopher Karl Popper in his 1935 book *Logik der Forschung* (later translated into English as *The Logic of Scientific Discovery*). One approach to emphasizing the data collection aspects of fieldwork is to encourage students to use field observations to test an idea or geological story you've told in class.

I (Russ) often begin my unit on the stories told by sedimentary rocks with consideration of the lake that once filled the Red River Valley. Most students here know about glacial Lake Agassiz, but my question for them is, "How do you know?" We launch into how sedimentary rocks tell us about past environments of deposition. With a knife and a bit of scraping, I show them the powdery particles of a shale and the sandy particles of a sandstone. We talk about where one sees mud and sand in modern environments. We break out the kitty litter basins, one for each group, half filled with water and half filled with various sizes of sediment. With a bit of sloshing to simulate waves, we watch the development of a sandy beach, a wave-cut scarp, and a muddy-bottomed lake—just like in the Red River Valley (see Figure 3.1, p. 40).

Figure 3.1

Experiment to create a wave-cut scarp, sandy beach, and muddy lake bottom in preparation for a field trip to see the field evidence for the former presence of glacial Lake Agassiz

Stretch Your Curriculum

Our goal in offering illustrations of field and experimental activities from our particular curriculum or region is not to suggest that this constitutes an appropriate curriculum for your classroom. Rather we offer these vignettes as inspiration and encouragement to go beyond your curriculum.

Often the curriculum inhibits the teacher from following up on student-initiated questions or questions of local significance—or even from offering personal experience or expressing interest in the material. Sherian Foster (1998, p. 2) wrote that "so called 'teacher-proof' curricula actually require the teacher to do little or no thinking about the ideas they are teaching."

That experiement sets up our field trip where we test *in the field* our emerging model for the story of the Red River Valley. We take samples from the former lake bottom with a small auger, and I use the traditional test for grittiness—grinding between my teeth—to confirm its fine-grained texture (I always express hope that the cows haven't been there recently). Then we visit the former beaches and discover the topographic scarps and sandy deposits at the edge of former Lake Agassiz, just like the ones developed in our experimental basin-lakes.

Lab, Field, and Model Inferences

The experimental activity in Figure 3.1 provides insight into how small-scale features form over time, but it provides no direct evidence that what we see in the kitty litter basin actually happens during real earth processes. In contrast, field observations reflect real earth processes, but often do not allow us to see features develop over time. It is by combining field and lab observations at different temporal and spatial scales that earth scientists construct models for earth processes.

Although the activity in Figure 3.1 contains elements of an experiment (simulating an erosional process and observing the features that develop), it also contains elements of a physical model (we aren't controlling and measuring variables so much as we are attempting to re-create natural processes and features at a small scale). Thus, modeling, which we talk about more in Chapters 6 and 7, is an important part of this activity. Experimental inferences, field inferences, and inferences from physical models often combine to inform earth scientists' understanding of earth processes and features.

The focus on testing an emerging idea allows students the opportunity to see field observations as data. Your region will have a different story and a different hypothesis to test. What are the key geological stories of your region? Take a key story from your region and ask, What prediction does it make? What observation will be a key confirmation of that story? Where are the rocks that tell that story?

Set up your students' understanding of the concepts through activities in the classroom, and then go test those ideas in the field.

Focused Field Discovery

If you have more time, it's worthwhile to go beyond simple field observation into field study. Most regions have the option for some type of field study that is accessible in a short field trip. For example, students can examine river meanders, point bars, river migration, oxbows, and erosion or deposition. They can study beach ridge and runnel formation, swash zone processes, tidal processes, wave patterns, ripple and dune formation, erosion, and deposition.

Transforming casual field observations into field experiments requires attention to observational sampling ("What do you notice?") and selection of sampling areas and sampling methods. As stated in the *Framework* (NRC 2012, p. 59),

> *Planning for controls is an important part of the design of an investigation. ... In many cases, particularly in the case of field observations, such planning involves deciding what can be controlled and how to collect different samples of data under different conditions.*

Encourage students to make observations and *document* those observations. Documentation is an important step in turning casual observation into a field study. Encourage them to consider what the observations tell us. But remember that during a short field trip, students will not be able to figure it all out, and they will need significant prompting in what to notice and what it might mean.

HOW MUCH GUIDANCE, HOW MUCH FREEDOM?

In fieldwork, as in classroom science, the teacher needs to find the balance between providing so much guidance that students don't have the opportunity to explore and providing so little guidance that the experience does not engage students in any meaningful process of science. How much guidance to provide in doing fieldwork is particularly challenging because (1) field opportunities are typically short and (2) there are a great deal more nonrelevant observations to compete for student attention in the field. The article "Field-Based Learning" (Kastens and Manduca 2009) on the SERC (Science Education Resource Center) website run by Carleton College in Minnesota emphasizes the difficulty of knowing what to notice in the field:

> *In a school laboratory, most objects on the lab table are relevant to the inquiry at hand, whereas for a field-based inquiry most objects in view are not relevant, and it is not obvious to the novice which are relevant and which are not.*

This complexity in doing field-based teaching might come into conflict with other educational directives that emphasize student discovery. For example:

> *Resist the urge to jump in and correct misconceptions or answer questions. The process of students doing their own thinking is what energizes this learning method (inquiry)! Allow students to work through and modify their ideas. (Richards, Johnson, and Nyeggen 2015, p. 57)*

Finding the right balance between guidance for the students and giving them freedom to explore might take some practice. We offer some of our experiences in the following paragraphs.

Around 2008, I (Mary) developed an interdisciplinary field activity along our local (Minnesota) Buffalo River under the theme "The Place We Call Home." In addition to photography exercises as part of their art class, my students made observations and took measurements along the river as part of their math and science classes.

Activities that resembled classroom problems worked quite well. Students drew topographic profiles across the Buffalo River valley, identifying the river channel, floodplain, and bluffs at the valley wall. Students took measurements of stream velocity, channel width and depth, and calculated stream discharge values.

What worked less well was the observational field data exercise. They were supposed to find and map locations of particular river features—point bars, cutbanks, and oxbows—and how those features relate to faster and slower water flow and to processes of erosion and deposition. But students were on their own in this exercise, and they didn't have a way to process or evaluate what they saw. None of them found the oxbow, and finding the other features was sporadic at best.

My takeaway message was that fieldwork is so unfamiliar to students that even when they have encountered all the concepts in the classroom, they need substantial guidance and prompting in the field to recognize key features and separate those from the blizzard of other observations crying out for their attention in the "great outdoors."[C8]

I (Russ) had a similar experience with my first college field trip along the Buffalo River. I took a different approach to engaging students in the study of sediments, focusing on the classic geology premise "the present is the key to the past." My students and I looked for features in the modern sediment that might become clues to how environmental signatures are recorded in the rocks. I expected them to find asymmetric ripple marks with finer particles on the downstream side, to discover differences in sediment size between the swift center of the stream and slower edges, to find cobble pavements on point bars, and perhaps to spot mud cracks or footprints along the shore—all things we'd talked about in class. I hoped a few of them might notice the orientation of shells in the sand along the shore. Shells tend to get turned with their concave side down in moving water, a more stable configuration. This can give an indicator of the original "up" direction in a rock layer.

No one even noticed that there were any shells in the sand until I pointed them out, and the other features were nearly as invisible. The world was too full of things to notice, and they had too little practice at noticing the little features in the sediment.

Now, when I try this exercise, I give them some initial prompts: "Notice the shells in the sand and how they are oriented. What do you think that means?" "Think about the things we've talked about in class, such as ripples or mud cracks, can you find some of them? What do they say about the environment they form in?"

Even with the prompts, we need to gather together to talk about what we've seen. Students need an expert mentor on the spot to point out what they are seeing and offer prompting observations and questions so they can learn what to pay attention to and how to assess whether their observations are relevant.

MAPPING THE WORLD

Ideally, fieldwork in geology should do more than reinforce learning, test hypotheses, or study modern processes.

The example activity design in this chapter includes a student mapping exercise. Before diving into that exercise, let's consider a classic study in geological mapping from the 1700s. Field studies from history can be a way to give students some experience with fieldwork inside the classroom.[C9]

A key feature of the geological map is the *contact*—the boundary between different types of rock. Recognizing the contacts between different lava flows allowed Nicolas Desmarest (the guy who said, "go and see") to prove that some rock comes from volcanoes.

Scientists had studied the volcanoes of central France for decades, and many scientists believed that some rocks were of volcanic origin even in regions where there were no active volcanoes. However, the matter remained unresolved until Desmarest's field investigation in the late 1700s (Adams 1938; Geikie 1962; Taylor 2009).

Desmarest carried out years of study, walking over the landscape west of Clermont, France, and mapping the relationships between what he believed were old lava flows and the extinct volcanos. He recognized that the relationships between flows—some on top of others, some flooding eroded valleys that cut through older flows—indicated that the flows came from different periods of time, which he called epochs. For each time period and flow, he mapped out the orientation and locations of rocks, demonstrating clear connections with particular extinct volcanoes west of Clermont. One of his maps,

Figure 3.2
Nicolas Desmarest's map

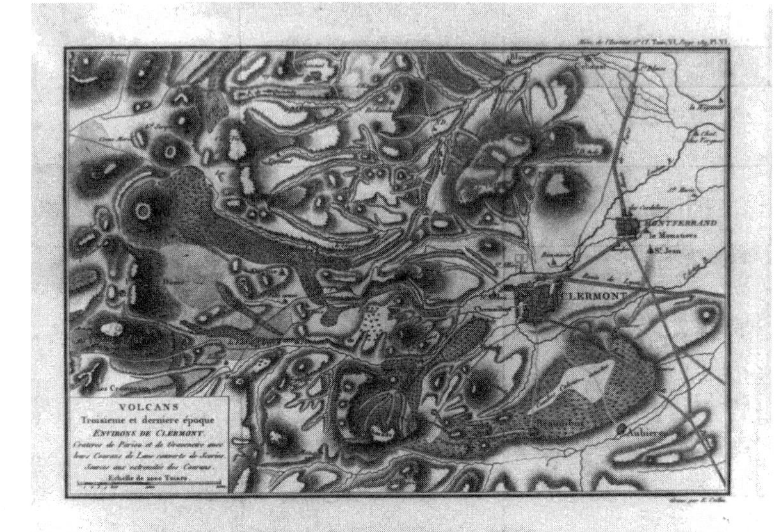

This map from Desmarest's 1806 publication *Mémoire sur la Détermination de Trois Époques de la Nature par les Produit des Volcans* shows some of his field evidence supporting the connection between rocks he believed were old lava flows and the extinct volcanoes west of Clermont, France. Lava flows are illustrated with a stippled pattern. One lava flow associated with the volcano in the left center is seen extending toward Clermont.

Source: Public domain. Available from the Library of Congress, *www.loc.gov/resource/cph.3b41449.*

showing the connection between a flow (stippled on left center of the map) and one volcano, is shown in Figure 3.2.

A Story of Black Stone

In 1774, after 11 years of field studies, Nicolas Desmarest published his observations and interpretations of the basalt layers and their association with volcanoes of the Auvergne district of central France. His words (quoted in Adams 1938) reflect the measured excitement of his discovery:

> *I followed the thin sheet of black stone and recognised in it the characters of a compact lava. Considering further the thinness of this crust of rock, with its underlying bed of scoriae, and the way in which it extended from the base of hills that were obviously once volcanoes, and spread out over the granite, I saw in it a true lava-stream which had issued from one of the neighbouring volcanoes. With this idea in my mind, I traced out the limits of the lava, and found again everywhere in its thickness the faces and angles of the columns, and on the top their cross-section, quite distinct from each other. I was thus led to believe that prismatic basalt belonged to the class of volcanic products.*

The map, however, was not his primary evidence, but rather a way to portray his evidence for others. In fact, he did not even publish the map until more than 25 years after the first report of his studies. Instead, his evidence was found in the field relationships among the rocks: relationships that established which flows were older and which were younger, how the lava flows followed valleys that existed at the time of the lava flows, and how flows could be traced back to particular volcanoes. He reported locations where the various relationships could be found in the rocks, allowing future investigators to check his work.

FINAL THOUGHTS

Field opportunities are hard to implement, and students struggle mightily to see features and understand concepts in the field. However, fieldwork provides insight into the practice of geological science that cannot be gained in any other way, exposing students to the blizzard of observations that scientists have used to understand our world and universe. If our goal were simply to convey factual information, talking in the classroom would work fine. But the goal is to understand the enterprise of science and participate in it. In geology, that requires field experiences.

Consider the different levels of field opportunity that we've presented—reinforcing classroom learning with field observations, testing a hypothesis with field observation, focused field study with documentation, and geological mapping—and do as much as your resources, energy, and time allow, remembering that no amount of classroom activity, however skillfully implemented, can completely replace field experiences. We include in this assertion many well-designed computer-assisted simulations of field experiences; although these simulations can be an important part of classroom activities, they offer only a controlled interaction with scientists' interpretation of data and not exposure to the uncontrolled complexity of true field study.

REFERENCES

Adams, F. D. 1938. *The birth and development of the geological sciences.* New York: Dover.

Foster, S. 1998. New perspective–new practice. *Principled Practice in Mathematics and Science Education* 2 (1): 1–10.

Geikie, A. 1962. *The founders of geology.* 2nd ed. New York: Dover.

Kastens, K. 2011. You map it, you own it. Earth and Mind: The Blog. *http://serc.carleton.edu/earthandmind/posts/youmapit.html.*

Kastens, K., and C. Manduca. 2009. Field-based learning. *http://serc.carleton.edu/research_on_learning/synthesis/field.html.*

National Research Council (NRC). 2012. *A framework for K–12 science education: Practices, crosscutting concepts, and core ideas.* Washington, DC: National Academies Press.

NGSS Lead States. 2013. *Next Generation Science Standards: For states, by states.* Washington, DC: National Academies Press. *www.nextgenscience.org/next-generation-science-standards.*

Patterson, C., G. Tilton, and M. Inghram. 1955. Age of the Earth. *Science* 121 (3134): 69–75.

Popper, K. 1935. *Logik der Forschung.* Vienna: Verlag von Julius Springer.

Read, H. H. 1957. *The granite controversy: Geological addresses illustrating the evolution of a disputant.* New York: Interscience.

Richards, J., A. Johnson, and C. G. Nyeggen. 2015. Inquiry-based science and the *Next Generation Science Standards*: A magnetic attraction. *Science and Children* 52 (6): 54–58.

Taylor, K. L. 2009. Desmarest's "Determination of some epochs of nature through volcanic products" 1775/1779. *Episodes* 32 (2): 114–124.

Twain, M. 1903/1962. Was the world made for man? In *Letters from the Earth: New uncensored writings by Mark Twain,* ed. B. DeVoto. Robbinsdale, MN: Fawcett.

Young, L. B. 1986. *The unfinished universe.* New York: Simon & Schuster.

EXAMPLE ACTIVITY DESIGN

FIELD OBSERVATION: LAYER-CAKE ROCKS

An Introduction to Geological Mapping

Evidence in layers of rock tells the story of how a place changed through time. This example activity design (EAD) involves mapping of horizontal rock layers, a first step in understanding how we know about the changes on Earth through time. Students will gain insight into how rock occurs in layers that can be traced laterally over a region. They will see how rock layers are stacked on top of each other stratigraphically. They will gain experience with three-dimensional visualization of rocks in the field. They will practice gathering and interpreting spatially associated data.

Some areas don't have layered rocks (such as our region in northwestern Minnesota). However, many places do have layered rocks, and if you live in such a place this EAD might work for you. Many other locations could have rocks that are layered but the layers are tilted. Other places might have nonlayered rocks, like metamorphic or igneous rocks. These areas might support a geological mapping project, but the specific prompts we offer below would need some modification.

This EAD presumes that students are already familiar with topographic maps.

TEACHER PREPARATION AND PLANNING

One challenge of this EAD is to find the right location for the project. You need to find a place where layers of rock are exposed in multiple locations, where at least some boundaries between rock layers are exposed, where students can have safe access, and which encompasses a small enough area so that students can explore it in the time available, which typically won't be more than a day or two and may be as short as two to three hours.

Field Activity Safety Planning

The National Science Teachers Association Safety Advisory Board has prepared a "Field Trip Safety" document to help teachers in planning field activities. Highlights of the document are given here; you can view the entire document at *www.nsta.org/docs/FieldTripSafety.pdf.*

Preparation of students: The teacher must identify—and communicate to students in advance—clear learning outcomes for the site and the strategies for accomplishing them as a result of the field trip. Issues include student behavioral expectations, addressing issues with special needs students, communications with parents prior to the trip, and planning for emergency first aid. Before the field trip takes place, the teacher should visit the destination site to determine if there are any safety or health hazards and how to address them. Students should never be taken on a field experience if the location has not been visited by the teacher ahead of time!

Permission: "All school and district policies and procedures must be followed when planning a field trip. A parent and/or guardian must provide written permission for students to participate in any educational experience off school grounds." This section notes that there may be additional safety concerns that need to be addressed, such as appropriate attire and risks involved with native plants and animals. It is important to be aware that written permission for a child to participate is not a waiver of liability for student safety. This section also brings up the issue of transportation to the site: "Teachers should not transport students in their own vehicles, even if parents or guardians have given written permission to do so."

Supervision: What should be the ratio of chaperones to students? How should chaperones be trained for the trip? What is the means of communications in the field between the teacher and chaperones? Who should be responsible for student medications and first aid equipment/supplies? Is there a listing of contacts for emergency response personnel should there be an accident?

Outdoor field experiences: There are a number of critical items that need to be planned for during a field experience. "Before the field trip, field trip supervisors should create a checklist of needs that may occur outdoors. These include, but are not limited to, parking, availability of drinking water, washing and lavatory facilities, trash disposal or recycling, and other needs. These needs can best be determined by a visit to the site prior to the field trip." Some other issues that may need to be addressed are changing weather conditions, removal of artifacts, cell phone reception, and the need to seek immediate shelter because of a safety or security issue.

First, read up on the geology of your area. Internet searches will yield information about almost any location. Local parks often have information about geology. You might acquire a geological map of your region to assess the possibilities. (The entire United States has been mapped at the scale of 1:24,000 in the 7.5 minute quadrangle series by the U.S. Geological Survey [USGS], which local libraries or universities might help you locate or, with more effort, might be found at USGS databases such as *http://ngmdb.usgs.gov/ngmdb/ngmdb_home.htm.)* The *Roadside Geology* series of books (Mountain Press) has information about the geology of many areas in most states.

Once you get enough information to ask good questions, talk with rangers at local parks or touch base with local or regional college geology faculty to gain more ideas.

Students will need topographic maps of the region (the geological map will be constructed on top of the topo map). They will also need compasses, GPS-capable units, and writing tools.

The geology of every region is different, so every field experience will be different, but to help you see how this EAD might evolve, we offer a simulated student map for an activity based on mapping of Quarry Hill Nature Center in Rochester, Minnesota (see Figure 3.3).

EXAMPLE PROMPTS AND LIMITING OPTIONS

- Make sure that students know the extent of the map area and that they have been coached in safety. Have students work in teams, and have someone present with all teams to coach on both geology and safety.

Figure 3.3

Topographic maps for recording initial observations and a simulated student geological map

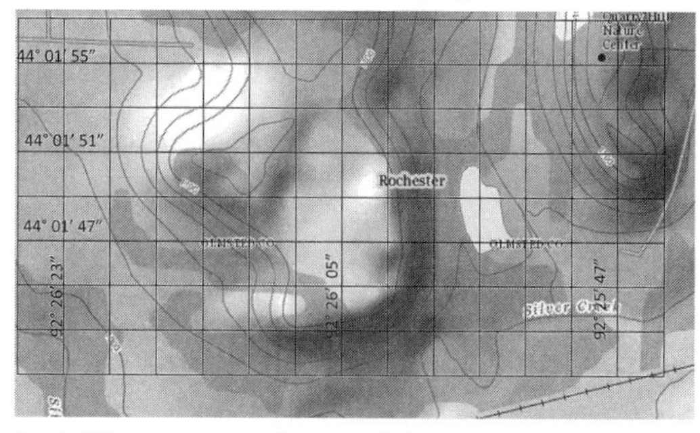

(a) USGS topo map of Quarry Hill Nature Center with overlain latitude and longitude grid (created from Google Maps locations using a PowerPoint table and a few scaling calculations).

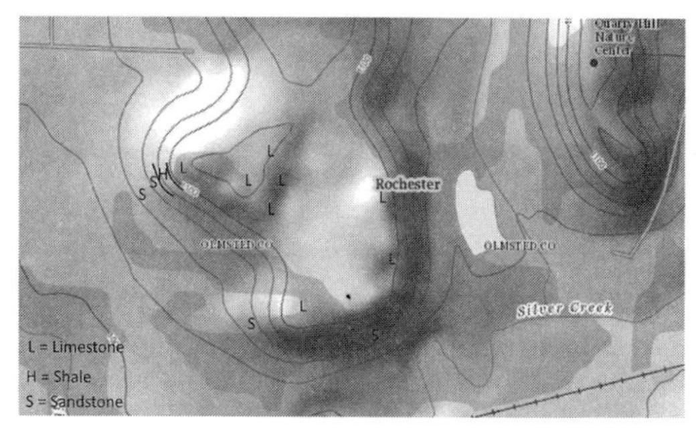

(b) USGS topo map of Quarry Hill Nature Center with locations and type of rock outcrop such as might be observed and measured by students. Observed contacts between rock layers are shown as short lines.

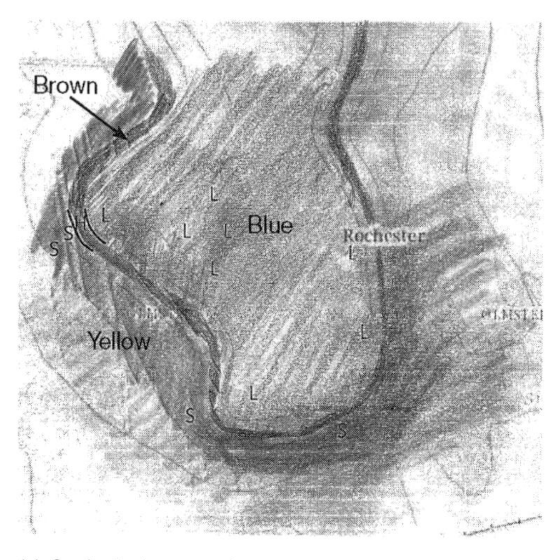

(c) Geological map such as your students might construct, based on outcrops and contacts observed in the field. Yellow = sandstone, brown = shale, blue = limestone.

- You might prompt students to find and mark on their map the location of each outcrop of rock. For example, if you find a sandstone outcrop at a particular location, mark the sandstone symbol at that location on the map. You might do the first outcrop with your students to show them how to find the location by GPS and how to plot it on your map. (If rocks are tilted, students might measure the angle and direction of tilt using a Brunton compass, marking the measurement with a strike and dip symbol on their map).
- Are students making sure that their outcrop is indeed an outcrop and not a large block of rock fallen from the hillside above?
- Pay particular attention to contacts between rock layers. This could be marked on the map with a short line marking the contact between layers with the appropriate rock type marked on either side of the line (e.g., sandstone on one side, shale on the other). Students need to be aware of direction (compass) and where they are on the map, to make this marking. This is an important part of their mental construction of a three-dimensional image of the geology.
- When the entire map region has been walked, with outcrops identified and coded onto the map, students might compare maps to see if they have differences. Any differences can be checked while still in the field. Was someone lost and therefore didn't record their data in the correct location? Did some people find outcrops that others missed?
- Maps can be completed in the classroom. Given where outcrops occur, where are those rocks likely to extend beneath cover of soil or vegetation? Where should you draw the contact between rock layers?
- One big clue for drawing in horizontal rock layers is that the contacts between rock layers must be parallel to topographic lines, which are also horizontal (being lines of equal elevation).

EXAMPLE INTERACTION

Student question: Is this an outcrop?

Teacher prompt: Well, what are the clues?

Student observation: It seems to go down into the earth.

Teacher prompt: Is the rock oriented like all the other rocks we think are outcrops?

Student observation: I think so.

Teacher prompt: Well, why don't you measure its dip? See if it's the same.

Student observation: I can tell that it's flat, the layers in it are horizontal.

Teacher prompt: Well, does it make sense that it should be an outcrop? Does it fit in with where you are mapping this kind of rock?

SUMMARY CHECKLIST FOR TEACHER AS PRACTITIONER OF SCIENCE

- Review field trip safety planning using the "Field Trip Safety" document from the National Science Teachers Association Safety Advisory Board.
- Let students explore for a while; they don't have to start measurements right away.
- Help them recognize the difference between float and bedrock—it's quite difficult for a beginner.
- Make sure students can tell the difference between the different kinds of rock in the region. Not all sandstones look the same, nor do shales or limestones.
- Do the first outcrop together—show how to find the location by GPS and how to plot it on your map.
- Give students the freedom to make some mistakes, but give them a chance to recognize those mistakes when the whole class comes together to compare results.

TEACHER REFLECTION

The students have had a chance to gather spatially related field data and see the patterns that result on their own geological maps. In drawing the contacts on the map, students have begun the process of interpreting their data. You might follow this field experience with classroom activities using maps, geological maps, geological cross-sections, stratigraphic columns, or activities with crosscutting relationships. You might include classroom activities with geographic information systems—which will now be more meaningful since they will understand better how the actual features in the field are reflected in the symbols shown on a map.

Mapping has implications in fields beyond geology or mapping of rocks. We will be considering geological maps and cross-sections in greater detail in Chapter 5 ("Analyzing and Interpreting Data, Part 2: Maps and Cross-Sections") and Chapter 11 ("Stories in Rock Layers").

ANALYZING AND INTERPRETING DATA, PART 1

GRAPHING

To understand a graph, it's important to remember that graphs represent real-world observations. Graphing is a language that turns an initially puzzling blizzard of real-world data into a coherent story.

Back before stories went viral on Twitter, anonymous jokes circulated by e-mail. I (Russ) received one such e-mail on March 20, 1999, that illustrated the complexity of crunching through a large body of observations to reach a synthesizing conclusion.

> *Sherlock Holmes and Dr. Watson went on a camping trip. After a good meal and a bottle of wine they lay down for the night, and went to sleep. Some hours later, Holmes awoke and nudged his faithful friend.*
>
> *"Watson, look up at the sky and tell me what you see."*
>
> *Watson replied, "I see millions and millions of stars."*
>
> *"What does that tell you?" says Holmes.*
>
> *Watson pondered for a minute. "Astronomically, it tells me that there are millions of galaxies and potentially billions of planets. Astrologically, I observe that Saturn is in Leo. Horologically, I deduce that the time is approximately a quarter past three. Theologically, I can see that God is all powerful and that we are small and insignificant. Meteorologically, I suspect that we will have a beautiful day tomorrow. What does it tell you, Holmes?"*
>
> *Holmes was silent for a minute, then spoke.*
>
> *"Watson, you idiot. Someone has stolen our tent."*

With the last sentence, obscure elements, such as why Holmes woke him up and how a unique solution to the question is possible, become clear. There is a point in an investigation when new observations are no longer bewildering but fit neatly into a growing understanding. At that point, we gain the ability to predict future observations.

ORGANIZING THE BEWILDERING

Graphs bring large numbers of bewildering observations together in a single, coherent picture. They provide a means to explain how variables are related to each other and can be a starting point for evaluating possible cause-and-effect relationships. They provide a way to make predictions based on correlations between variables, which can be used to test whether the correlations are causative or not. To develop those explanations and make those predictions, we need to understand the language of the graph.

In this chapter, we offer a quick journey through reading a graph. As you review the case study and the subsequent sections on working with graphs, pay attention to the habits of mind that you use and think about how you might guide students in developing those habits.[C10]

Considerations in Learning the Language of the Graph

- Does this graph plot variation in one parameter against another? Most graphs used in experimental science do, but bar graphs don't.
- What are the labels on the axes? How do the values apply to the real world? Do you need students to explore any aspect of the labels? For example, on the Keeling Curve discussed in this chapter, would your students understand the idea of *concentration in ppm?*
- Is there a correlation or not?
- If there is a correlation, is the relationship linear or not linear? What does that mean in terms of real-world behavior?
- Is the slope positive or negative? What does that mean in the real world?
- What is the value of the slope? What does that mean in the real world?
- If there is a correlation, is there a causal relationship? What is your evidence? What predictions can we make from that causal relationship?
- If there appears to be a correlation, is it real—that is, is it greater than the uncertainty in the measurements? Is the scatter in the data around a correlation trend smaller than the change in value due to the correlation?

A CASE STUDY IN GRAPH READING

Let's consider a classic graph from the earth sciences that continues to have a significant impact on national policy. The Keeling Curve shown in Figure 4.1 is a record of the concentration of carbon dioxide (CO_2) measured at the top of Mauna Loa since 1958.

Figure 4.1

The Keeling Curve charting the measured change in carbon dioxide in the atmosphere at the summit of Mauna Loa in Hawaii

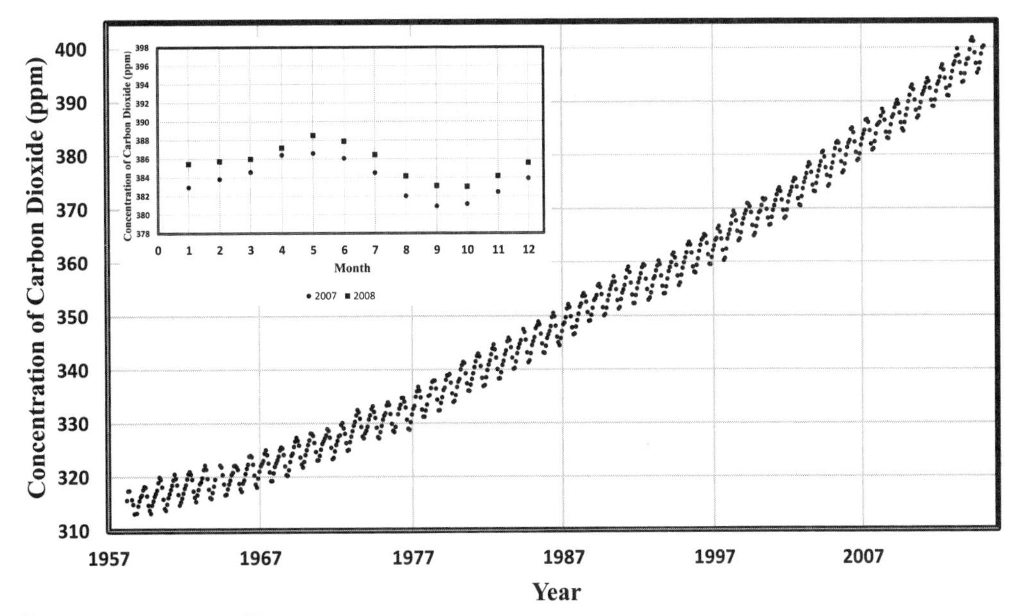

Note: ppm = parts per million.

Sources: Based on data from *ftp://aftp.cmdl.noaa.gov/products/trends/co2/co2_mm_mlo.txt.* Carbon dioxide data were collected by David Keeling of the Scripps Institution of Oceanography (SIO) before 1974 and by both the SIO and the Earth System Research Laboratory at the National Oceanic and Atmospheric Administration since then.

What does this graph tell us? We see that the amount of CO_2, in parts per million (ppm), is plotted on the *y*-axis and time is plotted on the *x*-axis. The concentration of CO_2 since 1958 is not constant but increases to the right, telling us that the concentration of CO_2 is increasing with time.

We can see that the trend is not linear, but rather the slope of the trend becomes steeper with time—notice that you can't draw a single straight line through data for all years. This means that the rate of increase in CO_2 has itself been increasing, which is consistent with humans burning more fossil fuels and making more cement in the 2000s than in the 1960s (although this correlation does not prove that human activity is the only cause of this increase). We can see that the trend in the data does not intersect the *y*-axis at zero on the left-hand side of the graph, meaning that some CO_2 was already present in the atmosphere in 1958.

We also notice a peculiar sawtooth pattern to the variation in CO_2 concentration. From the inset on the graph, we see that each rise and fall is exactly one year long. Each year, the concentration increases from about October (month 10) to May (month 5) and decreases from about May to October. Like Sherlock's missing tent, this observation makes sense when we realize that from May to October, plants in the Northern Hemisphere are consuming CO_2 during their growing season. The zigzag pattern of the Keeling Curve proves a strong *correlation* between variations in CO_2 and time of year. When combined with other evidence for seasonal variations in CO_2 due to plant respiration, this correlation provides a strong case that the zigzag variations are *caused by* seasonal variations in plant growth.

Discovery of the Carbon Dioxide Annual Cycle

In the mid-1950s, Charles David Keeling developed a system for measuring the carbon dioxide (CO_2) concentration in air. As a postdoctoral fellow at the California Institute of Technology, he measured the CO_2 concentration in forests and grasslands, places where the air would be more affected by natural biological activity than by human activity. He discovered that air impacted by the forests and grasslands showed a regular daily cycle in CO_2. He was able to relate the cycle to respiring plants giving off CO_2 at night.

In March 1958, Keeling and his team began making CO_2 measurements on Mauna Loa. Unexpectedly, he found that the concentration of CO_2 at Mauna Loa rose from March until May and then declined until October. The same pattern repeated in 1959. In Keeling's own words: "We were witnessing for the first time nature's withdrawing CO_2 from the air for plant growth during summer and returning it each succeeding winter" (Scripps Institution of Oceanography 2016a, 2016b).

The Keeling Curve is often combined with measurements of world air temperature to show a correlation between CO_2 concentration and average global temperature (see Figure 4.2). In this graph, we see a strong positive correlation between CO_2 and temperature. A positive correlation means that they are both increasing together. We might jump to the conclusion that the increase in CO_2 is *causing* the increase in temperature. However, since these data come from field measurements and not from a controlled experiment in which only one variable is changed and another variable responds, the correlation does not necessarily imply causation.

The distinction between *correlation* and *causation* is an important one for students to consider. I (Russ) often present the difference this way: "Every morning I get up, and also

Figure 4.2

Correlation between the average annual concentration of carbon dioxide measured at the summit of Mauna Loa and average annual world surface air temperatures (T)

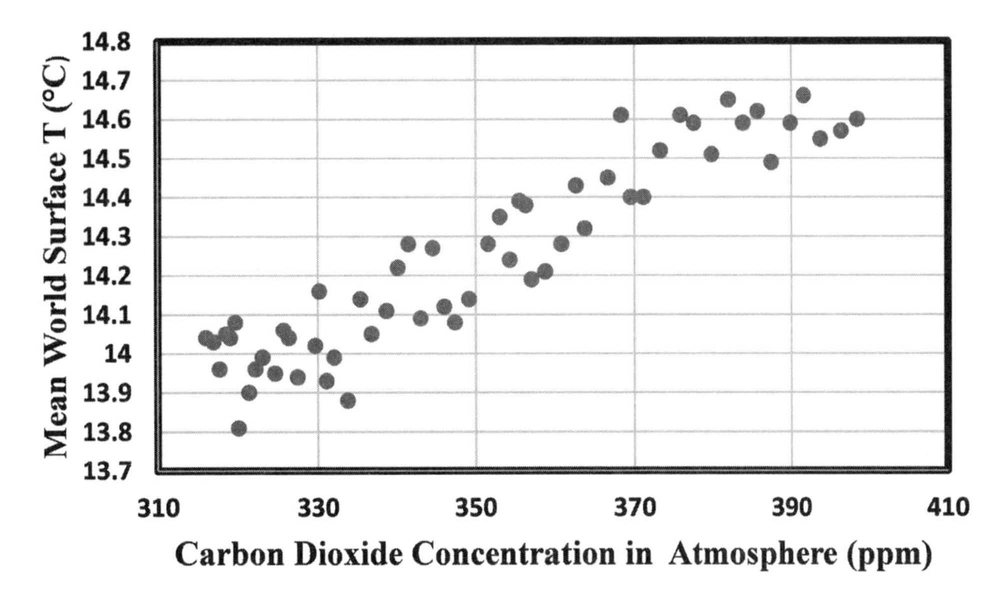

Sources: Based on data from *http://data.giss.nasa.gov/gistemp/graphs_v3/Fig.A2.txt* (for temperature) and *ftp://aftp.cmdl.noaa.gov/products/trends/co2/co2_annmean_mlo.txt* (for carbon dioxide). Carbon dioxide data were collected by David Keeling of the Scripps Institution of Oceanography (SIO) before 1974 and by both the SIO and the Earth System Research Laboratory at the National Oceanic and Atmospheric Administration since then.

every morning the Sun rises. Every morning! A clear correlation. Wow. I must cause the Sun to rise!"

Correlation simply means that two things happen together. However, a correlation does present the possibility that one change causes the other. The possibility that I cause the sunrise can be tested by making a prediction and seeing if the prediction is born out: If my getting up in the morning causes the sunrise, then we predict that if I get up at different times of the day the sunrise should follow (I have tried this, and, disappointingly, it doesn't work).

For the correlation shown in Figure 4.2, the case for causation is strengthened by other experiments showing that CO_2 absorbs infrared radiation—radiation that might otherwise escape into space—and so *could cause* the temperature of Earth to rise. Testing for causation by making a prediction such as we did with me and the sunrise would require that we produce different amounts of CO_2 over long periods of time and see how it affects world temperature. This is an experiment which, for better or worse, we are in the process of carrying out.

From Figure 4.2, we can also get a feel for uncertainty in the data and how that uncertainty affects our confidence in the correlation between temperature and CO_2. *Uncertainty* is one of the most important ideas of science. It's important for students to understand that correlations are never perfect, both because no measurement is ever exact and because real variation that is not explained by our correlation model might exist in natural data. Uncertainty is an estimate of how imperfect our measurements or correlation models might be.

As simplified in the following text, a graphical analysis of uncertainty can be done by students even without doing a lot of math. Although graphical statistics don't have the rigor of mathematical analysis, students get a feel for what uncertainty means and how it can be inferred from data.

Notice that values for temperature in Figure 4.2 vary from 13.8°C to 14.7°C and CO_2 concentrations vary from about 315 ppm to nearly 400 ppm. Although the variations in temperature and CO_2 are correlated, they are not perfectly correlated—that is, we can't draw a single smooth line or curve that passes through all the data points. We can think of the total variation as being the sum of the variation that goes along with the correlation trend (the variation explained by a curve through the data), plus any "extra" scatter around that trend (the amount of variation from the curve). This "extra" variation can be caused by many factors such as imprecisions in our measurements or real variation in the data that are correlated with other, unidentified causes. This extra variation gives us a feel for the uncertainty in the data.

You might have students consider the magnitude of the variation that is "explained" by the correlation and compare it with the magnitude of the remaining scatter around the trend. The smaller the magnitude of the scatter compared with the variation explained by the correlation, the more confidence we have that the correlation is real and not a random artifact of data variations. For example, in Figure 4.2 the total variation is a bit less than a degree, whereas the scatter is about 0.2 degrees. You might also have students draw a "maximum reasonable slope" line and a "minimum reasonable slope" line through the data. From these two trends, students can get a feel for how much the uncertainty might affect the slope of the trend.

NOT JUST A TECHNIQUE

Although students need to develop a number of technical skills to make graphs, such as proper scaling and proper plotting, the mechanics of drawing a graph are not the main point of data analysis, interpretation, and graphing as proposed by the *Next Generation Science Standards (NGSS).* For example, science and engineering practice 4 matrix for grades 6–8 in Appendix F of the *NGSS* includes this element: "Construct, analyze, and/or interpret graphical displays of data and/or large data sets to identify linear and nonlinear relationships" (NGSS Lead States 2013).

Nor do the technical mechanics of drawing a graph address the *NGSS* crosscutting concept of Cause and Effect: Mechanism and Explanation, as described in *NGSS* Appendix G for grades 9–12: "students understand that empirical evidence is required to differentiate between cause and correlation and to make claims about specific causes and effects."

In our experience, students are much better at plotting graphs than at understanding what the graphs mean. This struggle that students experience in connecting graphical data to lab and field observation has been reported by teachers and education researchers for many years, as pointed out, for example, by Robert Beichner in his 1994 paper "Testing Student Interpretation of Kinematics Graphs." Yet, deriving meaning from the graph is the main point of the graph. What does the slope represent? What does the y-intercept mean? How does the information on the graph relate to the natural system under study?

In 2014, I (Mary) had my students study graphs showing the rate of weathering of tombstones in Sydney, Australia (see Figure 4.3). I began with simple exercises such as considering whether the slope was positive or negative and what that might mean in terms of the rock changing through time. Then, I had students calculate the slope (millimeters per year). For technical reasons, some students ran into difficulties in calculating slope. In math, students sometimes determine slope by counting squares in a grid; some

Figure 4.3

Weathering rate of marble tombstones in Sydney, Australia

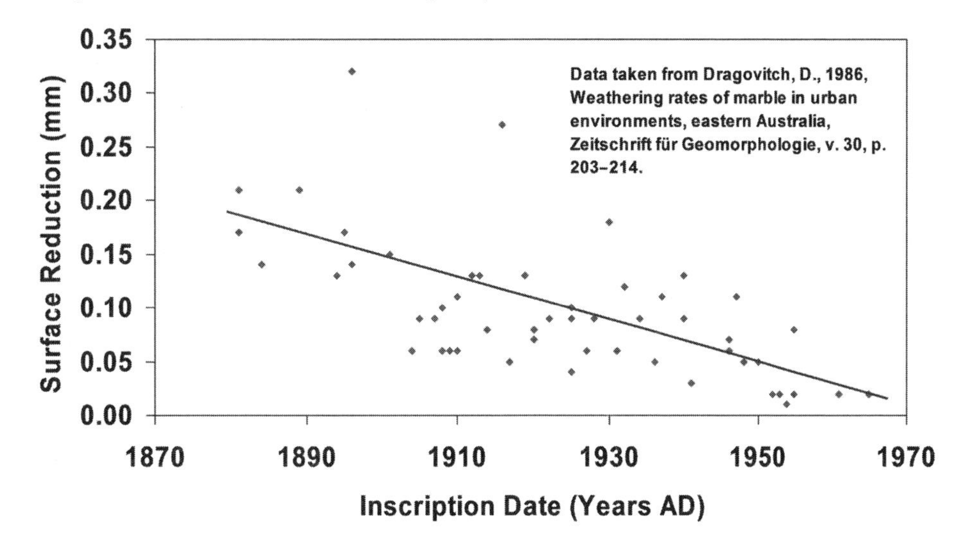

Source: Graph and activity developed by Rebecca Teed and published as "How Fast Do Materials Weather" in *Starting Point: Teaching Entry Level Geoscience*, available at *http://serc.carleton.edu/introgeo/interactive/examples/weatrate.html.*

students couldn't find slope because there was no grid on the graph. Other students tried to measure the rise and run with a ruler—not realizing that the scales of the two axes were quite different and plotted different types of values (millimeters vs. years). In science, graphs rarely plot dimensionless numbers against each other. To transfer math graphing skills to the science classroom, students need guidance in thinking about labels and scales on the axes.

Our conclusion is that students need to practice reading scientific graphs even if they know the mechanics for creating them. Students need practice thinking of numbers not as dimensionless values but as values connected to real-world measurements such as mass, temperature, and solubility. Students need practice relating trends on graphs to real-world processes and relationships.

We view reading a graph as a language skill, like learning to read a book, rather than a technical skill, like learning to write the letters of the alphabet. In the following section, we offer some ways to practice that language skill.

GRAPH-READING CHALLENGES

In learning to read and write in the language of the graph, students might practice translating graphs into real-world understanding or translating real-world data into graphs.[C11]

For example, in the activity with the tombstones, after students calculated the slope and associated that slope with a rate of weathering (about 2 mm/100 years), one student commented that it didn't seem like a very fast weathering rate. I (Mary) asked if the tombstone would still be there in a million years. That question launched the class into a calculation of how long it would take to weather the tombstone away completely, an activity that engaged them in translating the graphical information into a real-world application.

One way to practice going the other direction—translating observational data into a graph—is through the use of short graphing puzzles. I (Russ) often engage students in conceptual analysis of what observational data imply about the slope on a graph, or whether the observational data are most consistent with a positive or negative slope—basic concepts of graph reading that students often struggle with most. These puzzles are easy for teachers to create and adapt to a variety of lessons and topics.

For example, most students understand that the Earth gets hotter as one goes deeper. This increase in temperature with depth is called *geothermal gradient.* However, not all locations have the same geothermal gradient. Yellowstone National Park has magma near the surface, resulting in a higher geothermal gradient in that area. With a graphing puzzle such as that presented in Figure 4.4, you might prompt your students by asking which trend represents the geothermal gradient at Yellowstone (the one that's hottest close to the surface). For follow-up, you might ask, "What would be the temperature at the point the different trends converge?" and "What would be the physical meaning of the other trends?"

Figure 4.4

Conceptual graphing puzzle on the geothermal gradient at Yellowstone National Park

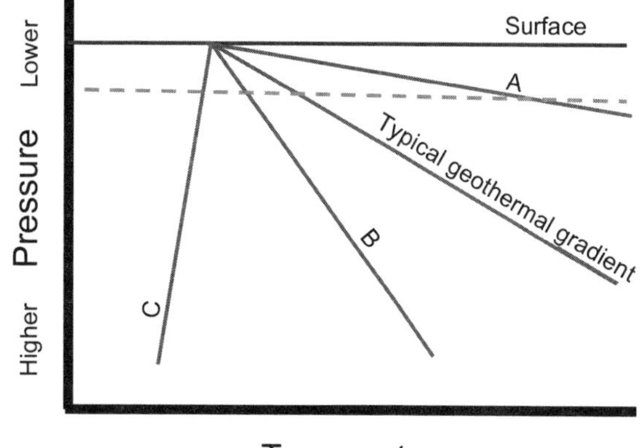

Prompt questions you might use with students include the following: Which line most closely portrays the high geothermal gradient present at Yellowstone National Park? What value for temperature or pressure do you expect where the lines converge? What is the real-world meaning of the other trends? Note that pressure is plotted on the *y*-axis—as you go deeper into the Earth, there is more rock above you and therefore greater pressure. The dashed line shows one particular depth, which can guide students in determining which trend shows the highest temperature at that depth.

When Experience Contradicts Learning

Many students have visited caves, usually on summer vacations with their families, and noticed that the air temperature *decreases* as they descend into the earth. It's cold in the cave, especially in contrast to the hot summer weather. And especially if one ignores the signs that say "Cave is cold, bring a coat!"

Students often ask how the Earth can get hotter with depth if caves are colder. Yeah. What's going on here? Which is it? Hotter or colder as you go down into the Earth?

Cave temperatures are often close to the average annual temperature outside the cave. They will indeed be cooler than the outside air in summer. But they are warmer in winter.

In Chapter 2, "The Controlled Experiment," we reported experiments for measuring the solubility of water vapor in air as a function of temperature. Suppose that you're working on a similar unit. You might prompt your students with the following puzzle (see Figure 4.5, p. 62): "Consider the observation that on a hot, muggy day, water condenses on a cold can of pop when you take it from the cooler out at the lake. If the water forms on the can of pop when the water condenses from the air, which trend line would represent how solubility of water vapor changes with temperature?" You might then ask these follow-up questions: "What would be the physical meaning of the other lines?" "Does a vertical line even make sense?" "Why or why not?"

Figure 4.5

Conceptual graphing puzzle on the temperature dependence of water vapor solubility in air

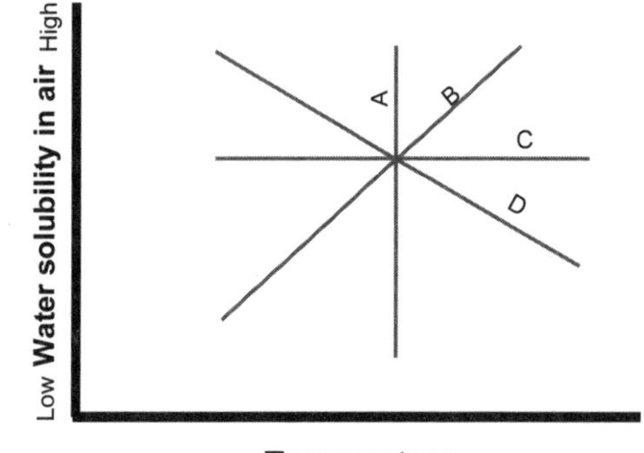

Prompt questions you might use with students include the following: Which of the lines best portrays the trend of solubility with temperature, given that water condenses on a cold can of pop on a hot, muggy day? What is the real-world meaning of the other trends (none of which actually occur in our dimensional reality)?

The vertical line (line A in Figure 4.5) doesn't make sense because it implies that only one temperature is possible and, at that temperature, water solubility in air simultaneously takes on all possible values. The horizontal line (line C in Figure 4.5) means that the water solubility does not change with temperature. Line D in Figure 4.5 implies that water vapor becomes more soluble with decreasing temperature, the opposite of the relationship observed.

PAYING ATTENTION TO AXIS LABELS

You can't know what a graph is telling you if you don't know what's been plotted on it. Students often automatically and wrongly assume that the axis labels will correspond to whatever data they are given in a classroom problem.

For example, finding the location of an earthquake epicenter is a common classroom activity in which data from the seismogram and the graph axis labels don't match up. The input data are typically arrival times—the times when the P and S waves arrived at a seismograph—but the graph plots travel times—time elapsed between earthquake and arrival of the seismic wave at the seismograph.

Seismic Travel Time Analogy

Measuring seismic arrival times is like noting that you saw a flash of lightning at 2:00 p.m. (the P wave) and then heard the sound of thunder at 2:00:25 p.m. (the S wave). To figure the distance to the lightning bolt, you use the difference in the arrival times of the light and the sound. Given that sound travels at about one-fifth mile per second through our atmosphere (1 mile every five seconds), the hypothetical lightning bolt was 5 miles away. From this, you can calculate the time of the lightning strike (because the speed of light is so fast, this is essentially the same as the arrival time of the flash, about 2:00 p.m.). The travel time for the thunder was thus about 25 seconds.

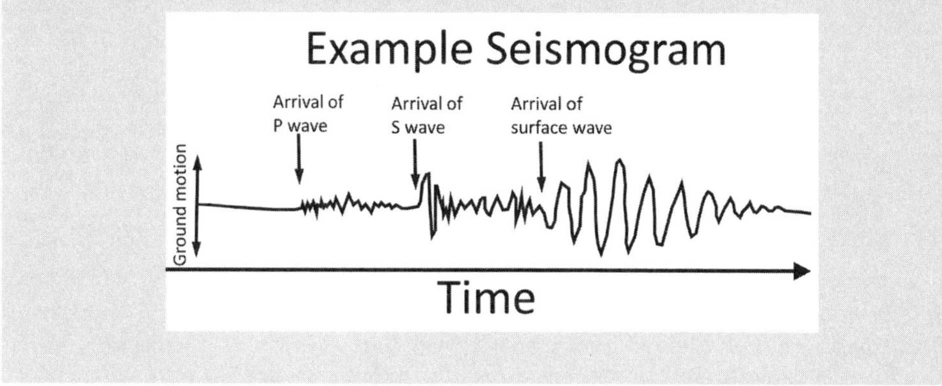

To determine how far each seismograph is from an earthquake, students need to figure out how the arrival times at three seismograph stations can be applied to a graph that plots travel time versus distance from earthquake (Figure 4.6, p. 64). They then draw a circle on a map around each of the three seismograph stations with the radius determined from the graph. The epicenter is located at the point where the circles intersect. Younger students might use a graph that plots the difference in P and S arrival times instead of the travel times, making for a simpler problem. However, in either case, the values in the data sets given to the students (arrival times) do not directly correspond to what's plotted on the graph (either travel time or difference in travel times). When students try to chart the arrival times on the axis labeled "travel time" (or "difference in travel times"), they get nonsensical results, which forces them to go back and rethink what the graph means.

Figure 4.6

Graph of seismic wave travel times versus surface travel distance

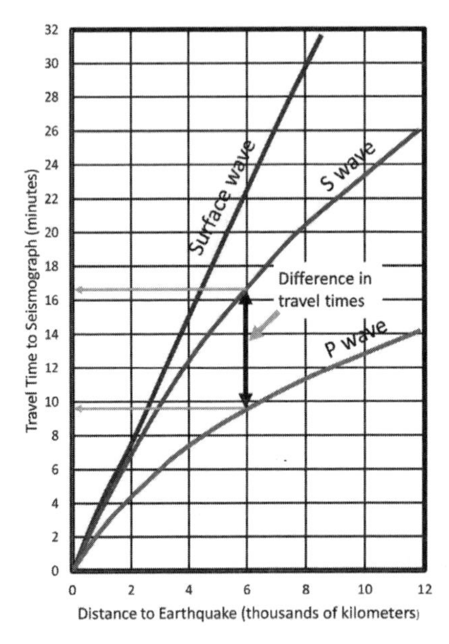

Seismograph data provide arrival times for P and S waves from an earthquake. In figuring out how far each seismograph was from the earthquake, students need to realize that this graph does not plot arrival times, and adjust their strategy accordingly. They also need to realize that the difference in the travel times for the P and S waves and the difference in arrival times will be the same value.

Another way to give students a chance to think about axis labels is to let them choose how to plot their own data. Unless specifically directed otherwise, students tend to gravitate toward histograms, rather than the *x-y* scatter plots that show the correlations between two experimental variables. Not specifically telling students what kind of graph to use can introduce an opportunity to talk about different kinds of graphs.

One year, I (Russ) had college students measure the effect of viscosity and eruption rate on the slope and diameter of sugar-water volcanoes on the distant (imaginary) planet of Lollipop. Students were asked to determine the relationship between volcano diameter and one of the following: composition of sugar water, eruption rate, or temperature.

Several groups chose to study the effect of melt composition (sugar acting as the proxy for silica concentration, which influences the viscosity of lava in volcanoes on Earth). Despite knowing the goal of the experiment, several groups didn't initially include with their report an *x-y* graph showing the variation in diameter with composition. One group plotted volcano diameter versus the trial number for three different trials of each of three compositions, making a line graph as in Figure 4.7. This graph provided useful information, just not the relationship between the dependent and independent variables the way a scatter plot would. For example, the graph showed the reproducibility of their results, giving a feel for experimental uncertainty, and offered a way to consider if there were any unexpected trends with time in their measurements. The freedom to plot a variety of graphs gave us the opportunity to talk about the value of different types of graphs.

Figure 4.7

Student graphs from an experimental investigation of the relationships between melt composition and volcano diameter (at fixed flow volumes, flow rates, and melt temperature) on the imaginary world of Lollipop

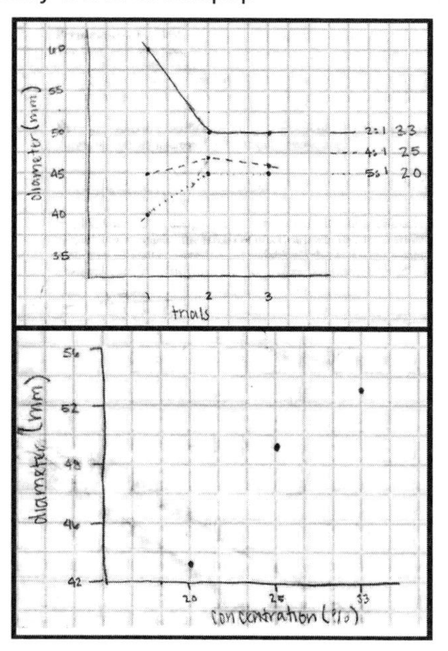

The results are reported for three melt compositions (by mass): 20% water and 80% sugar, 25% water and 75% sugar, and 33% water and 67% sugar. Volcano diameter was plotted against trial number in the upper graph—a type of line graph—rather than plotting melt composition versus diameter in an *x-y* scatter plot. This provided an opportunity to talk about the value of different types of graphs. After discussing, students plotted the average diameter against melt composition (lower graph), but the lower graph plots categorical data along the *x* axis, ordered by time of experiment, making it no different in concept from a histogram—notice that there is no meaningful scale on this graph; rather, the data are simply plotted against the three different compositions.

FINAL THOUGHTS

Reading a graph requires paying attention to the details of the graph, including what variables are plotted against each other, whether the variables are correlated, and the slope of that correlation. Most important, reading a graph requires paying attention to what the graph means in the real world. In science, graphs are not simply abstract numerical constructs. Graphs represent real phenomena, and understanding a graph requires visualizing the real-world meaning of it.

R. T. Rybak, a former mayor of Minneapolis, gave a presentation for the Minnesota Science Teachers Association in 2015. He recalled a story told by a colleague whose daughter taught math first in Los Angeles and then in Minnesota. She was surprised to find that her Minnesota students understood negative numbers much faster than her California students. Rybak associated that difference with Minnesota's cold winters, and the corresponding mental image Minnesota kids have of a thermometer scale that goes both above and below zero. In fair disclosure, we Minnesotans are always on the lookout for reasons to think well of our winters, so maybe we should take the story with a grain of salt, but the value in associating numbers with life experience remains sound.

We might call this life connection to numbers *experiential meaning*. Reading and constructing graphs requires that we connect the content of the graphs to experiential meaning. We need to think about how the graph—its slope, its values—relates to the world around us. We need to help our students do that.

REFERENCES

Beichner, R. J. 1994. Testing student interpretation of kinematics graphs. *American Journal of Physics* 62 (8): 750–762.

NGSS Lead States. 2013. *Next Generation Science Standards: For states, by states.* Washington, DC: National Academies Press. *www.nextgenscience.org/next-generation-science-standards.*

Scripps Institution of Oceanography. 2016a. Charles David Keeling biography. *http://scrippsco2.ucsd.edu/charles_david_keeling_biography.*

Scripps Institution of Oceanography. 2016b. The early Keeling Curve. *http://scrippsco2.ucsd.edu/history_legacy/early_keeling_curve.*

EXAMPLE ACTIVITY DESIGN

THE IMPACT-CRATERING EXPERIMENT

WHAT TO GRAPH, HOW TO GRAPH IT, AND HOW TO MAKE PREDICTIONS

The graphing activity we present here is a continuation of the experimental activity begun in Chapter 2. In Chapter 2, students identified questions to address, developed their experimental methods, made measurements, and organized their results into a table. Here we walk through how to coach students in analyzing and interpreting their experimental data using graphs.[C12]

TEACHER PREPARATION AND PLANNING

Look over the data your students derived from their cratering experiments. Make a graph or two of your own. Which are the independent and dependent variables? On which axes should they be plotted? Would it be better to combine some results onto a single plot? For example, you might plot drop height versus crater size for a variety of different masses or impactor size. Think about how you can encourage and guide your students as they construct their graphs without telling them exactly what to do. Think about any peculiarities in the students' data so you are ready to facilitate discussion when they discover differences between groups.

EXAMPLE PROMPTS AND LIMITING OPTIONS

Taking a Look at the Data

You might get students started analyzing results by asking them to look at their data, still in table form. For example, you might give the following prompts:

- Can you spot any consistent trends?
- What variables appear to affect crater size and in what way?
- Do the trends appear linear or nonlinear?
- Are the results reproducible within each group, meaning are data similar for multiple trials of the same variable?
- Are the results reproducible between the groups that tested the same variable?
- Are the results what you expected?

- Do the data make sense given the experimental observations and conditions?
- If there are differences in results between groups, can you identify possible reasons? For example, did someone measure in different units?

Preparing to Graph the Data

Talk about types of graphs and the kinds of data that are best plotted on them. In particular, talk about the difference between a bar graph (good for data that can be counted and put in bins) versus an *x-y* scatter plot (good for seeing correlations between two numerical variables). If students need a refresher on the mechanics of graphing, you can go through some simple examples. For example, you might ask students the following questions:

- Which axis do you want to plot the variables on?
- Do you want to do one graph with all your data, or more than one graph?
- What scale do you want to choose for each axis?

Note: We don't recommend using computer graphing routines, which choose the scales behind the scenes and make setting up the graph seem like magic, but if your goal is to also teach students how to use graphing software and you can spare the large upfront time to learn the technology, go for it.

Graphing the Data

Have students graph their own results first, then perhaps graph the data from another group that measured the same variables, plotting it on their own graph. This can help bring home the value of clear data table formulation that other people have to read. It also gives students an opportunity to recognize any differences in results between groups. You might also have groups plot the results for other variables measured by other groups. At this point, it's not necessary for students to draw any lines or "connect any dots" on their graphs.

Analyzing the Results

Remind students how to look for trends in their data. For example, you might ask the following questions:

- Does it make the most sense to "connect the dots" of the data, or does drawing a smooth line (or curve) of best fit through the data make the most sense?

- If you were to draw a line or curve through your data, would it be linear or nonlinear?
- Can you explain the meaning of the trends you see?
- What does the trend tell you about the relationship between crater size and the variables that affect it?

Help students explore the difference between one group's results and the results from another group who experimented with the same variables. Can they identify causes of the differences? Group members often respond to this kind of question by saying "we measured wrong." Encourage students to think beyond this simplistic response. Let groups show one another how they arrived at their measurements. Prompt students' discussions with these questions:

- Might someone have inadvertently introduced a new variable?
- Might differences in results be due to the coarseness of your measuring tools or differences in the way groups chose to do their measurements?

Help students develop a feel for the experimental uncertainty in their measurements. You might prompt them with these questions:

- Based on the scatter in your data, how reproducible are your results?
- Would a line or curve of best fit go exactly through each point, or would the line or curve take a path between points?
- What does this mean about the measured values?
- If there is a large amount of scatter compared with the overall trend, what does that mean for your confidence in the trend?
- If there is a lot of scatter in the data around a trend, what does that mean for your confidence in the exactness of the measured values?

The Big Challenge—Making a Prediction

Valid science makes predictions that can be shown to be either true or false. Discuss the idea that if their experiments show a causative relationship between variables, then they should be able to predict an outcome that was not directly tested in the experiments.

Tell students you are going to drop a ball of mass *X* from height *Y*. Give groups enough time to consider their graphs and data table and to write down their predictions for how big the new crater will be. You can do this for several different masses and drop heights.

We suggest considering only one variable at a time for any one prediction. For example, if a group varied the mass of impactors, not drop height, you would choose an impactor with a mass that is different from the ones the students used, but the same drop height. If a group varied drop height, but not mass, then choose a drop height different from their experimental ones but using the same mass. Maybe for one prediction you could mix it up by changing multiple variables and having them interpolate.

After each prediction, do the experiment two or three times and measure the crater diameter for each. Perhaps measure the crater diameter in two directions for each experiment. Calculate an average of the multiple measurements so students understand the idea that repeated measurements decrease uncertainty. Tell the students the results and direct them to compare the measured value with their predicted value. Discuss how close the predicted value needs to be to the measured value to still be counted as "right." For predictions that are way off, discuss possible explanations for the difference.

For scoring and grading, you might choose the "right" answer to be anything within 10% or even 20% of the predicted value. No one will ever predict the crater size exactly (you can't even measure it exactly). This provides another chance to discuss experimental uncertainty if you choose. You can give credit for the accuracy of their prediction, giving higher scores to those who are closer, say in increments of 10%, which gives the students some "skin in the game" when they make their predictions.

EXAMPLE INTERACTION

Student question: When I draw the line on the graph do I just connect the dots?

Teacher prompt: Do the dots fall exactly along a perfect line or curve?

Student observation: Well, sort of, but they kind of bounce around a bit. The dots don't really line up.

Teacher prompt: Do you think the not lining up reflects how velocity really affects the size of the crater?

Student interpretation: I think the bouncing around is probably because we didn't measure it exactly right.

Teacher prompt: Why do you think that?

Student reasoning: Well, we redid this one experiment and one measurement is higher and the other measurement is lower, so it doesn't seem like it's a real difference.

Teacher prompt: So do you think that connecting the dots gives you a better measure of how the real world behaves, or would a smooth curve do that?

Student conclusion: Probably the smooth curve.

SUMMARY CHECKLIST FOR TEACHER AS PRACTITIONER OF SCIENCE

- Plan time for students to discuss the meaning of their data *before* they start graphing.
- Review types of graphs and talk about how and what to plot, depending on students' background knowledge.
- After students graph their own data, plan for discussions on what the graphs and data mean in terms of what students experienced and observed during their experiments.
- Have students use graphed data to predict new impact crater sizes.
- Find some way to give students some skin in the game—for example, through grading or class challenges.

TEACHER REFLECTION

Consider ways to use your students' newly exercised graphing skills in future units and lessons. For example, you might have students sketch conceptual graphs given a basic understanding of natural variations. How will air volume change as the temperature of air increases; what will the graph look like, in concept? Or you might take graphical data from real research or something in the news and have students interpret what the graphed data are telling us.

Tie your students' experimental research to the big ideas and driving questions on which your curriculum is based. In what ways can your students' experimental analyses help them make sense of natural phenomena? For example, this cratering research project ties to our big idea that planets have a history of change; it also ties to the *NGSS* disciplinary core idea ESS1.C that we can learn about the history of our solar system by looking at asteroids, meteorites, and the cratered surfaces of other planetary bodies.

ANALYZING AND INTERPRETING DATA, PART 2

MAPS AND CROSS-SECTIONS

Visualizing Earth data in three dimensions has always been a key contribution of earth science to economic prosperity. The geologist William "Strata" Smith had to think spatially to know where to build the canals needed to transport coal and iron in 18th- and 19th-century England. In the 20th century, meteorologists had to think spatially to predict the presence of the jet stream in planning bomber routes in World War II. And in this century, geological engineers have had to think spatially to plan horizontal drilling for oil in the Bakken formation of North Dakota.

Back in the 1800s, English coal miners had to visualize rocks in three dimensions to follow a coal seam across a fault. Movement along the fault shifted the coal seam—perhaps up, perhaps down, perhaps sideways—and to find the continuation of the seam, miners needed to know which way to dig. Given the cost in money and lives, digging in the wrong direction was unacceptable. The miners' livelihood depended on a correct visualization of what they could not directly see.

The miners learned that with a normal fault, they needed to continue excavation in the same (normal) direction until they picked up the coal seam again. With a reverse fault, they had to reverse direction to pick up the coal seam (see Figure 5.1, p. 74). In fact, the terms *normal faulting* and *reverse faulting* were coined by the English coal miners who recognized these differences in how the two kinds of fault offset a horizontal coal seam.

Figure 5.1 ✪

Cross-sectional illustrations of normal and reverse faults

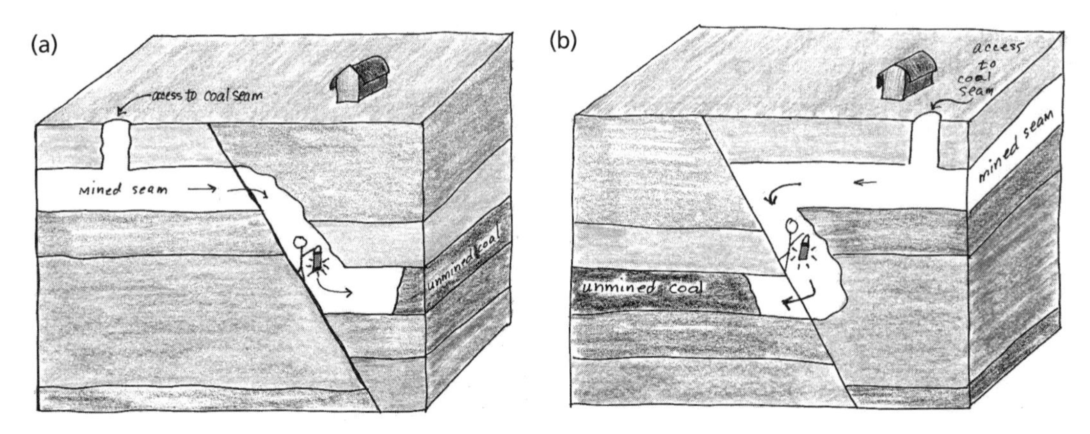

Spatial reasoning is important in mining because miners need to be able to visualize the location of ore bodies in three dimensions. Coal miners in the 1800s gave different kinds of faults the names *normal* and *reverse* depending on whether they had to reverse course to pick up the coal on the other side of the fault or whether they could continue in the same direction normally. Illustration (a) shows a normal fault, and illustration (b) shows a reverse fault.

THE IMPORTANCE OF SPATIAL REASONING

With the development and pervasive use of geographic information systems (GIS), spatial thinking is becoming even more important. According to an article in the journal *Nature*, geospatial technology is one of the three most important emerging fields in the 21st century, along with nanotechnology and biotechnology (Gewin 2004). Spatial reasoning affects almost every aspect of our economy, from identification of basic resources to the distribution of final products, and is critical to the public understanding of geopolitical issues such as the risks in fracking for oil.

The important educational contribution from earth science to geospatial technology is not found primarily in the mechanics of how to use the technology, but rather in the spatial-thinking skills required to make using the technology worthwhile. Understanding the Earth requires spatial thinking, and the tools for visualizing Earth's three-dimensional nature are the map and the cross-section. Together, these constitute a key science language of equal importance to the graph.[C13]

Data analysis of all sorts is a key objective in the *Next Generation Science Standards* (*NGSS;* NGSS Lead States 2013), and spatial reasoning shows up specifically in the *NGSS* and in *A Framework for K–12 Science Education: Practices, Crosscutting Concepts, and Core Ideas* (*Framework;* NRC 2012). For example, consider the following excerpt from the science and engineering practice 4 (Analyzing and Interpreting Data) matrix for grades 6–8 in Appendix F of the *NGSS:* "Use graphical displays (e.g., maps, charts, graphs, and/or tables) of large data sets to identify temporal and spatial relationships." The *Framework* says, "Once collected, data must be presented in a form that can reveal any patterns and relationships and that allows results to be communicated to others" (p. 62). Additionally, the *Framework* begins the discussion of disciplinary core idea ESS2 with the statement that "Earth's surface is a complex and dynamic set of interconnected systems … that interact over a wide range of temporal and spatial scales" (p. 179).

Spatial reasoning is a practice of science quite different from other academic skills such as math or language. In our experience, many students who struggle with other academic practices may excel at spatial reasoning. Unfortunately, few students are exposed to spatial-reasoning practices in school, at least not the kind of spatial skills used in earth science. According to a 2015 report by the Center for Geoscience and Society, only two states in the United States require a yearlong earth science course for high school graduation (compared with 30 and 20 states that require life and physical sciences, respectively).[C14]

AN INTRODUCTION TO SPATIAL REASONING

In one sense, maps and cross-sections are to field data what graphs are to lab data; both provide a synthesis and portrayal of important observations. However, maps and cross-sections deal with spatial data and spatially correlated numerical data. Often, a key goal in analyzing spatial data is to consider several different spatially correlated variables at the same time, looking for insight into the nature of the overarching system.

With this in mind, I (Mary) often begin my academic year with a two- to three-week activity examining two large wall maps that engage my eighth-grade students in thinking about how elevation, geomorphic features, rock types, and rock ages vary with location. One map is a topographic relief map of the United States, titled *Landforms and Drainage of the 48 States* (Allan Cartography 1992), and the other is a geological map showing rock ages, titled *A Tapestry of Time and Terrain* (Vigil, Pike, and Howell 2000).

Students begin the activity by considering how elevation and features change with location, as shown on the shaded relief map. The large amount of data stirs a lot of questions that form the foundation for exploring earth processes throughout the rest of the school year: "Why is the U.S. flat in the middle?" "Why are some elevations higher than others?" "Why are all the mountains in the west?" "Why do the rivers start where they do?"

Students identify regions that are alike or different and group those regions into geomorphic provinces. This starts them thinking about the possibility that there must have been some process that caused those regions to be different.

Never Underestimate the Power of Ownership to Motivate Learning

When I (Mary) was having my class work with the map *A Tapestry of Time and Terrain*, one of my students asked, "Can we use the scale to measure how wide the U.S. is?" When I said yes, the students got right to work with an urgency born from the excitement of being allowed to choose what they did.

Scale and spatial-thinking problem solving started right away. Many groups started with a ruler, then decided that a meterstick would be better, and finally one group asked for some string. One student asked, "Where do we measure from on the map?" "You decide," I said.

They had to evaluate the quality of their results. "I got 13 miles wide for the U.S., but that can't be right. I drive 75 miles to my grandmother's house!"

Each group posted their measurements. In considering the differences between groups, some students realized that they had used the scale for kilometers instead of miles.

"So what is the *right* answer?" one student asked. They checked the distance between western California and eastern Maine with Google maps, which led to another student-initiated discussion about how much variation in their measurements was reasonable.

Students tackled the challenge with a passion because the project was *theirs!* They initiated it, decided what to do, and evaluated the results. After the first day, Emma asked, "Can it just be *our* class that does this?" If the other classes did it, too, the activity would become just another teacher-initiated lesson.

In considering the geological map, students correlate age of rocks not only with geographic location but with geomorphic province. With the addition of rock ages to the data set, questions become more complex: "Why do the older rocks in the plains poke through the younger rocks in patterns that look like tree branches?" "What causes that spot in the Central Valley of California that has a different age?" Students often notice that rocks in the mountains have a larger range of ages than the rocks on the plains. This observation becomes a reference point later on when students learn how rocks occur in layers and why there might be a wider range of ages in the mountains.

After pursuing some of the student questions and promising to consider more throughout the school year, I push the class to examine a particular question. For example, in the fall of 2014 I picked up on my own interest in the older spot of rock in the midst of younger rocks in the Sutter Buttes of California's Central Valley. The feature shows up on both the geological map and the shaded relief map. What is that feature? What kind of rock is it made of?

I had students do a Google investigation to figure out what this feature was. Students didn't have the experience to fully understand the meaning of what they found—the Sutter Buttes are a volcanic dome surrounded by younger sediment eroded off the surrounding mountains. However, the exercise gave them practice thinking about multiple types of data (geomorphic, elevation, rock age, and rock type) as a function of location. They began thinking about geological process—every place is different, and there must be some cause of that difference. They began to visualize sequences of events—older rocks nestled among younger rocks.

Although this map-reading activity did not address a *specific* driving question, it introduced students to a variety of map types, developed a sense of where major landform regions occur and their correlation to rock ages, and gave students practice thinking about scale—all key aspects of spatial reasoning. It became a reference point for future activities with topographic maps, erosional processes, and history of change on Earth. The activity provided a visual framework for concepts throughout the rest of the academic year.

A CASE STUDY IN SPATIAL REASONING

Whereas graphs are great for portraying relationships between numerical variables, maps and cross-sections are great for showing spatial relationships. This makes them particularly important in portraying large-scale conceptual models of earth processes.

For example, understanding of plate tectonics dawned for students one year while interpreting the map and cross-sectional views of earthquake locations at a subduction zone. I (Mary) had my students exploring the driving question "How do we know that plate tectonics is happening on Earth?" when they made the connection between the three-dimensional data and the conceptual model of a subducting plate. Getting to that moment of insight took a six-week lesson—here's how it came about.

After considering some of Alfred Wegener's evidence for continental drift (Rogan-Klyve et al. 2015; Wegener 1929/1966), I had students engage with an online GIS-like activity developed by Lehigh University titled *Environmental Literacy and Inquiry: Tectonics (www.ei.lehigh.edu/eli/tectonics/index.html).* Students engaged with a variety of spatial data organized into map layers. They manipulated layers that showed epicenters and focal depths for earthquakes greater than magnitude 4, volcano type and location, seafloor bathymetry, earthquake hazard probability, historical earthquakes greater than magnitude 8, age of seafloor, surface heat flow, plate tectonic boundaries, and the like. In one investigation, students considered geohazards nearest to their school. Another investigation had students identify and draw plate boundaries. A third engaged students in thinking about seafloor spreading and the patterns in heat flow and the age of the ocean floor.

The spatial reasoning needed to use these maps involved more than understanding scale, understanding latitude and longitude, and paying attention to the map legend. Students needed to engage with maps long enough to learn their spatial symbology and translate that symbology into a three-dimensional mental image.

By the third investigation, relating the multiple lines of map data to plate tectonics had brought students to a raised level of understanding. Students commented that "it's starting to make sense now" and "all the pieces kind of fit together."

After students had worked with these maps in the computer lab for about a week, I brought them back to the classroom to focus on what kind of data maps portray and how we think about that data.[C15] I put a composite map of earthquakes and volcanoes from the Lehigh activity up on the screen (see Figure 5.2) and asked the students, "What information and data does this map show?"

Figure 5.2

Screen shot from the Lehigh University online instructional activity *Environmental Literacy and Inquiry: Tectonics*

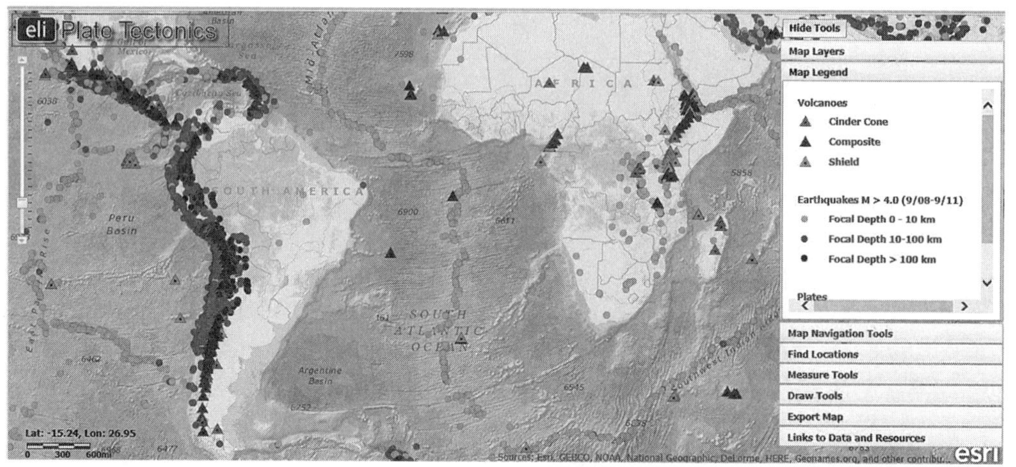

The map layers are from "Investigation 1: Geohazards and Me" and combine seismic and volcanic data on a sea floor topography base map. Types of volcanoes and depths of earthquakes provide information about type and placement of plate boundaries.

Source: Environmental Literacy and Inquiry Working Group at Lehigh University, *http://gisweb.cc.lehigh.edu/tectonics/investigation1.*

Students quickly jotted down "volcanoes and earthquakes" and "plate boundaries" and then start chatting with their friends about what they did over the weekend. To refocus their attention and get them to think past the first answer that popped into their head, I said, "If you wrote volcanoes and earthquakes, that's not good enough."

Murmurs and confusion—"Whatdyamean, not good enough?"

"Does the map just tell you that there are earthquakes, or do you know something else?" I asked.

"Well, we can see where the earthquakes happened."

"So, do the earthquakes happen randomly in different places, or do they happen in particular locations?"

"Particular locations."

They still weren't getting my point—that they needed to consider the maps more carefully—so I took a more direct approach. "Why are there three different symbol colors for the earthquakes?"

Aha. Students remembered that the colors represented different focal depths, with ranges of 0–10 km, 10–100 km, and >100 km. After considerable discussion, I summarized: "So the map doesn't just show earthquakes, it shows their spatial distribution and focal depth."

"And what about the volcanoes?" I asked. "Why are there three different colors for the volcanoes?"

"Oh, yeah. Composite volcanoes, shield volcanoes, and cinder cones."

I continued with another question: "How many wrote down that the map shows plate boundaries?" Hands went up. I asked, "What's the symbol on the map for plate boundaries?"

Students checked the map legend for the symbol—the legend being what tells you what is plotted, the equivalent to the axis labels on a graph. But of course it wasn't there, because this map didn't show plate boundaries. Those were plotted on a different map. "But I can see them!" one student said.

I took the opportunity to point out an important distinction between evidence and interpretation on maps: "We can see where a plate boundary goes because the map plots some of the evidence for the boundary, but it doesn't plot the boundary itself." We then explored the difference between plotting measured data on a map versus plotting one's model for what causes the distribution of the data. In a graph, data might be plotted as points, but when you draw a line or curve through those points you have constructed a model. Maps can also show both data and models. Data include things such as earthquake location and depth. Models are interpretations of that data, like plate boundaries. Keeping interpretations separate from data is critical to reading maps. It's also critical to arguing from evidence, which we talk about more in Chapters 8 and 9.

I asked another question: "Can you describe three patterns in the data? Start with 'I can see that … ' and fill in the blank. What can you see?"

Already alerted to my high expectations, students' answers became more sophisticated: "I see that a few earthquakes are scattered around, but most are grouped," and "I see that earthquakes follow the coastlines."

"Are they always on the coastline?" I prompted. "And do all coastlines have earthquakes?"

Well, no and no.

One student noticed that earthquake epicenters in the Atlantic Ocean mimic the curvature of the coastlines on either side. Another student made the connection—after checking out an ocean bathymetry map hanging in the room—that the mid-Atlantic earthquake locations match the location of a high ridge that runs down the center of the Atlantic. Another student noticed that the Atlantic Ocean earthquakes are shallower than many of the earthquakes on land. And another noticed that the deeper earthquakes correspond to the locations of the composite volcanoes, and that the volcanoes tend to be in chains rather than scattered around randomly.

One important pattern that none of the students noticed was how the depth of earthquakes increases inland from the coast of Peru and Chile. Because earthquake depth was a key evidence for plate tectonics that I was working toward, I zoomed the map to that feature and pointed it out (see Figure 5.3).

Figure 5.3

Detail of the Lehigh University earthquake and volcano map zoomed to the west coast of Peru and Chile, showing how the depth of earthquakes increases inland

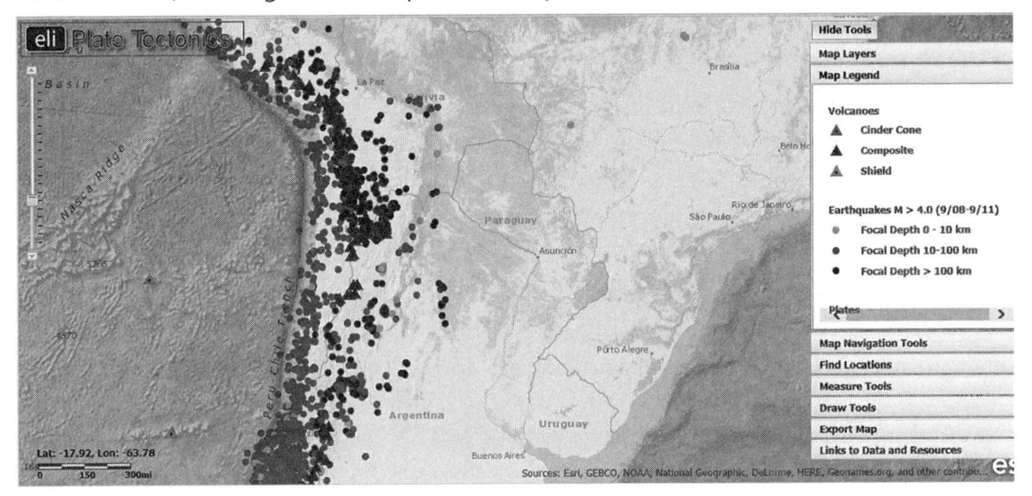

Source: Environmental Literacy and Inquiry Working Group at Lehigh University, *http://gisweb.cc.lehigh.edu/tectonics/investigation1*.

Then, we spent some time thinking and talking about the significance of narrow belts of earthquake epicenters and chains of volcanoes. For example, if an earthquake results from movement along a break in the Earth's crust, then the locations of the earthquakes should follow the trend of the break, occurring along narrow bands at the boundary between plates rather than randomly or in clumps or clusters.

Students had an illustration in their textbook showing plate boundaries drawn in cross-sectional views as well as map view (a block diagram). To get them to visualize plate tectonic theory in three dimensions, I had them reproduce this drawing by hand, encouraging them to think about how the drawing related to features in the real world.

After my students had drawn their pictures, I asked, "So, if earthquakes happen where rocks slide past each other, where in the cross-sectional view of a subducting plate do we expect earthquakes?" I gave students time to think about and discuss the question before plotting their thoughts for the whole class to consider (Figure 5.4).

Figure 5.4

Block diagram of a convergent plate tectonic boundary

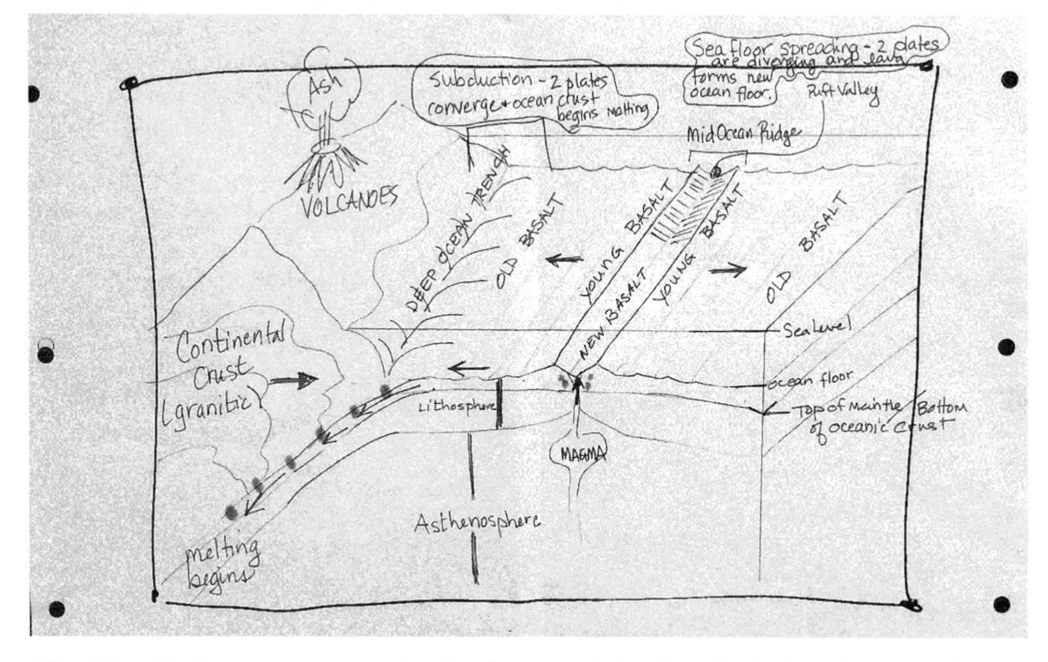

I (Mary) drew this diagram as a summary of student diagrams and class discussion. Regions where one might expect earthquakes are marked by the dots near the mid-ocean ridge and those that descend from the trench under the volcanoes.

Then, we looked at a map of earthquake focal depth, with the Peru-Chile subduction zone enlarged (see Figure 5.5, p. 82). This map was slightly different from the online version they'd been looking at before, with four rather than three focal depths plotted and the background image being plain black and white instead of a color-shaded image. Even though this was a new map, my students' previous experience with maps made it easy for them to interpret—their ability to generalize map-reading skills expanded by exposure to different maps.

Figure 5.5

Map showing earthquake location and depth for the Peru-Chile convergent boundary, 1990–1996 (magnitudes of 4 and greater)

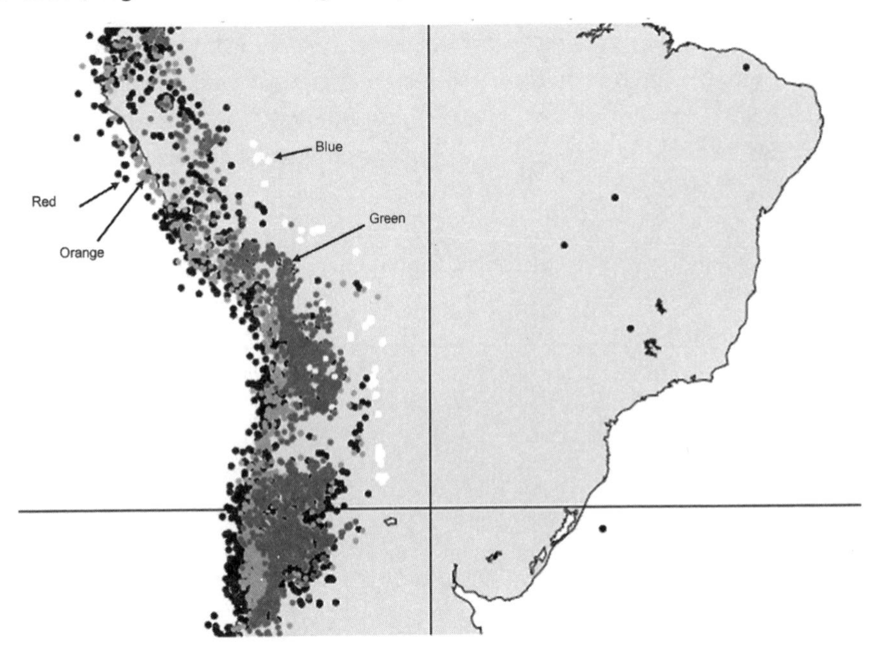

The addition of a fourth category of earthquake depths makes the increase in depth of earthquakes as one goes inland more striking in this map than in Figures 5.2 and 5.3.

Note: Color indicates depth: red = 0–33 km, orange = 33–70 km, green = 70–300 km, and blue = 300–700 km

Source: This map is part of "Discovering Plate Boundaries," a classroom exercised developed by Dale S. Sawyer at Rice University *(dale@rice.edu)*. Additional information about this exercise can be found at *http://plateboundary.rice.edu.*

Two hands shot up. "Is that really right? Do earthquakes get deeper that way?" The lightbulb switched on for Taylor and Kayla. They recognized that the cross-sectional image was a theoretical model of a subduction boundary. They recognized that the model predicted that earthquakes should get deeper inland from the boundary. They realized that the map of earthquake focal depth was observational data, not theory. And they realized that the data supported the theoretical model for plate tectonics illustrated in the cross-sectional view. For the first time, they were able to visualize in three dimensions what the maps and cross-sections portrayed and how all the pieces fit together into a coherent whole.

Their insight into the three-dimensional picture and the link between observation and theory was reached through spatial reasoning, considering many different maps and cross-sections over a period of six weeks. The takeaway message for me was that the six weeks were time well spent in laying the foundation for spatial reasoning that would be used in other areas of earth science as well.

ADDITIONAL THOUGHTS ON MAPS AND CROSS-SECTIONS

Relating maps and cross-sections to real-world features is as important to spatial thinking as relating graphs to real-world processes is to graphing. Although experts look at maps and cross-sections and see important patterns that reflect the real world, students often see lines and curves to memorize for a test. They have not yet developed the spatial-reasoning skills that allow them to connect the abstract portrayals on maps and cross-sections to real phenomena.[C7]

I (Russ) sometimes have students draw the cross-sectional view of plate boundaries in college exams. Student versions often show ocean crust towering above the mountains or volcanoes deep within the Earth's mantle. These errors reveal that students aren't always making the connection between the cross-sectional illustration and physical reality (see Figure 5.6). I even warn my students before the exam that they need to think about the physical reality that the cross-section is illustrating and to make sure that they don't have an ocean crust poking up so high that it might get hit by the Moon! Even knowing that

Figure 5.6

Cross-sectional illustration of a convergent plate boundary with subduction

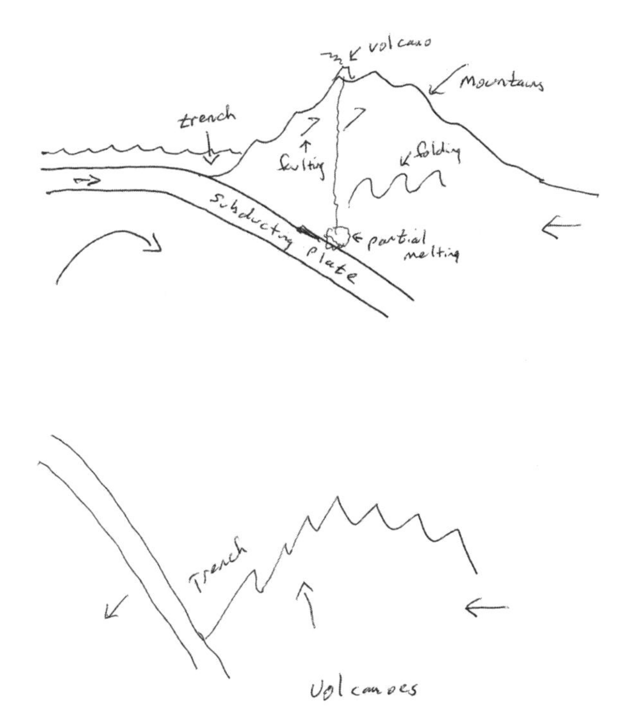

The top illustration shows the kind of cross-sectional drawing that I (Russ) hope to see on exams. The lower illustration shows an example with three-dimensional visualization errors that are common with 5–10% of college students. These students are not making the connection between the cross-sectional illustration and the physical world. Middle school students need explicit guidance in how to construct a drawing—it's helpful to use sea level as a line of reference.

ahead of time doesn't eliminate the nonsensical illustrations if students haven't had enough practice connecting drawings to real-world information.

As Kim Kastens, Thomas Shipley, and Alex Boone noted in the abstract for an invited talk at the American Geophysical Union fall 2012 meeting, "When geoscience experts look at data visualizations, they can 'see' structures, and processes and traces of Earth's history. When students look at those same visualizations, they may see only blotches of color, dots or squiggles." Turning those splotches and squiggles into real-world spatial understanding is more like learning a language than it is like memorizing a technique. Plate tectonics can probably be taught without engaging students in the complex skills involved in reading map layers and cross-sections. However, reading maps and cross-sections is a practice that is critical to science literacy and arguably more important to education than merely learning the theory of plate tectonics.

The Most Important Spatial-Thinking Skills

I (Russ) asked Dr. Kirk Stueve, one of my colleagues who teaches the Spatial Reasoning course at the Center for Geospatial Studies at Minnesota State University Moorhead, "What are the three most important practices or skills in spatial thinking?" Simplified, these were his responses:

- Recognizing patterns in mapped data at multiple scales
- Visualizing the third dimension—cross-sectional profiles—from two-dimensional maps
- Thinking how features represented on a map change through time

Given the time commitment needed to correlate cross-sections and map data and gain a working mastery of spatial reasoning, it's not possible to go through extensive map and cross-section correlation activities for every unit. But it's a good idea to dive in deep in at least one. It doesn't have to be plate tectonics.

One year, instead of looking at plate tectonics, I (Mary) had my students looking at maps and cross-sections of global circulation patterns like those shown in Figure 5.7. Again, the students had to analyze surface data, in this case wind direction patterns, and integrate that data with a conceptual model of convection in the atmosphere.

Figure 5.7
Illustration of global circulation in map and cross-sectional views (vertical scale is exaggerated)

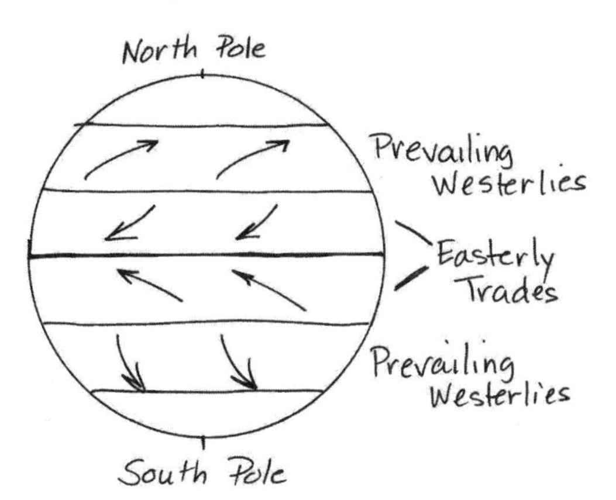

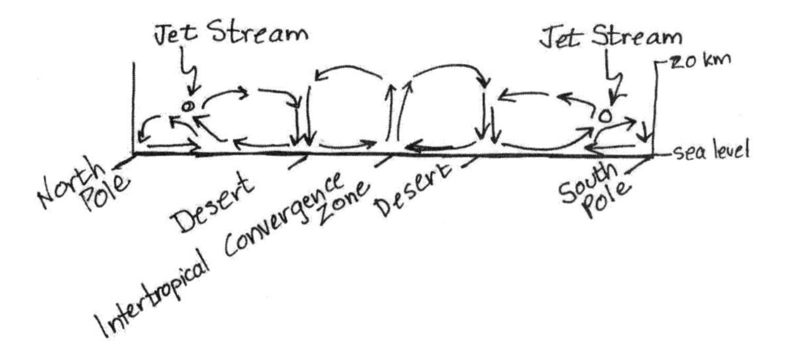

CONSTRUCTING TOPOGRAPHIC PROFILES AND GEOLOGICAL CROSS-SECTIONS

Chapter 3 briefly described a field activity in which I (Mary) engaged my students in constructing a topographic profile across the Buffalo River. Drawing this profile encouraged students to think about the topographic expression of river channel, floodplains, and surrounding bluffs. Based on that profile, students might address driving questions such as "Was the valley formed by glaciers or river erosion?" Students might also draw a topographic profile of the same section of river based on a topographic map (see Figure 5.8, p. 86). They could then compare the classroom profile with their conceptual profile drawn in the field, helping them connect real-world observation with map data.

Figure 5.8

Buffalo River (Minnesota) topographic map with hand-drawn topographic profile

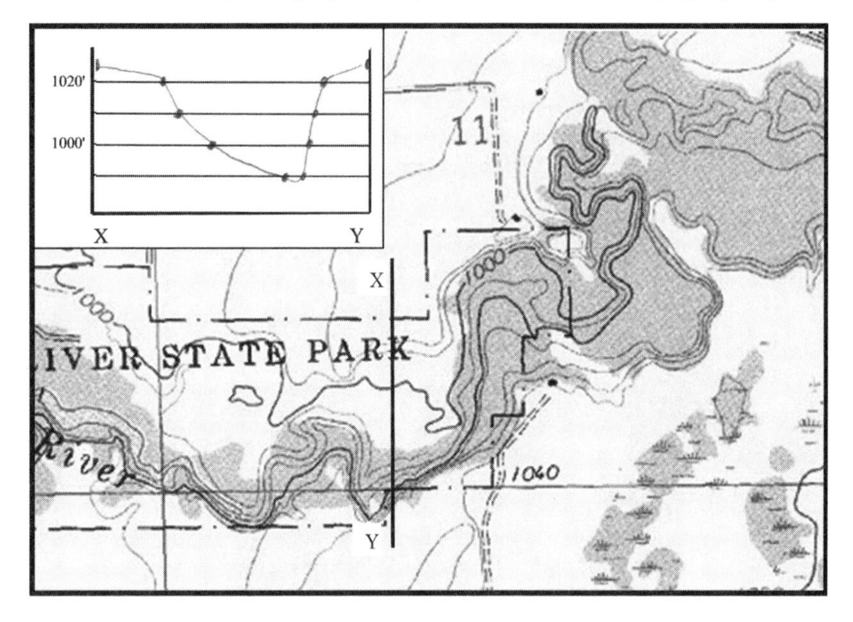

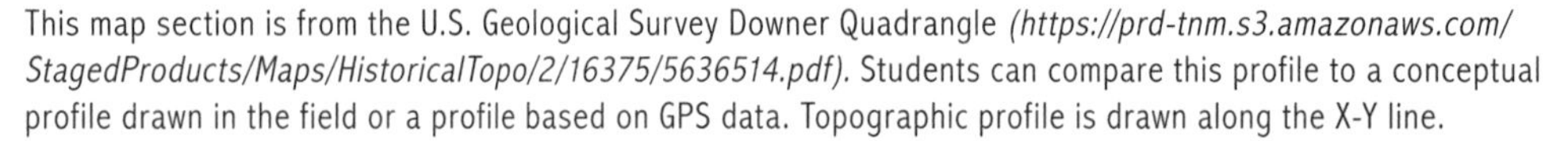

This map section is from the U.S. Geological Survey Downer Quadrangle *(https://prd-tnm.s3.amazonaws.com/StagedProducts/Maps/HistoricalTopo/2/16375/5636514.pdf).* Students can compare this profile to a conceptual profile drawn in the field or a profile based on GPS data. Topographic profile is drawn along the X-Y line.

Another approach could be to have students take GPS location and elevation measurements across the valley and draw a profile based on that data. This would give them less experience with conceptualizing the profile, but more experience with geospatial technology, plotting measured data, and thinking about how one represents spatially distributed data.

Teachers need to choose how much time to devote to the practice of spatial reasoning and how much time to devote to learning a technological tool. Our approach is often to try to make using the technology brief and seamless so as to maximize student exposure to spatial-reasoning problems. However, experience with technology is also important.

Construction of geological cross-sections usually requires prior experience with both topographic and geological maps. If students only get a single earth science class, typically in middle school, it may be tough for a teacher to work geological cross-sections into the class without compromising the students' experience with other practices of science. However, the ideal educational scenario should include encountering earth science a second time (in high school perhaps), and geological cross-sections would be appropriate there.

Incoming college students typically have no understanding of the arrangement of rocks beneath the surface. Given a geological map of rock layers and asked to draw a cross-section, students will often draw the rocks straight down into the ground, even for a simple case such as the layer-cake geology of the Grand Canyon (see Figure 5.9). This is sometimes true even for students who have been to the Grand Canyon and seen the horizontally layered rocks there. The takeaway message is that (1) noticing rocks in the field and connecting those observations to a three-dimensional mental image takes practice and (2) students have typically not received that practice in high school.

Figure 5.9

Simplified conceptual geological cross-sections of the Grand Canyon

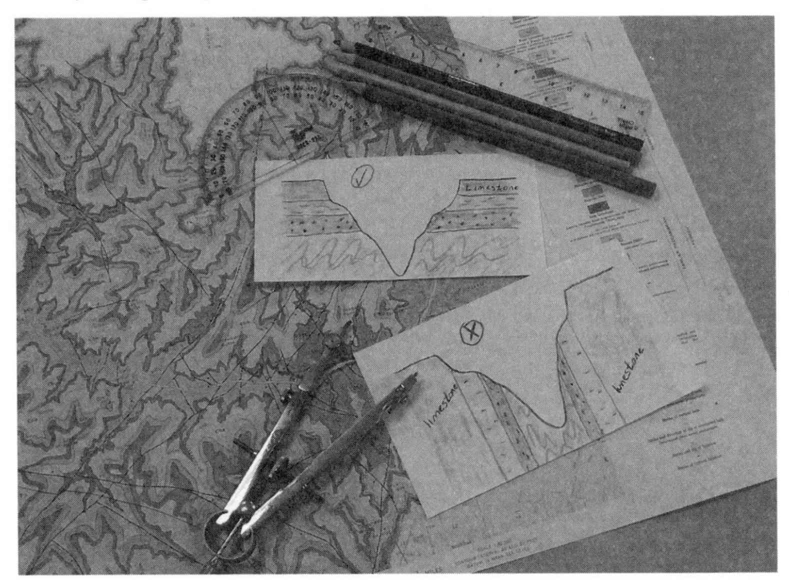

A correct conceptual cross-section of three rock layers and the underlying Vishnu Schist is marked by the circled check mark. A common student spatial visualization error—drawing rocks straight down into the earth—is designated by the circled *X*.

We offer a geological cross-section activity in this chapter's "Example Activity Design" section. In the "Digging Deeper" section at the end of this chapter, we offer some brief ideas for how you and your students might "dig deeper" into three-dimensional thinking by learning about contour maps and applying mapping concepts and cross-sections to areas of earth science other than geology.

REFERENCES

Allan Cartography. 1992. *Landforms and drainages of the 48 states.* Medford, OR: Raven Maps and Images.

Center for Geoscience and Society. 2015. *Earth and space sciences education in U.S. secondary schools: Key indicators and trends.* Earth and Space Science Report No. 2.1. Alexandria, VA: American Geosciences Institute. *www.americangeosciences.org/sites/default/files/education-reports-SecondaryES_Report.pdf.*

Gewin, V. 2004. Careers and recruitment: Mapping opportunities. *Nature* 427: 376–377.

Kastens, K. A., T. F. Shipley, and A. Boone. 2012. What do geoscience experts and novices look at and what do they see when viewing and interpreting data visualizations? Presentation made at the American Geophysical Union fall meeting, San Francisco. Abstract ED11B-0731. *http://adsabs.harvard.edu/abs/2012AGUFMED11B0731K.*

National Research Council (NRC). 2012. *A framework for K–12 science education: Practices, crosscutting concepts, and core ideas.* Washington, DC: National Academies Press.

NGSS Lead States. 2013. *Next Generation Science Standards: For states, by states.* Washington, DC: National Academies Press. *www.nextgenscience.org/next-generation-science-standards.*

Rogan-Klyve, A., M. H. Randall, T. St. Clair, and R. Gray. 2015. Bringing historical scientific arguments back to life: The case of continental drift. *Science Scope* 38 (7): 25–33.

Vigil, J., R. Pick, and D. Howell. 2000. *A tapestry of time and terrain.* Geologic Investigations Series 2720. Reston, VA: U.S. Geological Survey. *http://pubs.usgs.gov/imap/i2720.*

Wegener, A. 1929/1966. *The origin of continents and oceans.* Translated from the 4th revised German edition by J. Biram. New York: Dover.

EXAMPLE ACTIVITY DESIGN

MAPS AND CROSS-SECTIONS

An Introduction to Geological Cross-Sections

If you are able to develop a field geological mapping activity such as that presented in Chapter 3, then a logical extension of that activity is to construct a geological cross-section based on that map. We present an example of that in this example activity design (EAD). You can also engage students in constructing cross-sectional views of published geological maps. We give an example from a published map of folded rocks in Pennsylvania.

This EAD engages students in thinking about topographic and geological information and gives them practice *constructing* profiles and cross-sections. By making the connection between raw information and constructed visualization, students have the opportunity to practice their spatial-reasoning skills, particularly three-dimensional thinking. This EAD could fit within a unit exploring the evidence for how Earth has changed through time.

TEACHER PREPARATION AND PLANNING

- Review the field mapping activity you did with your students previously (such as the EAD in Chapter 3, pp. 47–51).
- Acquire a geological map of a folded region. Our example (see panel b in Figure 5.10, p. 90) is from the Somerset County, Pennsylvania, geological map, but you can also get geological maps for most areas from the U.S. Geological Survey. Geological map–teaching packets can be purchased from various educational outlets.
- Get familiar with the maps and go through the exercise ahead of time, taking note of your own errors and challenges in constructing the topographic profile and geological cross-section. This can help you know how to prompt your students while still giving them room to explore and interpret the maps and cross-sections.

Figure 5.10

Geological maps and cross-sections of Quarry Hill Nature Center in Minnesota and Somerset County in Pennsylvania

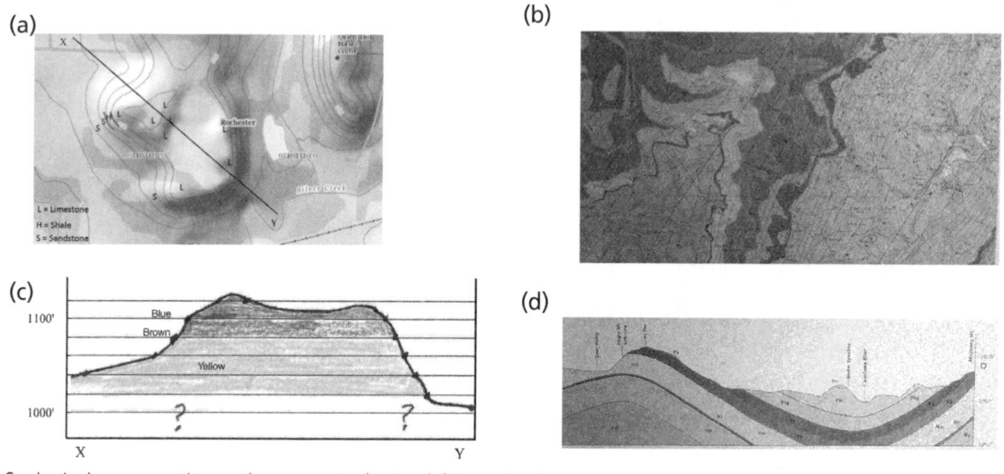

Geological cross-section such as your students might construct, based on the map from the activity in Chapter 3 (see Figure 3.3, p. 49). Yellow = sandstone, brown = shale, blue = limestone.

Source: Flint, N. K. 1965. Geologic map of the southern half of Somerset County, Pennsylvania. In *Geology and mineral resources of Somerset County,* C56a. Harrisburg: Pennsylvania Bureau of Topographic and Geological Survey. Available at *http://collection1.libraries.psu.edu/cdm/ref/collection/pageol/id/52067.*

EXAMPLE PROMPTS AND LIMITING OPTIONS

- Review the mechanics of topographic profiles with your students. You might have the students sketch the profile conceptually to start. Usually, for constructing a formal cross-section, students are encouraged to mark specific elevations on the topographic profile, then draw in the topography, and finally draw in the subsurface geology. Methodologies for marking cross-sections based on a topographic map are explained in most physical geology lab books or in many online sites.

- Choose horizontal and vertical scales. The vertical axis should go a bit higher than the highest elevation along the line of profile and lower than the lowest point by enough to allow room for rock layers beneath the surface. The vertical scale can be the same as the horizontal scale or different, depending on whether you want to emphasize what the slope of the rocks truly looks like below the surface or whether you want to expand the scale so the rocks are easier to see. Although it's possible to choose a horizontal scale for the cross-section that is different from the horizontal scale on the map, we recommend keeping the same scale, unless your primary objective is to give your students a lesson in mathematical conversions.

- As students draw their topographic profiles, take note if there are any missed ideas that you might prompt them to think about. Does the students' placement of rivers or valleys match up with low places on the profile? Are the tops of hills the high points on the profile? Are they in the correct locations? This is a good opportunity to point out the connection between the maps/cross-sections and the real world. If they've put a river at the top of a hill, that might be a problem!
- Go through a few basic concepts for drawing a geological cross-section. Unlike the topographic profile, the geological cross-section cannot be analytically determined in every aspect. It is an interpretation of the data present at the surface of the Earth as represented on the geological map. More than one interpretation is possible, but some are better than others, and some interpretations are clearly wrong. The key constraints are that (1) the top of the cross-section must match up with what is observed at the surface of Earth and (2) a particular rock layer in one location must be related to that same rock layer in another location by some reasonable geological process.
- Be ready to prompt your students with helpful reminders, such as starting with what they know—the rocks at the surface—and then expanding their thinking to construct an interpretation of what lies below the surface. You might use pieces of paper or modeling clay to show what layers or folds look like in map and cross-sectional view to help students visualize the third dimension.
- Don't simply tell your students that a particular interpretation is "wrong" and give them the "right" interpretation. Challenge their ideas with questions that cause them to consider alternative subsurface interpretations that are better supported by the evidence. You might ask questions such as these:
    - Does the geological cross-section match with rocks at the surface?
    - Does the slope of the rocks correspond to the strike and dip of rocks as shown at the surface?
    - Can pieces of a particular layer of rock be related to each other through reasonable geological processes such as folding and faulting?
    - Do the rock layers remain parallel to each other unless a geological process (such as formation of an unconformity) made them not do so?
    - Do the rock layers maintain more or less constant thickness (unless you have reason to believe they shouldn't)?

EXAMPLE INTERACTION

Student question: How can I possibly know what the rocks look like underground? Where do I start?

Teacher prompt: Well, what is it that we know? What's our data?

Student question: I'm not sure ... what kind of rocks are there?

Teacher prompt (clarifying the question): What kind of rocks are where?

Student response (clarifying the question): On the map.

Teacher prompt: Does the map show what's underground?

Student observation: No, only what's at the surface.

Teacher prompt: So, what you know is the rocks at the surface. You can start by marking what's at the surface on your cross-section.

SUMMARY CHECKLIST FOR TEACHER AS PRACTITIONER OF SCIENCE

- Review or introduce the mechanics of making topographic profiles.
- Offer a few pointers on how to start a geological cross-section.
- Remind students that their geological data are for what exists at the surface and that their interpretation of the subsurface must relate to the surface data.
- Remind students to think about how their constructed drawings relate to the real world.
- Think ahead and be ready with prompts for how to visualize the third dimension when constructing the geological cross-section.

TEACHER REFLECTION

Think about how the students' hard-won three-dimensional thinking might be used in future lessons. For example, it might provide the foundation for future class exploration of natural resources (where we get oil or coal) or questions about water quality (how and where water moves in the subsurface). Does it bear on any concerns in the news, such as the potential for fracking to contaminate ground water? Ideally, relating this activity to a real subsurface data problem in your hometown will provide a context for these spatial-thinking skills.

DIGGING DEEPER

CONTOUR MAPPING WITH WEATHER

The following information was given in "Captain Favorite's Weekend Fishing Forecast" in the May 21, 2015, edition of *The Bradenton Times* (Bradenton, Florida): "A tightening pressure gradient later in the period could result in evening easterly surges that may create caution or advisory conditions at times." If you want to fish for red grouper along the Florida Gulf Coast, you need to keep an eye on the pressure gradients. If you want to fish for walleye in one of the 10,000 lakes of Minnesota, you need to know the water depth. Both require an understanding of contour mapping.

We would be remiss if we didn't mention the importance of the contour map. This type of map portrays data that vary continuously as a function of location. Contour maps are useful in showing both three-dimensional spatial data (such as topography and water depth) and nonspatial but spatially correlated numerical data (such as barometric pressure and growing degree days [GDD]).

Traditionally, contour mapping has been introduced in the classroom as part of a topographic mapping unit. Many classroom topo map activities engage students in creating a contour map based on a plastic mountain model sitting in a basin. Students fill the basin with different levels of water and then map the contours formed where different water levels intersect the model mountain. This gives students a feel for what the contours mean in the physical world.

Contour maps can also be studied in association with other types of data. For example, contour maps are used to represent depth of overburden (important for assessing how much rock will need to be removed in extracting an ore body); depth and flow direction of groundwater; and agriculturally significant information such as mean precipitation, evaporation, or GDD.

A particularly important use of contour maps is in portraying atmospheric pressure data. From a contoured pressure map, forecasters and red grouper fishermen can infer wind speed and direction.

By taking measured pressure data, constructing a contour map, and then drawing a profile across the map, students can connect the map to real-world phenomena and get a feel for what these maps tell us. For example, Figure 5.11 (p. 94) shows measured barometric pressure data, contours interpolated and drawn from that data, and a pressure profile based on the contours.

In drawing a contour map, students need to learn some plotting techniques (e.g., how to choose a contour interval and how to interpolate), but the broader educational goal is to learn to visualize the three-dimensional character of the plotted contours and to connect map data to real-world experience.

Figure 5.11

Illustration of a contouring and spatial reasoning activity with barometric pressure data

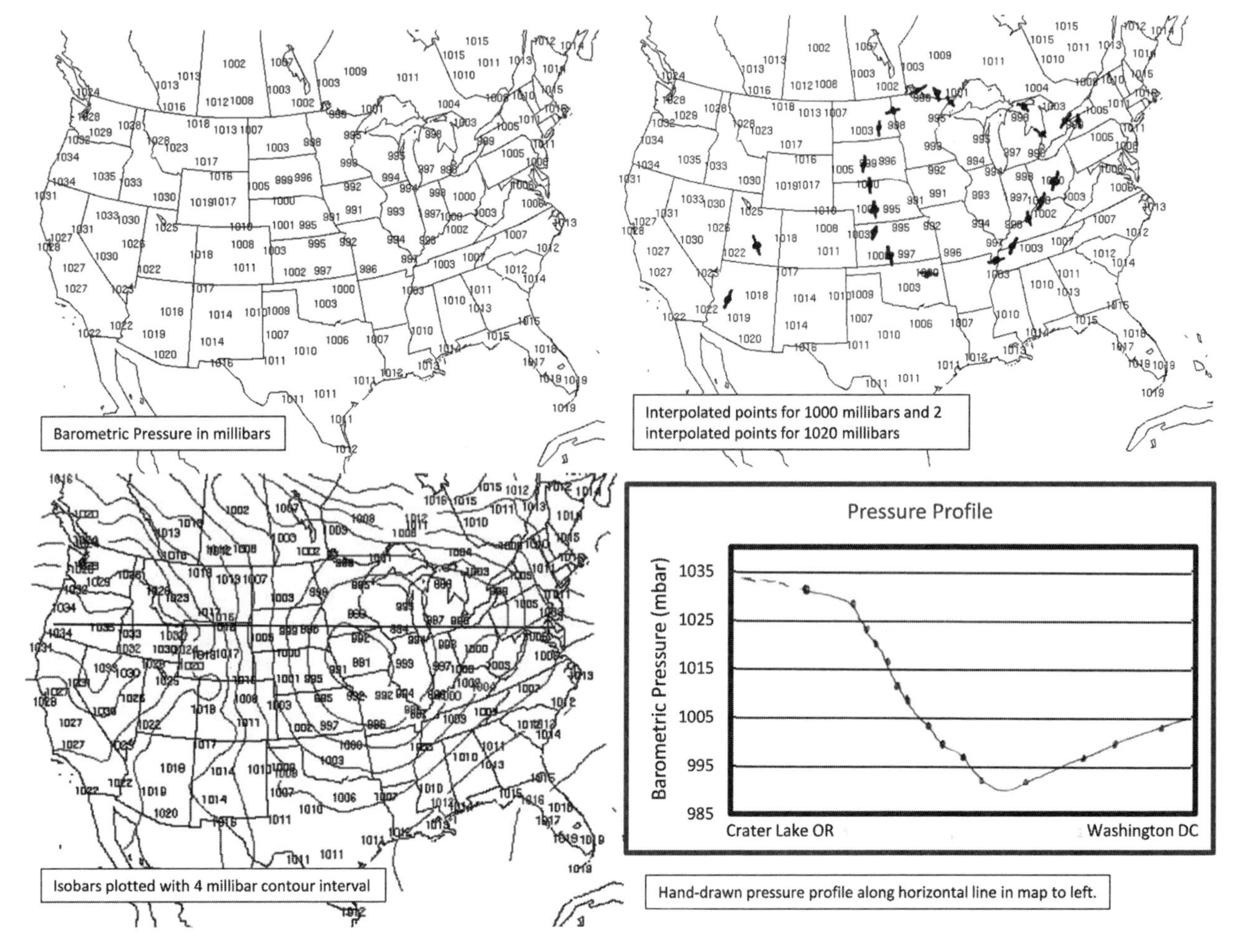

Starting with the data presented in the map on the upper left, a student can interpolate points (examples for interpolating 1000 millibars are shown in upper right), and then draw in pressure contours (lower left). This solution to the activity defines a contour interval of 4 millibars, making contour lines at 992, 996, 1000, 1004, 1008, 1012, 1016, 1020, 1024, 1028, and 1032 millibars. From the contour map, students can draw pressure profiles analogous to topographic profiles (lower right). The activity can be linked to a unit on weather concepts. Winds are expected to be strongest where the pressure gradient is steepest. Wind direction can also be inferred from the isobar map.

MODELING, PART 1

CONCEPTUAL MODELING

The meaning of the word *model* is confused by its wide variety of uses: Teachers model good practices, students build models of rivers using downspouts and sand, scientists construct numerical models of weather systems. Or—here's a classroom favorite—students build model volcanoes with plaster of Paris. In the *Next Generation Science Standards* (*NGSS;* NGSS Lead States 2013), modeling is much more than building objects that simply look like natural systems—such as a plaster of Paris volcano. A model is a conceptual understanding of how the universe, or some part of it, works. A model provides both the foundation for understanding the universe and a framework for making predictions about its future.

THE PURPOSE OF MODELING IN TEACHING AND RESEARCH

What Is a Model?

One definition for a model, useful in both the classroom and research, might be that it's the pattern we make in our head as we construct understanding, envisioning a picture of how things work, telling a story of how things work. In some sense, drawing a line through data on a graph is a model of how the variables relate to each other. Drawing a cross-section of a subducting plate is a model of how plates behave at a convergent boundary. Imagining molecules of air colliding with a tree is a model for wind. It is our conceptual models for the Earth that allow students and teachers to understand the answer to questions such as "Can you drive a submarine under an island?" or "What would happen to the water if you drilled a hole in the bottom of the sea?" Conceptual models are also the key to answering curricular questions such as "How does buoyancy of warm air produce rising air and wind?"

Models Are Constructed, Not Merely Used

Consider the parallels between constructing models in research and constructivist learning in the classroom (see the section "Constructivist Learning and the Practice of Science," p. 98). In both cases, learners are trying to make sense—to tell a story—of how the universe works. In both cases, learners are engaged with open-ended questions that require creative interpretations of complex constraining observations. In both cases, constructing the model is an iterative and ongoing process.[C16]

A couple of decades ago, I (Russ) attended a workshop on changing teacher licensure standards in the state of Minnesota. The presenter told a story to illustrate the idea that models change as we gain new experiences:

> *Two donkeys, one with a heavy load of grain, the other with a lighter load of wool, came to a river. The first crossed the river and found that, while crossing, the weight of the grain seemed less, and once across, the weight was little changed from its original weight. The donkey with the wool saw this and reasoned that, because its load was lighter, the river would make its load lighter still. So that donkey crossed as well. Of course, once the water saturated the wool, it became much heavier than either the dry wool or the grain. The second donkey continued on, chagrined at its much heavier load.*

The presenter's summary of the story was that, after crossing the river, the two donkeys made different assumptions—one assumed that the river lightens loads, the other that the river makes loads heavier—and the assumptions directed their future models of reality. I had a somewhat different interpretation of the story. Before crossing the river, the donkeys believed either that they were carrying the same type of material or that the two types of material would have similar properties in the river, a mental model that lacked any observational support and therefore might be called an assumption. Once they crossed the river, they could conclude that their original assumption was wrong! They were not carrying materials with identical properties. In both interpretations, the donkeys' mental model of reality changed in response to a new experience.

The takeaway message is not that we must choose which model is right (my interpretation of the story or the presenter's, for example) but rather that constructing understanding is an ongoing process and something that each learner has to do on his or her own.

The teacher's goal is not to impose models of reality on the students, nor is that the purpose of textbooks. In 2013, the National Science Teachers Association (NSTA) selected several dozen educators to find and curate lessons that address the *NGSS* performance expectations. During discussion of models at a workshop for the curators in Charlotte, North Carolina, one of the curators commented, "Textbooks are full of models—all those diagrams. What's to stop the students from just memorizing the diagrams?"

Well, nothing probably. Learning a diagram of how things work that someone else has created is a place for students to start in constructing understanding. But a memorized diagram is a bunch of colors, lines, and squiggles—not a model. Students need to make the model their own, and that means interacting with it, trying it out, talking about it, modifying it, and using it to make and test predictions.

The goal of science education is not for the teacher or the textbook to give the students a model which they can then "learn and use." The goal is to create experiences that help students construct their own mental models, a process that mimics the scientific practice of model construction.

Modeling Modeling

In keeping with our philosophy that the teacher should be a practitioner of science and a fellow scholar along with his or her students (see Box 6.1; this idea is discussed in more detail in Chapter 17), we suggest that the teacher also needs to be about the business of model construction. Students look to you, the teacher, to showcase, or model, how you revise and expand your own understanding. You need to model modeling. None of us has a perfect and complete mental model of reality. The ongoing modeling of reality is what keeps science alive for us. Constructing understanding is an ongoing process, even when we become experts.

Box 6.1
Teacher as Mentor and Practitioner of Science

In some sense, the teacher-student relationship is like the relationship between master and apprentice. The apprentice can learn from the master's greater experience, but both are engaged in practicing a skill together, through which practice they both learn and grow.

Our goal in this book is to inspire teachers to engage in learning with their students, and not to simply convey knowledge and be the keeper of the answer key. Therefore, we offer ideas and examples for activities and interactions—not a detailed curriculum—which teachers can adapt to their own classroom explorations.

Albert Einstein found the inspiration for much of his life's work in a wondering effort to construct mental models of how light works. Early on, he wondered what light waves would look like if he moved as fast as they did. Later, in 1917, he said "For the rest of my life, I will reflect on what light is" (quoted in Perkowitz 1998, p. 69). Thirty-four years after that, toward the end of his illustrious career, Einstein (quoted in Lam 1998, p. 1) said, "All the fifty years of conscious brooding have brought me no closer to answer the question, 'What are light quanta?' Of course, today every rascal thinks he knows the answer, but he is deluding himself."

The message is that all of us are learners together, engaged in constructing models of the universe—researcher, teacher, and student.

CONSTRUCTING UNDERSTANDING THROUGH MODELING

Constructivist Learning and the Practice of Science

We've drawn a parallel between constructing models (either in the classroom or in research) and constructivism as an educational philosophy. Constructivist learning is iterative and ongoing, like scientific research. In a book chapter titled "Constructivism: A Psychological Theory of Learning," Catherine T. Fosnot and Randall Perry (2005) proposed that learning "requires invention and self-organization on the part of the learner" and that "as learners struggle to make meaning, progressive structural shifts in perspectives are constructed—in a sense, big ideas."

In her book *Learner-Centered Teaching: Five Key Changes to Practice,* Maryellen Weimer (2013, p. 24) noted that constructivism "proposes that students must be interacting with the content." Scientists also construct meaning by interacting with the content, that is, their new observations and preexisting understandings.

One difference between the scientist and the classroom student is that the frontier of the scientists' construction of understanding, and thus the kinds of things that they call models, might be well beyond where students can go. However, students can participate in authentic construction of models, even if that construction is not at the current frontiers of science. They work at the frontiers of their own understanding.

Another difference between research lab and classroom is that the construction of meaning in the classroom must be somewhat accelerated—the classroom student doesn't have time to study problems for a year, or a decade, or a lifetime. It's the teacher's job to give students an authentic experience constructing models while providing enough guidance for them to make faster progress than if they simply recreated scientific discoveries on their own. In the following text, we offer a few examples of how to do that in the classroom.

Starting With the "Wrong" Model

As discussed in the preceding sections, students need to interact with a model to make it their own. Students can't create a model or make adjustments to it if they are simply given an already-perfected model to memorize. The *NGSS* note the importance of student involvement in model creation and revision; for example, the Appendix F matrix for science and engineering practice 2 (Developing and Using Models) includes "Develop or modify a model—based on evidence—to match what happens if a variable or component of a system is changed" for grades 6–8 and "Develop, revise, and/or use a model based on evidence to illustrate and/or predict the relationships between systems or between components of a system" for grades 9–12 (NGSS Lead States 2013).

To develop, revise, and modify a model, students need to start with something that is not yet perfected—the "wrong model," if you will. We suggest a couple of ways to do this in the classroom.

Building on Students' Existing Notions

One way might be to start with the students' existing incomplete or incorrect notions of how things work and use experiments and discussions to prompt modification of their models. Students often have notions—beginning models—which can provide a starting place for model development. In Chapter 2, "The Controlled Experiment," we considered the temperature difference between a tray of water left out overnight and the air in the classroom. In a classroom follow-up, I (Mary) and my students explored how evaporation of water is an important player in moving heat on Earth. Warmer temperatures produce more evaporation, which both cools the water and moves heat into the air.

One student explained the process this way: "Water evaporating into the air warms the air because it's the warm water that evaporates." This explanation implied a mental model in which both warm and cold molecules existed in the water and only the warm molecules evaporated. This "not-quite-there" model provided a starting place to engage students with thinking about temperature and thermal energy. I had students heat various volumes of water, measure their temperatures, and note how long it took each to reach boiling. Students discovered that each batch of water boiled at the same temperature, but that the time it took to reach boiling varied according to how much water they heated. The amount of heat needed depended on *all* of the water in the beaker, not just some of it.

Starting with a beginning notion, students used their experimental investigation to come to an improved understanding of the relationships among heat, temperature, and evaporation.

Safety Notes

1. Make sure students wear indirectly vented chemical-splash goggles, aprons, and nonlatex gloves during all aspects of the activity—setup, hands-on piece, and takedown.
2. Caution students to work carefully with heated liquids to prevent burns.

I (Russ) sometimes have students in introductory college courses do a lab on isostatic equilibrium and buoyancy. As one part of this activity, I give students modeling clay and ask them to make it float. The goal is for students to recognize that shaping the clay like a boat will displace more water, decrease its effective density, and allow it to float.

Students often begin with the wrong model, thinking that the modeling clay will get lighter if they make it very thin and that is what causes it to float. In fact, they can often make this work. If made thin enough, the clay layer will stay on the surface of the water by surface tension. No isostasy needed. However, when I ask them to weigh the clay, they discover that its mass wasn't changed by making it thin. This brings them to realize that their model needs some adjustment. Thinness doesn't make clay lighter. There must be another explanation for why boats float. This leads them to think deeper about the idea of buoyancy.

Building on a Seed Model

A second way to engage your students in model revision and development might be to offer a "seed" model that includes some of the key ideas for a model, but not all of them. This is not unlike the progress of scientific discovery. Consider the following model for why the ocean is saltier at the surface than at depth, from *Natural History*, an encyclopedia written in Latin by Pliny the Elder beginning in AD 77. For comparison, the modern model is that water evaporates from the surface, leaving behind a saltier layer (see Box 6.2).

> *Consequently liquid is dried by the heat of the sun, and we are taught that this is the male star, which scorches and sucks up everything; and that in this way the flavour of salt is boiled into the wide expanse of the sea, either because the sweet and liquid, which is easily attracted by fiery force, is drawn out of it, but all the harsher and denser portion is left. (Pliny 77/1938, pp. 349–350)*

Notice that the key idea behind our modern model is present in Pliny's analyses from 2,000 years ago: The sun dries the water and the salt is left behind. The *story* of how it works has changed considerably in our modern model—no flavors boiled into the sea or sweet liquid attracted by fiery force.

The key point is that you can start your students out with a "seed" model, which they can modify and improve based on new observations and reasoning, just like modern understandings of evaporation were built up from previous foundational models. Glenn Dolphin offers some examples of this approach using real models from the past in his 2009 article in *Science & Education*, "Evolution of the Theory of the Earth."

Box 6.2
Implications of the Salty Surface of the Sea

But wait! If evaporation makes the surface layer saltier, why doesn't that salty layer sink? Isn't the salty water denser than the less salty water?

It does sink!

As warm surface water moves toward the poles, evaporation makes the water saltier. However, the density of the salty water is balanced by the density of colder water at depth so that the warm, salty water stays buoyant. Once the salty water reaches polar regions and gets both cold and salty, the water sinks.

The sinking water in the Labrador and Greenland Seas is a key driver for the global circulation of ocean water that moves tropical heat to the poles and then recycles the cold arctic waters back to the equator. This movement of heat is a major part of Earth's climatic balance.

Example of Building on a Seed Model

In 2014, the exploration of Gale Crater on Mars by the Curiosity rover was in the news and I (Mary) wanted to include some of its discoveries in my classroom. The rover had discovered cross-bedded sandstone in Gale Crater, and this sandstone was interpreted to be composed of deltaic sediments deposited where rivers flowed into the crater when a lake existed there long ago. Farther out in the crater, the cross-bedded sandstone gave way to laminated mudstones of the lake bottom.

We had just completed an experimental unit in which students used stream tables to study erosion in rivers and deposition where the rivers flowed into a standing body of water. Students observed cross-bedded coarse sediments deposited at the mouth of the river and finer sediments deposited farther out in the stream table lake.

With this backdrop, I had the students read a simplified version of a *Los Angeles Times* news article about the Curiosity's discoveries (Khan 2014). I had them read the original *Times* article and watch several associated online videos (e.g., "The Making of Mt. Sharp" [linked from the Khan 2014 article] and "28 Months on Mars" [Bostock et al. 2014]). Then, I worked through my own model about what happened at Gale Crater in the past (see Figure 6.1, p. 102). This "thinking-out-loud" gave students a seed model against which to compare new and more detailed observations from Curiosity.

Figure 6.1

"Seed" model for the interpretation of features observed in Gale Crater on Mars by the Curiosity rover

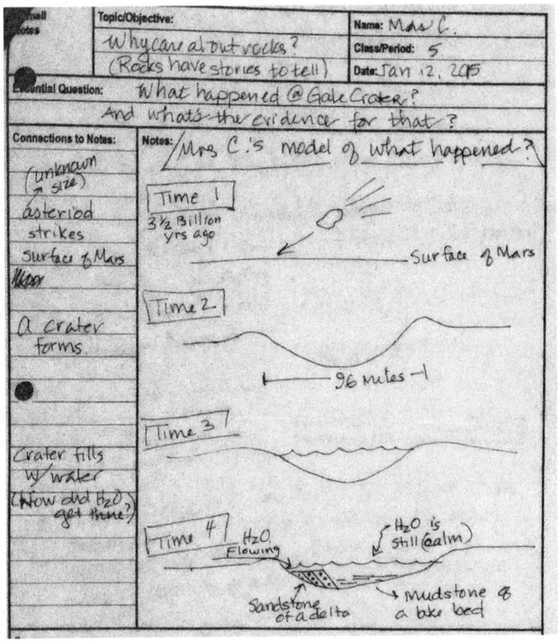

Topic/Objective: Why care about rocks? (Rocks have stories to tell)
Name: Mrs. C.
Class/Period: 5
Date: Jan 12, 2015
Essential Question: What happened @ Gale Crater? And what's the evidence for that?
Connections to Notes:
(unknown size) asteriod strikes surface of Mars
a crater forms
Crater fills w/ water (How did H_2O get there?)
Notes: Mrs C.'s model of what happened?
Time 1 — 3½ Billion yrs ago — Surface of Mars
Time 2 — 96 miles
Time 3
Time 4 — H_2O Flowing — H_2O is still (calm) — Sandstone of a delta — Mudstone of a lake bed

Next, my students next read a more comprehensive article from the *New York Times* (Chang 2014) that also examined the features in Gale Crater. I gave them the following assignment: "Given the new information from the *New York Times* about how features and structures formed in Gale Crater, make a suggestion for how to revise Mrs. Colson's model. Describe your suggested change below."

After some initial conversation about what constitutes new information or evidence (when reading news articles, middle-grade students tend to look for evidence in textual clues, such as language and tone, rather than in the rocks that the text talks about), students began making their suggested modification to my model:

- "I would've added more information and labeling like adding 3.4 billion years to Time 1."
- "I think you should make a new Time 5 and label it 'clay' for after the water leaves and the rock formations are left behind."

- "I think at the end of Mrs. Colson's model she should add another step. She should add that Gale Crater may have drained out, and water went back in many, many times over the years."
- "I think that instead of a single asteroid that there were plenty of them, all striking like a meteoroid storm."

One example student model is shown in Figure 6.2.

Figure 6.2

Student-enhanced model for the interpretation of features observed in Gale Crater on Mars by the Curiosity rover

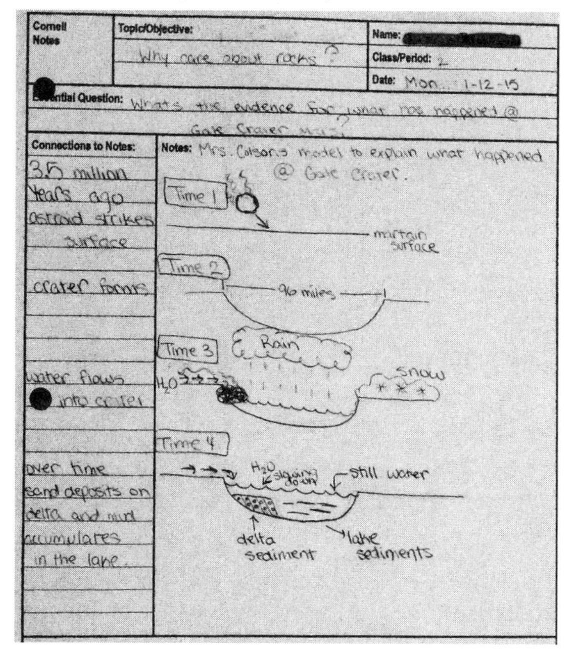

The student expanded on processes of precipitation and runoff in Time (step) 3.

Another student picked up on a key concept in the *New York Times* article that I hadn't included in my model. There is a large mountain in the middle of Gale Crater whose formation is not well understood. One model is that the mountain may have formed when surrounding sediment was removed from the crater by wind. This student said, "Instead of having water build up the mountain, have the crater fill up with sediment and have the wind blow some sediment away to form the mountain."

By engaging my students in the exercise, I showed them my own interest in learning—incorporating something new into the curriculum based on my own curiosity about Curiosity. I linked an exercise in reading comprehension to their experimental lesson. I encouraged them to think about the connection between evidence and model. I also gave them a chance to modify a model of an important geological process and make that model their own.

Constructing Understanding of What We Can't See or Measure

Models often deal with things that we can't see or measure. In a report on geysers, one of my (Russ's) students wrote, "Since scientists cannot see what happens beneath a geyser, they have formed theories to explain what causes them." The student certainly captured the idea of a model as something we can't see, if not quite getting the idea that models still must be based on observation.

It's important that teachers remember that models are mental images—stories—of things we can't see, and take the time to make tangible connections with the observable world. I (Mary) learned this lesson early in my career from a special student who suffered from a medical condition that severely limited his life expectancy. We were studying air pressure, considering how atoms of air press on everything around us. The student put his finger on a simple question that neither I nor the other students had quite identified: "How can we measure something we can't see?"

I was delighted with my student's question but had no ready response. I offered a few ideas such as "Well, we can measure what air does, like how it moves the trees." Over the next few years, I developed better answers to the question, exploring with experiments how we learn about air pressure and construct models of things we can't see.

The deeper significance of my student's question, and the part that took me several years to sort out, is that there are some aspects of models that not only can't be seen, they can't even be measured—at least not in the classroom. For example, observing and graphing the rise of mercury in a barometer with increasing pressure—something we can measure—is not the same thing as constructing a mental model of how air molecules press against the mercury and hold it up against the force of gravity. Scientists constructed models of atoms and the force they exert when no one had observed an atom pressing on the world around. Likewise, in the classroom, we construct models based on observational evidence, but we can't actually see or measure an atom pressing on the world around. Pressure and gravity are abstract concepts—models—by which we come to understand and organize the experimental results, making sense of the world through pictures and stories of things we can't directly see or measure.

Sadly, my student passed away in his early 20s, many years ago now. But his question has lived on with me and affected my entire career. In addition to giving me insight into the nature of models, the question taught me that teachers can learn from their students, even over an extended period of time, and that it's important to value everyone's contributions.

If models can't be seen or measured, it follows that a model is more than a physical representation or simple description of what we observe. Stephen Pruitt, who has led the development of the *NGSS*, said at a 2015 NSTA talk on science education in Chicago, "If you can eat it, it's probably not a scientific model!" We expand that a bit: If you can build it, it's not a scientific model (although physical models have a place in science and science education—see Box 6.3). A scientific model is a mental concept that gives us a way to explain observations. To understand models as a practice of science, students need to *use* models to make predictions and explain results.

Box 6.3
A Place for Physical Models

Building scale models of volcanoes, solar systems, and rivers, or using physical models to teach information or terminology from the textbook, does not provide the opportunity for investigation or development of a model in the sense envisioned by the *NGSS*. Even so, physical models have a place in science and science education. For example, a physical model built in the classroom, or an illustration of such a model, can engage students and help them understand or communicate a scientific idea. Also, physical models that involve experimentation with analog materials or at a scaled size are important investigative tools in both the classroom and research lab. We give examples of how physical models can be used in these ways in this chapter and in Chapters 2, 13, and 16.

Additional examples of classroom uses for physical models can be found in the paper "Students' Use of Physical Models to Experience Key Aspects of Scientists' Knowledge-Creation Process" (Kastens et al. 2013).

In Chapter 2, we talked about activities with cups, water, and air that I (Mary) do to engage students in thinking about air as an actual substance that can exert an actual force. After similar and additional experimental activities in your own classroom, you might engage students in thinking about models for air pressure.

The following exercise involves asking students questions about a classroom barometer and having them sketch pictures of their working model. This exercise gives the teacher a good feel for how students are constructing their understanding and gives students an opportunity to discuss differences between their models and come to understand the link between the model of things they can't see and the observation of things they can see.

Figure 6.3 (p. 106) shows a simple classroom water barometer in which air pressure inside the tube is equal to air pressure in the room. Students might sketch a model for

Figure 6.3 ✪
A simple classroom barometer

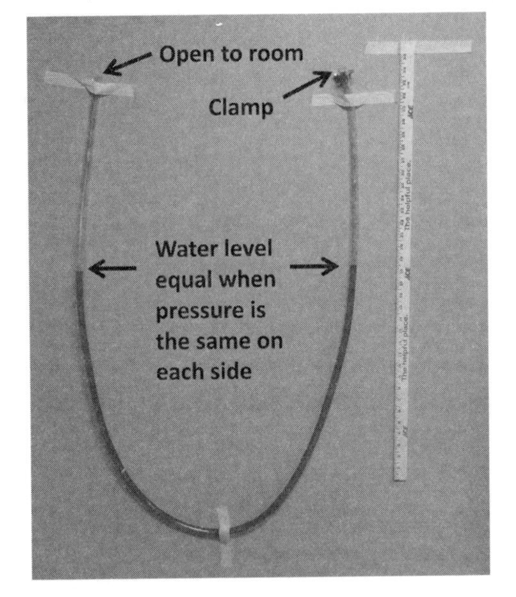

This design has the air pressure on the clamp side of the *U* equal to the pressure on the open-to-room side. Clamping the hose so that no air leaks through is essential to make this barometer work (an experimental challenge that I [Russ] often leave for my students to discover and solve). In the picture, the seal on the clamp is aided by pressing plastic modeling clay into the end of the hose. The liquid in the tube is water with red food coloring added.

this system, including illustration of air molecules both inside the tube and inside in the room. There are a couple of factors to consider in drawing the molecules: (1) Are the number of molecules inside the tube and inside the room equal or different? (2) Is the density of molecules—how close they are together—different or the same?

Students might discuss their pictures (see Figure 6.4 for an example).[C17] Questions to be discussed may include the following: How do the components of the model correspond to real air? What does drawing more molecules mean? What does drawing the molecules closer together mean?

Now, suppose that air pressure in the room increases. In terms of our model, how will the conditions in the barometer system change? Students can again draw and label an illustration, perhaps like Figure 6.5. There are at least three factors that students might include in their drawings: How will the water levels in the tube shift? How does the spacing of the air molecules in the room change? How does the spacing of the molecules in the clamp side of the tube change?

The modeling activity in the preceding paragraph produces a prediction—that the water will shift in a particular way in the tube. This prediction can be tested by setting up a barometer like this in the classroom. Attach an air bulb (e.g., an air bulb duster or a turkey baster) to the "room" side of the tube and squeeze to simulate higher room pressure. Does the water shift as students predicted?

As you plan your activity, you might consider having students revise their models on the basis of this experiment. You can prompt your students with questions to help them focus on key ideas: Does the pressure change on the open-to-room side of the barometer? Does the pressure change inside the tube on the clamp side? If the pressure changes, does it increase or decrease? How is that portrayed by the closeness of molecules in the model? Does the number of molecules of air inside the tube on the clamp side change?

You might have students explain to a classmate, in terms of their model, what is holding the water higher on one side of the *U* than on the other side. Their explanation of the model might include pressure forces and weights of the water if you have introduced those concepts to your students in previous lessons. You might also have students

Figure 6.4

Example conceptual model of air molecules for the barometer shown in Figure 6.3

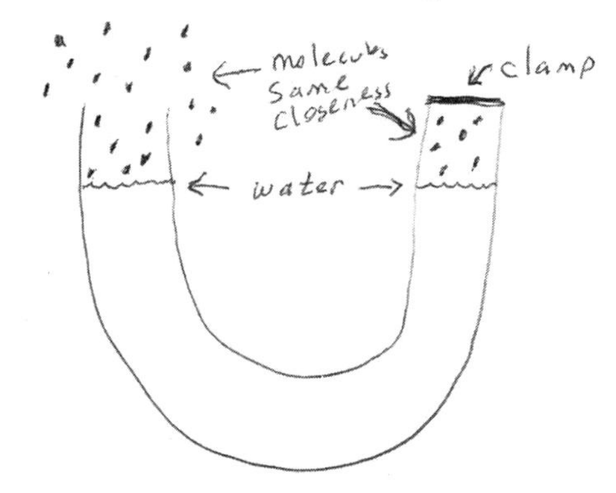

The student has correctly shown the compression of air molecules being the same on both sides of the *U*.

Figure 6.5

Example conceptual model for the case in which air pressure in the room increases relative to the case shown in Figures 6.3 and 6.4

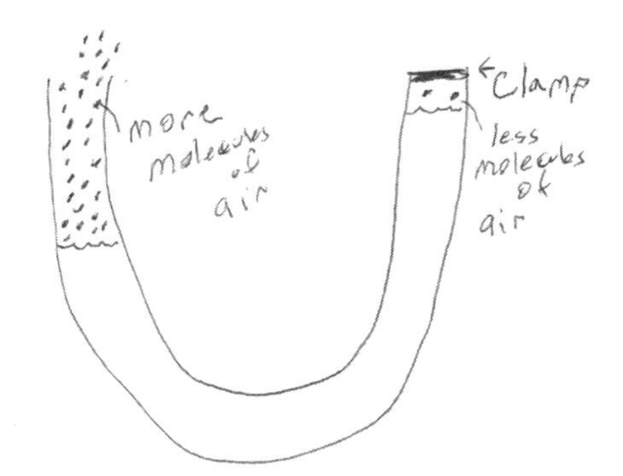

Here, the student has correctly shown that the compression of the molecules must be greater in the room than in the clamp side of the tube (air pressure is higher in the room than the tube) and that the water level will be higher on the right than the left. However, the number of molecules of air in the closed end of the tube should stay the same as what was shown in Figure 6.4 because that end of the hose is clamped shut and molecules can't leak in or out. There is some ambiguity about whether "more molecules of air" means more closely packed molecules (the key idea for modeling higher pressure) or literally more molecules (which might simply be related to higher volume).

explain to a classmate, in terms of their model, what *compression* or *decompression* means.

After students have discussed and explained, they can go through a third round of revising their models. As students continue to revise their models, they might converge toward something like Figure 6.6 (p. 108).

The physical barometer shown in Figure 6.3 works only so-so in measuring changes in atmospheric pressure because the small volume of air in the clamp side of the tube quickly compresses or decompresses to balance with the changed pressure in the room, keeping the change in water level on either side of the *U* small. This kind of classroom water barometer works better if you put a large flask on the tube side (see Figure 6.7, p. 109). Students can explain to a classmate, in terms of their model, the effect of adding a large flask to the closed end of the tube. Why is there a bigger change in the water level with a given change in room pressure? How does this new setup make the measurement of atmospheric pressure more sensitive?

> **Safety Note**
>
> When working with glassware and metersticks, make sure students are wearing eye protection—safety glasses or safety goggles.

Figure 6.6

Example conceptual model for the simple classroom barometer before (left) and after (right) the simulated increase in room pressure

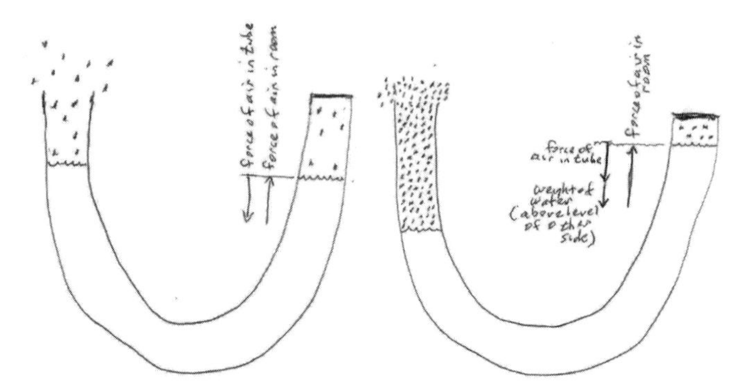

Forces acting on the water are modeled with vectors. The density of molecules on the open-to-room side increases when room pressure increases. The number of molecules on the closed end of the tube stays (symbolically) the same. The density of molecules on the closed end increases because of decreased volume, but the density remains below the room density, as it must for the water level to rise on the right side. Arrows show the balance of forces, conceptually.

FINAL THOUGHTS ON MODELING AND THE *NGSS* DISCIPLINARY CORE IDEAS

In 2013, I (Mary) attended a workshop on models offered by Joe Krajcik (*NGSS* writing team leader) at NSTA's national conference in San Antonio, Texas. One of the key ideas that I came away with, and wrote down in my notes from the workshop, was that there is a deep connection between the disciplinary core ideas (DCIs) in Earth and Space Sciences (ESS) and the practice of modeling. I wrote that the "value of the core ideas is that they can help students develop an integrated understanding of how ideas are connected" and "the core ideas are a mental framework to organize all the information."

It occurs to us that, in some ways, a DCI in ESS can be thought of as a model of models—a system that integrates the workings of many smaller-scale systems into ever larger processes and cycles. For example, consider these two sentences from DCI ESS1, Earth's Place in the Universe, in *A Framework for K–12 Science Education: Practices, Crosscutting Concepts, and Core Ideas* (NRC 2012, p. 176):

> *The solar system consists of the sun and a collection of objects, including planets, their moons, and asteroids that are held in orbit around the sun by its gravitational pull on them. This model of the solar system can explain tides, eclipses of the sun and the moon and the motion of the planets in the sky relative to the stars.*

No one person, using only his or her own data and observation, looked at the night sky and concluded that the Sun is the center of our solar system. Many observations, and many models, contributed to our understanding that the planets orbit the Sun. For example, Galileo was not able to observe Jupiter orbiting the Sun, but his observation and modeling of moons orbiting Jupiter proved that not everything orbited the Earth. Other observations and models also became part of our model of the solar system, making it a model of models.

This connection between modeling and the ESS DCIs is also apparent in the performance expectations. All 10 of the performance expectations dealing with the practice of modeling in middle and high school earth science (MS-ESS1-1, 1-2, 2-1, 2-4, 2-6; HS-ESS1-1, 2-1, 2-3, 2-4, 2-6) address the large, system-scale concepts of the DCIs. They address how giant systems are put together, not necessarily what you can do at the classroom level. For example, two of the middle school and high school *NGSS* performance expectations for Earth's Systems are, respectively, "Develop a model to describe the cycling of Earth's materials and the flow of energy that drives this process" (MS-ESS2-1) and "Develop a model to illustrate how Earth's internal and surface processes operate at different spatial and temporal scales to form continental and ocean-floor features" (HS-ESS2-1).

Figure 6.7

Modification of a simple classroom barometer to improve performance

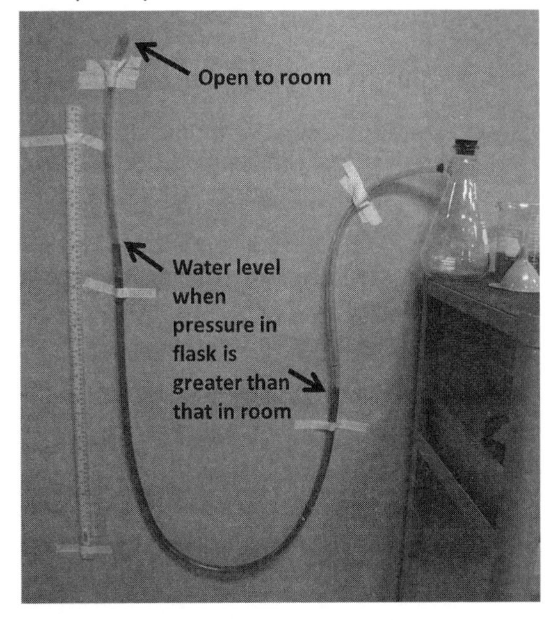

The performance of the simple classroom barometer shown in Figure 6.3 can be improved by putting a large flask on the closed end (the flask must be sealed from the room air). The larger volume of air allows for larger movements of water level for given changes in room pressure (because the larger volume of air in the closed end of the tube will compress and decompress by a larger volume as pressure changes). This allows greater sensitivity in measuring small changes in room pressure. For the situation shown, the room pressure is lower than the pressure inside the flask.

The ESS DCIs are very large-scale, system-level concepts that are quite difficult to teach in the classroom. In fact, there is some risk that if curriculum writers focus on the DCIs in implementing the *NGSS*, earth science courses will devolve into teaching big diagrams with lots of arrows—which is boring for students and doesn't really capture the spirit of the practices of science or the story-telling nature of earth science. In both the classroom lab and the science lab, students and scientists address small systems but fit the results into big systems. I (Russ) do research on geochemical processes, but I don't try to get my brain around "describe the cycling of Earth's materials" all at once. Instead, I focus on small parts of that large-scale system in my research.

Given that the ESS DCIs are a model of models—systems that integrate many smaller-scale systems into ever-larger processes and cycles—we propose that the place to begin in the classroom is where we also begin in research, at the small-scale end of that spectrum. Teachers and their students can work toward integrating classroom experiences into a bigger and bigger picture as students' understanding progresses.[C18]

REFERENCES

Bostock, M. S. Carter, J. Corum, and J. White. *New York Times.* 2014. 28 Months on Mars [video]. *www.nytimes.com/interactive/2014/12/09/science/space/curiosity-rover-28-months-on-mars.html?_r=0.*

Chang, K. *New York Times.* 2014. Curiosity Rover's Quest for Clues on Mars. December 8. *www.nytimes.com/2014/12/09/science/curiosity-rovers-quest-for-clues-on-mars.html.*

Dolphin, G. 2009. Evolution of the theory of the Earth: A contextualized approach for teaching the history of the theory of plate tectonics to ninth grade students. *Science & Education* 18 (3): 425–441.

Fosnot, C. T., and R. Perry. 2005. Constructivism: A psychological theory of learning. In *Constructivism: Theory, perspectives, and practice,* ed. C. T. Fosnot. 2nd ed. New York: Teachers College Press.

Kastens, K. A., A. Rivet, C. Lyons, and A. R. Miller. 2013. Students' use of physical models to experience key aspects of scientists' knowledge-creation process. Paper presented at the National Association for Research in Science Teaching annual international conference, Rio Grande, Puerto Rico. *http://ltd.edc.org/sites/ltd.edc.org/files/Kastens_NARST2013.pdf.*

Khan, A. *Los Angeles Times.* 2014. Mars Apparently Had Massive Lake, NASA's Curiosity Rover Finds. December 8. *www.latimes.com/science/sciencenow/la-sci-sn-curiosity-water-mars-mt-sharp-lakes-20141208-story.html.*

Lam, R. W., ed. 1998. *Seasonal affective disorder and beyond: Light treatment for SAD and non-SAD conditions.* Washington, DC: American Psychiatric Press.

National Research Council (NRC). 2012. *A framework for K–12 science education: Practices, crosscutting concepts, and core ideas.* Washington, DC: National Academies Press.

NGSS Lead States. 2013. *Next Generation Science Standards: For states, by states.* Washington, DC: National Academies Press. *www.nextgenscience.org/next-generation-science-standards.*

Perkowitz, S. 1998. *Empire of light: A history of discovery in science and art.* Washington, DC: Joseph Henry Press.

Pliny. 77/1938. *Natural history, Volume 1: Books 1–2.* Translated by H. Rackham. Loeb Classical Library 330. Cambridge, MA: Harvard University Press.

Weimer, M. 2013. *Learner-centered teaching: Five key changes to practice.* 2nd ed. San Francisco: Jossey-Bass.

EXAMPLE ACTIVITY DESIGN

MODELING ISOSTATIC EQUILIBRIUM

The modeling activity we present here began over a breakfast conversation in a pancake house on a road trip through South Dakota in 2013. We watched the ice cubes floating in our water glasses, and—nerd alert—discussed isostatic equilibrium.

Isostatic equilibrium models Earth's most significant topographic feature: the topographic contrast between high continental crust and deep ocean basins. Height of continents and ocean basins can shift when mass shifts. For residents of Scandinavia, isostatic rebound is occurring now. With the weight of the continental glaciers gone, the crust there is rising about a centimeter per year as it isostatically equilibrates. The modern harbor of Luleå, Sweden, is becoming too shallow for ships to use as a result.

This example activity design (EAD) addresses the driving question, "Why does Earth have continents and ocean basins?" It allows students to explore their mental conceptions of forces that act on a floating material. It gives them an opportunity to test their model experimentally and quantitatively. It gives them a chance to apply their lab experiments to big earth science questions.

Remember to make discoveries with your students, taking time to think out loud as you struggle to make sense of your observations. Remember to be imaginative and creative with your students—and don't just follow this EAD like a recipe.

TEACHER PREPARATION AND PLANNING

Before starting this activity with your students, try it out yourself, taking note of your preconceptions and misunderstandings. The first time I (Mary) tried this, I discovered I didn't understand buoyancy nearly as well as I thought I did!

Your students will need to have a familiarity with the ideas of density, mass, volume, and force. Students need to understand the idea of a vector as a way to portray the direction and magnitude of force. They will need access to ice cubes of different sizes and to wooden blocks of different densities, along with beakers of water, rulers, and balances.

Safety Notes

1. Have students wear indirectly vented chemical-splash goggles, and aprons during the whole activity, including setup, hands-on piece, and takedown.
2. Remind students to immediately wipe up any water spilled on the floor to avoid a slip and fall hazard.
3. Caution students in handling wood—slivers can puncture skin.

EXAMPLE PROMPTS AND LIMITING OPTIONS

Playing With Ice Cubes

To get your students started exploring the buoyancy behaviors of two materials with different densities, let them play with ice cubes floating in a beaker of water. As they explore, you might offer the following prompts:

- What's keeping the ice cube from sinking in water? (You might mention that we call that force *buoyancy.*)
- Does the ice have mass and weight?
- Does the ice still have mass and weight when it is floating?
- The ice doesn't sink in water, but would it sink in air?

Prompt students to experiment a bit:

- If you push the ice cube down into the water and then release it, what happens?
- If you lift it up a bit in the water and then release it, what happens? Why?

Help your students see what is hard to see: Does the level of the water change as the ice cube melts? Why or why not?

Modeling Ice Cube Movement and Balance of Forces

Given the idea that the ice cube remains static if there are no net forces acting on it but moves if there is a net force, have your students draw a model to show, using vectors, the balance of forces that are acting on the ice cubes for each of the cases examined while playing with ice cubes. For example, why does the ice cube bounce back up after you've pushed it down with a finger? There must be a net force upward. Why does the ice cube fall back down after you've pulled it up with your fingers? There must be a net force downward. Example correct vectors are shown in Figure 6.8 for the moment that you remove your fingers. Mass and weight of the ice cube are assumed to stay constant in these illustrations, which is true if there is no melting.

Figure 6.8

Example illustrations with correct vectors for modeling ice cube movement and balance of forces

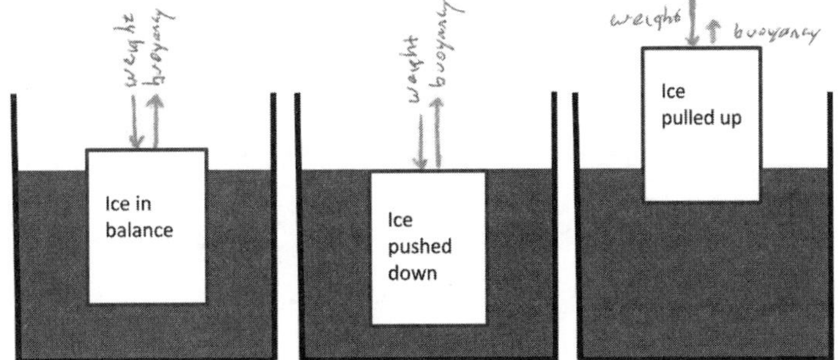

Measuring Ice Cubes at Isostatic Equilibrium

When there is no up or down motion, the ice cube is in isostatic equilibrium. Guide your students to develop a proportional relationship that describes their observations:

- How much of the ice cube is above the water line, and how much is below the water line?
- Does the larger ice cube, with more mass, sink deeper into the water?
- Does the same proportion hold true for a much larger ice cube?
- Are the above and below the water line proportions still the same regardless of the size of the ice cube?

Do a thought experiment with an ice cube–size batch of liquid water, which, like the ice cube, does not sink:

- Does the batch of water have weight?
- Why doesn't it either sink below or rise above the rest of the water in the beaker?
- Under what conditions are both the ice cube and the ice cube–size mass of liquid water in isostatic equilibrium with the surrounding water?

Point out to your students that the density of ice is about 90% that of liquid water (0.92 g/cm^3 vs. 1 g/cm^3). Give them time to consider the teaser question: "Might density have something to do with how high the ice cube floats above the water line?" If necessary, give this prompt: "Can you come up with an explanation for how much of the ice cube is below water and how much above based on your measurements, these density numbers, and your model of forces acting on the floating ice cube?"

Remind your students that the imaginary ice cube–size batch of water had a density that is 100% that of liquid water:

- How much of that water is at or below the surface?
- Does this thought experiment support your model?

Making Predictions

The test of a model is its ability to accurately predict new situations. Give your students a chance to make predictions based on their model:

- Suppose that you floated a block of wood in the water and the density of the wood is half the density of water. Based on your model from above, how much of the wood would be below the water line? How much above? Draw a picture of your prediction.
- Now suppose that the wood is only one-fourth the density of water. How much would be below the water line? Again, draw a picture to represent your prediction.

Testing Models

Have blocks of wood (e.g., pine) for students to float in water. The blocks should be at least ¾ inch thick and wider and longer than they are thick; otherwise, the block tips sideways. Students can measure the density of the wood (mass/volume). Have them predict how much of the block they expect to be below the water line. Then they can measure the proportion of the wood block that is below and above the water line and compare with their prediction. Encourage scientific argumentation within your students' lab teams:

- Did your prediction match the observation? If not, can you figure out why not?
- Try revising your model to match all of your data.
- Draw your revised model.

Try a different kind of wood with a different density (e.g., a low-density wood such as balsa or a higher-density wood such as zebrawood or teak), again making predictions followed by measurements. Are student predictions getting better?

Note: Students can base their models on conceptual ideas of proportionality and ratios (if the wood has a density 70% of the density of water, then 70% of the block of wood will be beneath the water). If you want to make the exercise more mathematical, you could give—or have your older students derive—the equation for isostatic equilibrium: *height of the block below water level = total thickness of block × (density of wood/density of water).*

Applying the Model: Rocks, Ocean Basins, and Continents

Because of the common misconception that continents "float on a liquid mantle," it might be worthwhile to emphasize to students that Earth's mantle is mostly solid, not liquid. However, over long periods of time, hot rock deforms viscously, so we can think of the crust of the Earth as rising or sinking in the mantle to achieve isostatic equilibrium. For example, in our region of northwestern Minnesota, the crust is slowly rebounding after being "pushed down" by the great mountains of ice that once were here. We can also explain big features of the Earth—continents and ocean basins—through isostatic equilibrium.

The Earth's mantle has a density of about 3.3 g/cm^3. The density of ocean crust (basaltic) is about 3.1 g/cm^3, with an average thickness of about 5 km. (One of the sources for density estimates in this EAD is the *Laboratory Manual in Physical Geology* [Busch 2008].) You can give students this assignment: "Using your models for isostasy, estimate how far the crust will extend below the top of the mantle and how much it will extend above the top of the mantle."

The continental crust (granitic) has a density of about 2.8 g/cm^3, less than either mantle or ocean crust. On average, continental crust is about 30 km thick. Analogous to the estimate for oceanic crust, you might have your students estimate how much of the continental crust extends below the top of the mantle and how much is above.

From the two estimates obtained in these assignments, students can calculate how much higher in elevation the average continent will be than the average ocean crust. For comparison with actual measurements of elevation, the continental crust rises 4.5 km higher than the ocean crust on average—how close can your students get?

Conceiving how to set up this calculation requires accessing and using the mental model for isostasy. If your students don't know where to begin, you may need to prompt them to think about how to apply their previous model to this new situation:

- Sketch a drawing of the ocean crust in isostatic equilibrium with the mantle and the continent crust in similar equilibrium.

- No, you don't actually have to draw mountains in your model, just represent density and relative buoyancy in your model.
- Which one will extend higher above the mantle and by how much?

Making Improvements to Models for Isostatic Equilibrium on Earth

Figure 6.9 illustrates two models developed in the mid-1800s to explain the presence of continents, mountains, and ocean basins on Earth (Watts 2001). Each model has one characteristic that matches the model for isostasy/buoyancy that you and your students have been developing. You might prompt your students to identify these characteristics for each model. Then, have your students combine these characteristics into a single improved model. The example student model shown in Figure 6.10 incorporates both the differences in density and the isostatic equilibrium depth into the model.

Figure 6.9

Illustration of two models developed in the mid-1800s to explain the presence of continents, mountains, and ocean basins on Earth

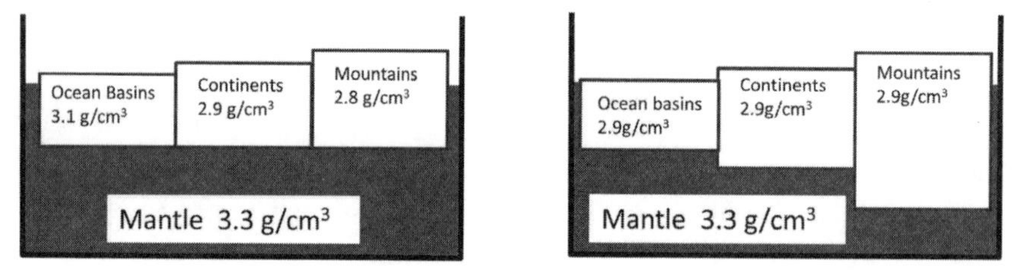

Figure 6.10

Simulated student model to improve on the Pratt and Airy models

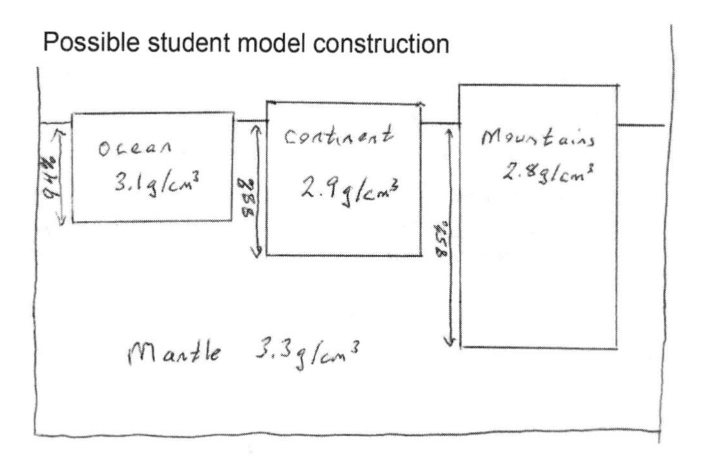

EXAMPLE INTERACTION

Note: Students might get bogged down when they apply the ice and wood models to understanding density differences in Earth's crust.

Student question: I don't know what to do.

Teacher prompt: Tell me what you're thinking.

Student question: What do ice cubes and wood have to do with continents?

Teacher prompt: Well, what were you testing when you floated ice and wood?

Student question: They were different densities and they floated differently.

Teacher prompt: What do you mean by "floated differently"?

Student question: The wood stuck up more above the water level.

Teacher prompt: So, which sticks up higher above the mantle, continents or ocean floors?

Student question: Well, continents.

Teacher prompt: What might explain that difference in "stick-up-ness" between continental and ocean crust?

Student question: Continents have mountains so they are higher and stick up more.

Teacher prompt: That's true, but what might cause this, given our ice/wood models? What about densities or thicknesses might be different?

Student question: So, I can think of continents like wood and ocean basins like ice, even though they don't look like each other?

Teacher prompt: Yep. It's OK that your model just represents density differences and not all the other aspects of Earth's crust.

SUMMARY CHECKLIST FOR TEACHER AS PRACTITIONER OF SCIENCE

- Have students play with ice cubes in water and think about net forces acting on them.
- Have student illustrate their model for a variety of scenarios discovered during their play with ice cubes in water. Include forces, conceptually, in the model (e.g., by using vectors).

- Plan how to give students opportunity to make predictions and then test those predictions.
- Provide an opportunity for more precise measurements than is possible with melting ice cubes. For example, make measurements of the buoyancy of blocks of wood and use the new data to refine the model for buoyancy/isostasy.
- Give students the opportunity to apply their model for buoyancy/isostasy to an earth science problem, such as explaining elevations of continents and ocean basins.
- Provide an opportunity for students to evaluate and improve on a "seed" isostasy model, using concepts developed in the lab.
- Watch for ways to prompt the student toward better results and practices without taking choice, exploration, and discovery away from them by telling them the "right" answers.

TEACHER REFLECTION

How well did your students do in developing, using, and revising their models? Did anything in particular arise as a stumbling block for your students? We have noticed that students often try to represent a system in all its complexity and end up with a mural or diorama, not a model.

How much guidance did you give to help students key in on important components of the model? How much help will you give the next time your students engage in modeling?

This work with isostasy and density could fit into an instructional sequence addressing (1) understanding how plate tectonics shapes the surface of Earth, (2) understanding the evidence for Earth's layered interior, or (3) understanding the role of gravitational and thermal energy in motions of the mantle and tectonic plates. This activity also has application to weather and oceanography. For example, differences in density and buoyancy cause warm air to rise. Likewise, differences in the density of seawater drive oceanic circulation.

REFERENCES

Busch, R. M., ed. 2008. *Laboratory manual in physical geology.* 8th ed. American Geological Institute and National Association of Geoscience Teachers. Upper Saddle River, NJ: Prentice Hall.

Watts, A. B. 2001. Isostasy and flexure of the lithosphere. Cambridge, U.K.: Cambridge University Press.

MODELING, PART 2

MATHEMATICAL MODELING

Students and teachers often have a special place in their heart for mathematics, either good or bad! Perhaps the strongly symbolic nature of math sets it apart from the other practices of science. The most basic elements of math are symbolic and contrast with the underpinning observational and real-world character of much of the rest of earth science. Even a number is only a symbol of quantity. This contrast between symbolic and real world is highlighted in a widely circulated math joke, first shown to us by a high school–age friend who later went on to complete degrees in math and physics. The joke is in the form of an SAT-like question. Given a picture of a triangle, and the length of two sides, find the third side, x. The joke is shown in Figure 7.1.

Figure 7.1
Math joke illustrating the tension between symbolic and concrete thinking

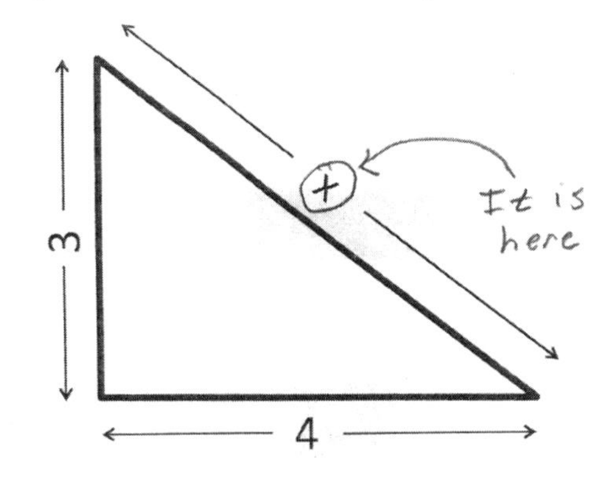

Given the information in the triangle above, find *x*.

"It is here" is a perfectly valid earth science mapping answer, spatially oriented and observationally sound.

Given that math is so deeply symbolic and that models are symbolic representations of reality, it follows that mathematical models invoke symbols within symbols. The exact nature of what constitutes a mathematical model is open for reasonable debate. Our son was defending his dissertation proposal on numerical modeling of molecular migration through nanotubes when a debate broke out among his PhD committee members about whether the work represented a true mathematical model (a mathematical description of the system) or only a better means to compute properties of the system. Our son explained the distinction to us by comparing two common equations in physics. The formula that calculates Earth's escape velocity gives us a calculated value but does not allow us to predict different states of a system as a function of time or input values. However, the sinusoidal expression for the motion of a ball on a spring models the position and velocity of the ball as a function of time, strength of the spring, and how far the spring was stretched; it has predictive capability. The latter is a model, whereas the former is not.

For earth science, we propose that a mathematical model can be understood as the mathematical portion of a conceptual understanding of how our universe works. As stated by mathematics professor John A. Adam (2003, p. 6), "The aim of mathematical modeling is the practical application of mathematics to help unravel the underlying mechanisms involved in, for example, economic, physical, biological, or other systems and processes." There is some gray area between a mathematical expression that calculates values in a system and a true mathematical model of a system, but one thing is certain: Having students plug numbers into equations that someone else has given them is not the *Next Generation Science Standards (NGSS)* idea of mathematical and computational thinking (NGSS Lead States 2013). Instead, as stated in *A Framework for K–12 Science Education: Practices, Crosscutting Concepts, and Core Ideas* (NRC 2012, p. 66), "students should have opportunities to explore how such symbolic representations can be used to represent data, to predict outcomes, and eventually to derive further relationships using mathematics."

In this chapter, we offer examples of how you and your students can infer mathematical relationships from experimental data, judge the quality of different mathematical models based on data, and use mathematical models to gain a deeper understanding of how the universe works.

Math, Like Graphs, Relates to the Real World

Unless encouraged otherwise, students often use memorized formulas like magic black boxes—if they put in the numbers in the right order and push the right buttons they get the right answer. Ta-da!

However, using math as a practice of science requires understanding how a mathematical expression relates to the real world. In earth science, knowing whether to multiply or divide is almost never a matter of memorizing a formula but rather a matter of understanding what the numbers represent in the real world and doing the operation that makes physical sense.

As much as possible, have your students think about what makes sense rather than plugging numbers into a memorized formula. The mathematical modeling activities in this chapter should not be viewed as a curriculum to follow but rather as inspiration to develop mathematical modeling activities of your own. Mathematical modeling can be applied to nearly every activity in science. Choose a few, and invent some lessons.

"DO I MULTIPLY OR DIVIDE?"

Earth science is not as math centered as physics, but there are still many opportunities for students to apply mathematical thinking to problems in understanding nature. For example, in association with a unit on isostasy, like that in Chapter 6 ("Modeling, Part 1"), students might make measurements of density.

In our experience, students often measure the volume and mass of a mineral or block of wood and then ask, "So, am I supposed to multiply the numbers, or divide?" How do you answer?

Not wanting to preempt student thinking by simply giving the answer, we might ask, "Well, what does the formula for density tell you to do?" Or maybe, "What are the units of density?" Students might answer "grams per cubic centimeter" meaning grams divided by cubic centimeter. You might follow with, "Given the units, which goes on the bottom and which goes on the top—mass in grams or volume in cubic centimeters?"

These are perfectly good responses—the first one encourages students to think about the formula and what it means, whereas the second one encourages them to think about units and how the units are tied to both the real world and to the equations. However, rather than tell students whether to multiply or divide, or even say "check the equation and units," try an answer that encourages students to consider the connection between the mathematical expression and their conceptual models. You might ask them, "What makes sense?"

Suppose that you imagine a situation in which the volume becomes larger but the mass stays the same. If you expect the density to be less, then do you need to multiply or divide? Which operation makes the proportion of mass to volume smaller? *Which makes sense?*

In this way, confusion with the math can be turned into an opportunity to build a stronger conceptual model in the student's mind and to connect that conceptual model with the mathematical expressions. Students can learn to see math as something that makes sense—something they can understand and check against the real world.

MATH-READING CHALLENGES: CONNECTING MATH TO NATURE

Connecting mathematical expressions to real phenomena is the key to reading math, in the same way that connecting graphs to the real world is the key to reading graphs and connecting maps to the real world is the key to reading maps.[C19] One way to teach the connection between math and nature is through equation-reading challenges that present students with opportunities to connect abstract equations to the natural world or design problems.

These math-reading challenges are easy to write for any mathematical expression or model that involves multiple variables—simply ask how the response variable will change if one of the input variables changes. These challenges provide a way to help students connect mathematical expressions to real-world consequences of changing values of variables; the more realistic the problem scenarios, the stronger the connection.

The Reynolds number example challenges included in this chapter use hydrology equations with application across multiple disciplines (earth science, physics, and engineering) and topics (flow in rivers, lava, air, oceans, or any other fluid). We envision these conceptual activities tucked into a unit on flooding (for example), providing a connection between math and the real world but not attempting more comprehensive fluid flow modeling.

Reynolds Number Challenges

One important equation in surface water systems is the equation for the Reynolds number (Easterbrook 1993). A high value for the Reynolds number correlates to turbulent flow; a low value correlates to laminar flow. The Reynolds number applies to problems as diverse as flow of water in rivers or through hydroelectric plants, flow of oil in pipelines, flow of lava from volcanoes, and flow of gas from a jet engine or rocket (I [Russ] like to tell my students that they really are doing rocket science!).

The Reynolds number is a dimensionless quantity (all the units cancel), making it useful for comparing processes at different size scales. For example, researchers might want to compare experiments in a small laboratory stream flume to a larger river. Or they might want to compare a river or laboratory flume to a flood so large that it's never been observed, such as the megaflood that produced the Channeled Scablands of eastern Washington.

Turbulence has implications for the behavior of a variety of natural and engineered systems. In rivers, turbulence can determine how vigorously the stream erodes its channel—high turbulence results in greater erosion. In baseball, turbulence determines how far the ball will travel when hit. In hydroelectric plants, turbulence can result in lower velocities through the turbines and therefore lower energy production.

Thus, the opportunity to think mathematically with the Reynolds number might fit into a unit on stream processes or a unit on a wide range of engineering problems. For example, units of study might address an *NGSS* performance expectations such as HS-ESS2-5, "Plan and conduct an investigation of the properties of water and its effects on Earth materials and surface processes," and HS-ESS3-2, "Evaluate competing design solutions for developing, managing, and utilizing energy and mineral resources based on cost-benefit ratios."

The Reynolds number is expressed by the equation $R = \rho dv/\mu$, where ρ = density; μ = viscosity; d = a characteristic dimension, which might be simplified as stream depth when applied to rivers; and v = velocity. You can think of the Reynolds number as a ratio of inertial forces (higher density, depth, and velocity) to forces resisting flow (viscosity).

Working in groups, students can puzzle through a series of challenges aimed at understanding the relationship between the mathematical expression and the real-world phenomena it describes, in this case the effect of several variables on the nature of the flowing water.

Challenge 1: Other things being equal, will the water be more turbulent in the middle of the stream where water is deeper and faster, or on the edge of the stream where the water is shallower and slower? *(middle; both greater depth and a faster stream increase the value of the Reynolds number, consistent with greater turbulence)*

Challenge 2: Suppose that a river is in flood. The flooding river is flowing faster and deeper than the preflood river, making it more turbulent and resulting in greater erosion of the banks. This produces a lot of sediment in suspension in the water. Sediment in the water increases both the density of the water and its viscosity, but we know from laboratory measurements that viscosity generally increases more than density. Will the eroded sediment cause even more erosion, or will it tend to decrease the amount of erosion? *(decrease; if a higher sediment load increases the viscosity more than the density, then a higher viscosity will decrease the Reynolds number, making the flow less turbulent than it would be without the sediment)*

Students often question the answer to the first challenge: "But the water looks more turbulent at the edge of the stream where it has to flow around the rocks." This gives us a chance to talk about the effects of roughness of the river channel or obstacles in a stream, neither of which is represented by a variable in the Reynolds number equation. In other words, we can use the Reynolds number to model the effect of stream velocity, depth, viscosity, and density on turbulence, but the Reynolds number doesn't give us a model for the effect of obstacles. Understanding what an equation takes into consideration and what it doesn't is an important part of understanding the value and limitations of the mathematical model, as noted in *NGSS* Appendix G (NGSS Lead States 2013) in the table on the crosscutting concept Systems and System Models: "models are limited in that they only represent certain aspects of the system under study."

Additional examples of math-reading challenges are presented in "Digging Deeper" (pp. 143–148).

DERIVING MATHEMATICAL EXPRESSIONS

Deriving mathematical equations can seem like a mystical art. I (Russ) remember reading through mathematical derivations of basic equations in introductory courses in physics and chemistry and being a bit befuddled. Often, the derivations would go through some fairly logical and straightforward steps and then say "at this point substitute (expression a) for (expression b)." Whoa! How do you know to do that?

For me, the substitution seemed to come out of left field, even though making the substitution led to a tidy solution. Perhaps the substitution wasn't straightforward for the mathematician who thought of it either. Somewhere along the line, the mathematician had a moment of insight, or maybe tried out lots of approaches and eventually came upon the solution. It may have taken days, or years, or a career.

To derive mathematical expressions in the classroom, students need some limiting guidelines—such as the "now substitute this here" guideline in my introductory class. Providing limiting guidelines so that students can authentically apply mathematics to their conceptual understanding within the time frame of an instructional unit is one of the most important tasks of a teacher. In the next two subsections, we offer a couple of ways to provide those limitations in deriving math expressions.

Prompting With a Starting Relationship

In Chapter 6, we examined a conceptual model for buoyancy and isostasy. We gave the equation for isostatic equilibrium of a wood block as *height of the block below water level = total thickness of block × (density of wood/density of water).* Given a starting point related to a conceptual understanding of the system, students can derive this mathematical description of the system (i.e., derive the model). To do this, students need to apply algebraic thinking to a scientific problem; this kind of thinking is a key *NGSS* science and engineering practice for both grades 6–8 and grades 9–12 (see practice 5, Using Mathematics and Computational Thinking, in Appendix F [NGSS Lead States 2013]).

The key constraint for buoyancy—one place for students to begin their derivation—is that the mass of the floating object is equal to the mass of the displaced fluid in which it is floating: *mass of displaced fluid = mass of floating block.*

It may be worth having students confirm this relationship by taking measurements in the lab (such as the measurements in the Chapter 6 "Example Activity Design: Modeling Isostatic Equilibrium," pp. 111–118). I (Mary) initially found this relationship somewhat unintuitive. I didn't quite believe it until I measured the relationship with a block of wood floating in a beaker of water. Even then, facing the data that clearly matched the mathematical model, I wondered "How can this be?"

Depending on how much time your students have, and how much of the problem that you want them to figure out, you might offer some or all of the following additional prompts:

- Your goal is to get an expression for the height of the block below the water level, in other words, *height of the block below water level = [expression].*
- *mass = density × volume*
- *volume of floating block = length × width × total height* (limiting students to a rectangular block)
- *height above water level + height below water level = total height of block*
- *volume of displaced water = height of block below water level × width × length*

From the key constraint *mass displaced water = mass of floating block* and from the relationships given in the bulleted list (either figured out by students or given as prompts), students can infer the following expression:

height of block below water level × length × width × density of fluid =
total height of block × length × width × density of block

Proceeding from the key constraint to this equation is likely to be the hardest step for students and take the most time to figure out. From this equation, solving for the height below the water level is a simple mathematical exercise:

height of block below water level × density of fluid = total height of block × density of block
and
height of block below the water level = total height of block × density of block/density of fluid

Realizing that they can derive the mathematical equation from a basic starting idea and that the mathematical expression matches their observational results can have a profound impact on students' connection between mathematical modeling and the real world.

Prompting With a List of Variables

Another approach to limiting options for students in deriving equations is to provide your students with the variables on which a particular parameter depends, along with some constraining information about the nature of the dependence. I (Russ) use this approach when students derive the expression for discharge in groundwater aquifers. This derivation follows an experimental activity in which students measure the permeability in

2-foot sections of piping filled with different sizes of marbles (the marbles represent particles of sediment in groundwater aquifers). Measuring permeability in this way is within the reach of most high school students, although extending the lesson to provide a connection to mathematical modeling like that described in the following text might require several additional class periods to give students time to process the ideas.

I give students the key variables on which discharge—the rate at which water flows through the pipes—depends: the cross-sectional area through which water is moving *(A)*, the distance of sediment through which it has to move *(d)*, the force per unit area that is pushing the water through the sediment *(f)*, and permeability *(P,* the ease of moving a fluid through rock). I also tell them that for each variable the dependence is either direct or inverse; that is, there are no exponential relationships. I then have students complete the following equation:

$$Discharge = Q =$$

Even before doing the experiments, students usually realize that increasing the size of the pipe *(A)* or the pressure per unit area on the water *(f)* will result in greater water flow, whereas increasing the length of the pipe *(d)* will decrease the water flow. Since *permeability (P)* is defined as the ease that water will flow, they can infer that greater ease—higher P—will result in greater water flow. Thus, from common experience, they can construct the following equation:

$$Q = P \times A \times f/d$$

Hydraulic gradient (I) is defined as the pressure difference divided by the distance of pressure drop *(f/d)*, so this expression reduces to $Q = PAI$, which is Darcy's law, the key mathematical model for groundwater movement. This equation is useful in problems as diverse as calculating rates of well water recharge and predicting migration of water-soluble pollutants underground. The equation was originally developed by Henry Darcy in the mid-1800s to understand movement of water through water purification systems (Freeze 1994).

This exercise provides another way for students to connect real-world processes to variables in a mathematical expression. If groundwater hydrology is not part of your curriculum, you can use a similar approach to mathematically model other types of experimentally determined rate or flow problems.

IMPLICIT MATH

Understanding math isn't always about interpreting, deriving, or writing equations. In Chapter 6, we mentioned an experimental activity that I (Mary) did with my eighth graders in response to a question about the difference between temperature and heat. To address the question, students put beakers with different amounts of water on a hot plate and measured how long it took for the water to reach the boiling temperature. Roughly, twice as much water took twice as long to reach the boiling temperature.

Safety Notes

1. Have students wear indirectly vented chemical-splash goggles and aprons during the whole activity, including setup, hands-on piece, and takedown.
2. Remind students to use caution in working with hot plates and heated liquids. They can seriously burn skin!
3. Remind students to immediately wipe up any water spilled on the floor to avoid a slip and fall hazard.

This activity was a side trip based on a question from a student, so I didn't take class time to do math calculations. However, interpreting the results involved significant mathematical thinking. Students had to recognize a one-to-one correspondence between time-to-boiling and amount of water. To relate the time-to-boiling to the amount of heat added, they had to recognize that the heat provided by the hot plate per unit time stayed roughly constant and that the total heat increased linearly with time. The total final heat corresponded to the heating rate times the time. They then had to recognize that a set amount of heat increased the temperature of a set amount of water by a set amount, the definition of *heat capacity.*

In the process, they also recognized that temperature is different from heat, the original goal of the activity. Of course, the math that my students used intuitively would have taken much longer to translate into mathematical expressions. In this case, I decided that converting our implicit math into mathematical equations was not the key goal of the activity.

THE REALLY COMPLICATED STUFF

There are a couple of *NGSS* PEs addressing computational simulations and representations that are so broad in scope that we feel we should offer our interpretation of their educational goal and how that goal can be reached. These PEs deal with relationships among interacting systems at the highest level and thus fall into the category of "model of models" that we talked about in Chapter 6.

HS-ESS3-3. Create a computational simulation to illustrate the relationships among management of natural resources, the sustainability of human populations, and biodiversity. *[Clarification statement: Examples of factors that affect the management of natural resources include costs of resource extraction and waste management, per-capita consumption, and the development of new technologies. Examples of factors that affect human sustainability include agricultural efficiency, levels of conservation, and urban planning.]*

HS-ESS3-6. Use a computational representation to illustrate the relationships among Earth systems and how those relationships are being modified due to human activity. *[Clarification statement: Examples of Earth systems to be considered are the hydrosphere, atmosphere, cryosphere, geosphere, and/or biosphere. An example of the far-reaching impacts from a human activity is how an increase in atmospheric carbon dioxide results in an increase in photosynthetic biomass on land and an increase in ocean acidification, with resulting impacts on sea organism health and marine populations.]*

The mind-boggling implications of these standards becomes apparent when you consider that the factors involved are not additive but are interactive, and that each interacting factor is in itself dependent on a complex interplay of subfactors, identification of which requires input from disciplines outside earth science. For example, the clarification statement of HS-ESS3-3 includes seven interacting factors, and the subfactors that affect just one of these factors—for example, cost of extracting a resource—are also many and complex.

The mind boggles.

Unless the entire year of study is devoted to this one PE, how can students create a meaningful computational simulation that includes even one or two of these factors? What earth science teacher has sufficient breadth of knowledge of economics, politics, and engineering, in addition to earth science, to provide guidance in developing the simulation?

That the writers of the PEs were aware of this problem is reflected in the *NGSS* assessment boundaries: for HS-ESS3-3, "Assessment for computational simulations is limited to using provided multi-parameter programs or constructing simplified spreadsheet calculations" and for HS-ESS3-6, "Assessment does not include running computational representations but is limited to using the published results of scientific computational models."

To interpret the intent of these PEs, we need to ask ourselves a couple of hard questions. Why ask students to create a complex, computational simulation of interacting systems at the global scale, when that activity can't be assessed and students almost certainly can't understand all the underpinning math and concepts involved? Why ask students to use a prepackaged simulation program when they have neither the experience to evaluate the results nor the means in the classroom to test them? Such activities seem a bit like computer games that purport to model the real world but leave students to either accept or reject the result on faith.

We suggest that these two PEs are written at the global (mind-boggling) scale not to imply that students should jump to that level in their classroom activity, but rather so that students and teachers have the big-picture context in which to fit a classroom-size study and so that the scope of classroom investigation is not unduly limited. We suggest that computational simulations and representations created and used by the students be of a scale that fits one of the following guidelines:

1. Students identify conceptual elements of a system that they put into an algorithmic sequence without necessarily specifying all the mathematics at each step. Some mathematical thinking should be included so the exercise doesn't devolve into drawing big pictures with lots of arrows.

2. Students create or use a simulation in the classroom that is sufficiently limited so that they can work through and understand the calculations involved and how those calculations relate to interacting components of the system.

Putting Math Into the Context of a Bigger Story

Math exercises such as those presented in this chapter should be integrated into instructional units that tell an earth science story. For example, suppose that you are doing a unit on the formation of the Channeled Scablands of Washington, formed by an ancient megaflood. While figuring out what processes shaped the Scablands, students can explore characteristics of flow (Reynolds number, Manning equation, discharge), experimentally model erosion and deposition (see "Suggestions for Designing an Activity: Experimental Petrology With Downspouts" in Chapter 10, "Stories in Rocks"), and mathematically model variables that have an impact on runoff (such as the runoff activity in the "Digging Deeper" section in this chapter). This type of integrated unit engages students in a variety of *NGSS* science and engineering practices, crosscutting concepts, and disciplinary core ideas and supports several high school PEs for Earth and Space Sciences (ESS).

The connection between classroom investigations and a bigger picture can arise out of driving questions that create a sense of story and give students ownership in those investigations. Consider the following two instructional starting points:

1. We are going to learn about megafloods; they are important in shaping the landscape of Earth and Mars.
2. Why do the Channeled Scablands look so different from other parts of the Columbia Plateau of Washington? What's going on here?

The first states the objective and why that objective is important. The second places the unit within the context of a specific earth science story through a driving question. It stirs curiosity and illustrates more authentically how the practice of science proceeds.

We talk further about putting small-scale studies into a bigger context in Chapter 15, "Teacher as Curriculum Narrator," and provide a list of connections between the *NGSS* and our example activities in Appendixes A and B.

Creating and Using Algorithms

Mathematical models don't have to be complex to be effective. M. King Hubbert, a geophysicist, proposed that the life cycle of a *wasting asset* (economic term for a nonrenewable resource that gets used up) will follow a simple bell curve, with an initial exponential increase in production, a peak, and then a decline that is more or less symmetrical with the increase (Hubbert 1956). He based this model on the observation that the production of a wasting asset divided by total cumulative production decreases linearly to zero when plotted against total cumulative production. This trend implies that we can predict the amount of material that will be discovered in the future based on how much has been discovered so far, assuming a more or less constant rate of technological advancement.

Hubbert's model has had a number of striking successes. For example, back in the 1950s, he predicted that U.S. oil production would peak between 1965 and 1970 (Hubbert 1956). The actual peak, at least so far, was around 1973, coinciding with the oil crisis and the rise of OPEC (Organization of the Petroleum Exporting Countries)—close enough to count as a success.

However, recent new technologies, sharp changes in oil price, and political factors have led to a steep rise in U.S. oil production beginning around 2006 that deviates from Hubbert's tidy bell curve (see Figure 7.2). So, was Hubbert's model wrong? Well, it didn't predict a new source of oil accessible by new technology (fracking and lateral drilling). His model is quite simplistic given that oil production depends on complex factors such as price of oil, tax policy, technology available for extracting oil, the amount of undiscovered oil still in the ground, and how much environmental risk the public is willing to

Figure 7.2

Data for annual U.S. petroleum production and the Hubbert model

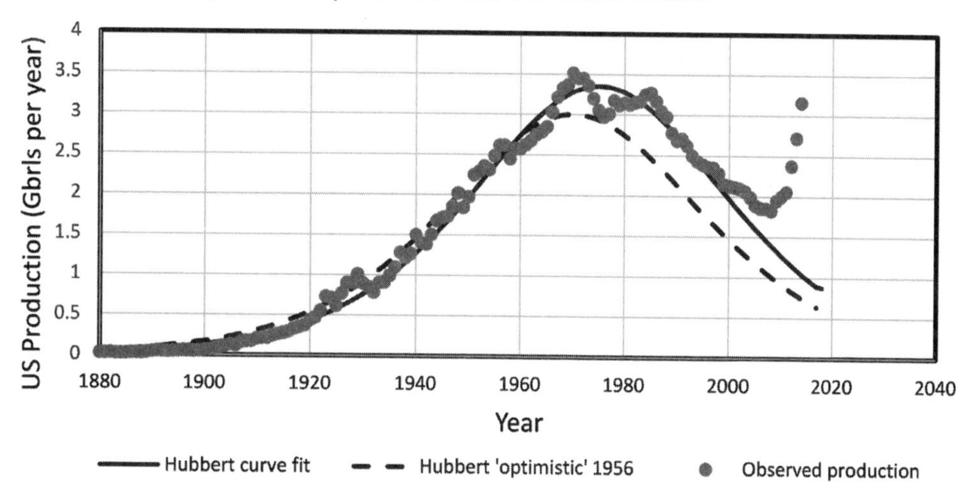

Data for annual U.S. petroleum production compared with the most optimistic of Hubbert's model curves published in 1956. We also provide a fit to Hubbert's model using modern production data, not available to Hubbert in 1956. Hubbert's most optimistic model curve predicted a peak near 1970, close to the observed peak, although his most likely model curve (not shown) predicted the peak in 1965. The recent deviation from the curve provides an opportunity to discuss how we interpret deviations from a simulated trend. Gbrls = billions of barrels.

Source: Data for U.S. production are from the U.S. Energy Information Administration website, *www.eia.gov/dnav/pet/hist/LeafHandler.ashx?n=PET&s=MCRFPUS1&f=A.*

shoulder to get the oil out of the ground. However, the simple model has worked for a number of situations.

Perhaps a better way to think about Hubbert's model is that it gives us a guideline for how the life cycle of a wasting asset might proceed given smooth progressions in technological development and limited swings in politics or economic well-being. Alternatively, perhaps the new oil available through new technologies should be viewed as the start of a new bell curve with its own life cycle.

The takeaway message is that mathematical simulations are not laws of nature, and they don't have to be exactly correct to be useful. They are often intended to provide predictive capability or insight into a complex interacting system only over a limited range of conditions. They give us a simplified way to understand how various factors might influence future outcomes. At this level of understanding, students can participate in constructing simple models of even quite complex systems. However, even when simplified, activities like the one described in the rest of this subsection should probably be part of an extended "special project" that you and your students have several weeks to prepare for and implement.

Consider how you might guide your students in simulating a small part of a complex system. You might pick one factor, let's say cost of resource extraction as proposed by the clarification statement for HS-ESS3-3, and have students brainstorm what the conceptual subfactors are that have an impact on the cost of resource extraction. This kind of initial brainstorming activity typically yields a variety of ideas and possibilities, but it doesn't necessarily lead to an understanding of how those ideas can be put into the form of an algorithm to calculate cost. Thus, it's important to push your students to the next step—considering what variables affect each subfactor. Finally, students need to consider the mathematical relationships between those variables and the subfactors that control the cost of resource extraction. In the following text, and in Figure 7.3, we offer an example of how this three-step process might proceed.

In earth science, assessment of resource cost often begins with a geologist constructing a contour map to show an ore body's depth below the surface (this type of contour map is called an *isopach map*). Greater depth corresponds to greater cost to extract the ore. You might have your students work in groups to come up with additional subfactors that affect cost. For example, students might identify (1) cost to remove overburden, (2) cost to bring the ore up to the surface, (3) cost to transport the ore to a processing facility, (4) cost to process the ore, (5) intangible environmental cost, (6) cost of lost alternative use of the land or materials, (7) cost of initial investment capital, (8) legal and political costs, and so on (see Figure 7.3).

Students might then create a list of variables that have an impact on each of the subfactors; for example, the cost of removing overburden would depend on the cost of fuel, labor, infrastructure, and distance between surface and the ore body. The cost of transporting raw ore to a processing facility would depend on the cost of fuel, labor, and infrastructure associated with transport.

Quantitative modeling of the numerical cost associated with resource extraction is probably beyond the scope of a high school classroom. However, students can place constraints on mathematical relationships by thinking through how the cost of each subfactor would change given various changes in the variables that have an impact on it (see Figure 7.3). For example, the cost of removing the overburden will be a fixed upfront cost but will be greater for deeper-seated ore bodies. Will the cost increase linearly or exponentially with an increase in depth to the ore body?

Students can reason that removing the top layer will have some fixed cost, let's say C_0 (for cost of the 0th layer). The next layer down (the first layer below the top, 0th, layer) will require bringing the same mass of material up from greater depth, and so the cost will be C_0 plus some percentage of C_0, say 10% ($cost = C_0 \times 1.1$). The second layer down will have an incrementally greater cost than layers 0 and 1—mathematically, its cost might be represented by $C_0 \times 1.1 \times 1.1$. In this way, students can realize that the cost will increase exponentially with increasing depth of overburden ($cost = C_0 \times 1.1^d$, where d =

Figure 7.3

Example algorithmic thinking exercise addressing the cost of resource extraction included in the clarification statement for *NGSS* performance expectation HS-ESS3-3

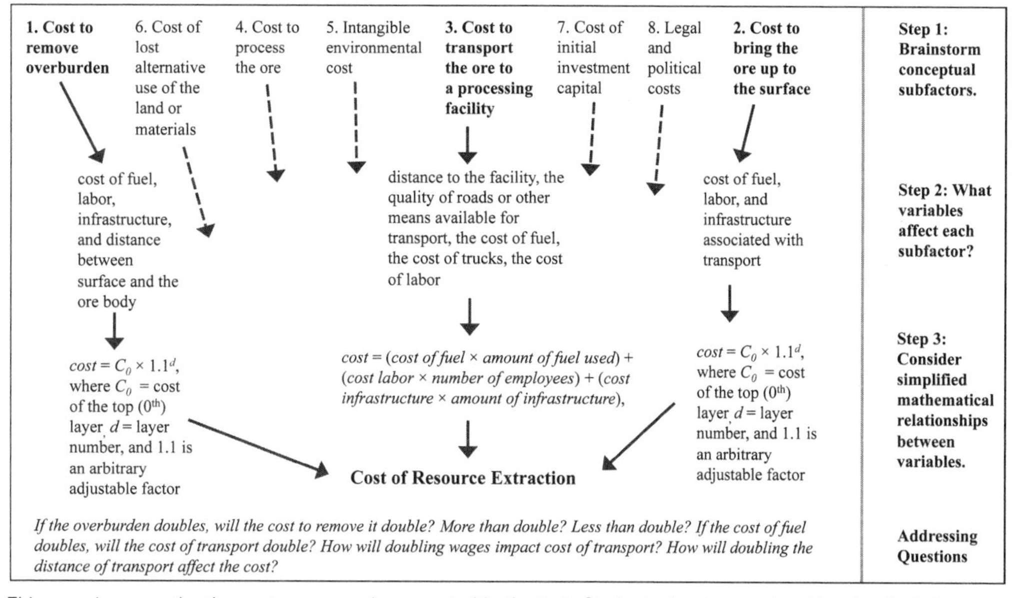

This exercise uses the three-step approach suggested in the text. Students develop an algorithm that includes some mathematical architecture, but they do not carry the problem to rigorous mathematical completion. In real-world application, this alogrithm could provide constraints on the cost structure and address real economic questions such as those shown in the figure, even without the full mathematical solution. Although this activity is more economics than earth science, the *NGSS* encourage this overlap between earth science and human activity. If you choose a similar activity in your own classroom, you might look for a problem of local interest or importance.

layer number). You might have them try out different exponential models for that cost. A similar analysis would be appropriate in thinking about the cost to bring the ore to the surface.

Cost to transport the ore to a processing facility depends on the distance to the facility, the quality of roads or other means available for transport, the cost of fuel, the cost of trucks, the cost of labor, and other subfactors. The total cost of transport might be represented mathematically as *(cost of fuel × amount of fuel used) + (cost labor × number of employees) + (cost infrastructure × amount of infrastructure),* plus expressions representing additional subfactors. From this expression, students can answer questions such as "If the cost of fuel doubles, will the cost of transport double?" and "How will doubling wages impact the cost of transport?" You might encourage students to think of other questions and how to address them mathematically, such as "How will doubling the distance of transport affect the cost?"

Students can go through similar analyses of other subfactors and try to come up with variables and mathematical relationships that model the cost for each. Although a detailed mathematical representation of the costs of these steps is out of reach for your students (unless you provide them with published mathematical models as in our activity on runoff in the "Digging Deeper" section at the end of this chapter), students should be able to come up with simple conceptual mathematical relationships, similar to the examples already given.

This activity gives students an opportunity to consider the factors important in a simulation, engage in algorithmic thinking, and construct simple mathematical expressions to predict outcomes given a variety of input variables.

Creating and Using Computational Simulations

Although computer-assisted calculations have expanded the scope and speed of computational simulations, computational simulations were done before we had modern computers. For example, in the early 1900s, Milutin Milankovitch simulated variations in solar radiation as a function of orbital mechanics and showed that the ice ages might have resulted from changes in Earth's orbit (Berger 2012).

Computer-Assisted Modeling

It's been estimated that weather and climate forecasting models alone, a type of computer-assisted simulation, have used more than a quarter of the world's super computing power to do calculations that in the past would have been impossible to do by hand fast enough to be meaningful. In 2015, the U.S. version of the world weather model broke the atmosphere into approximately 180 million blocks of air. In each block of air and for each small increment of time, the computer calculates the results of all thermodynamic processes and force vectors acting on the air. Each simulation in each incremental block of air depends on all the increments preceding it in that block and on the simulation in all the surrounding blocks. These simulations are carried out to the duration of the forecast, say 10 days. With present technology, these blocks of air are about 13 km wide by 13 km long (National Weather Service 2016)—a big chunk of air to treat as though it were all the same! To get better forecasts, the resolution of the model needs to be significantly increased. In 2005, Erik DeBenedictis estimated that an end target resolution would require one million petaflops calculation power, or 29,000 times more power than the fastest supercomputer on Earth in 2015 (see *www.top500.org/lists/2015/06*).

While on his honeymoon in 1914, Milankovitch was arrested for being a Serbian citizen in the Austro-Hungarian Empire. During his captivity, he devised mathematical methodologies to calculate how the bodies in our solar system interact with each other. After his captivity, he used those formulations to simulate variations in Earth's orbital parameters (eccentricity of orbit, angle of tilt, direction of tilt). He then calculated how those variations in orbital parameters affected the amount of sunlight reaching different places on

Earth in the past. He ran this hand-calculated simulation back for hundreds of thousands of years. These results were compared with the geological record of ice ages, revealing correlations that suggested variations in orbital parameters had altered climate.

The real science work of a computational simulation, whether done by hand or by a computer, is in setting up a sequence of calculations to match the processes that take place in the real world. For this reason, we suggest that the educational benefit of computational simulation is to be found in student engagement in creating and using the simulation rather than simply running through a computer program to reinforce classroom "learned facts."

In the "Digging Deeper" section, we offer an example simulation that estimates storm runoff in a small watershed given a set of initial conditions. Students can create the algorithm for this model, use and interpret model results, or both.

TESTING MATHEMATICAL MODELS AGAINST OBSERVATIONAL DATA

Mathematical models are sometimes derived theoretically from basic principles and other times might be derived empirically by fitting a curve to data. In either case, a mathematical model enters the realm of science when it is tested against observation. In the example activity design in this chapter, we offer an example of how students can test a variety of possible mathematical models against their own data.

REFERENCES

Adam, J. A. 2003. *Mathematics in nature: Modeling patterns in the natural world.* Princeton, NJ: Princeton University Press.

Berger, A. 2012. A brief history of the astronomical theories of paleoclimates. In *Climate change: Inferences from paleoclimate and regional aspects,* ed. A. Berger, F. Mesinger, and D. Sijacki, 107–129. Vienna: Springer-Verlag.

DeBenedictis, E. P. 2005. Reversible logic for supercomputing. In *Proceedings of the 2nd conference on computing frontiers,* pp. 391–402. New York: Association for Computing Machinery.

Easterbrook, D. J. 1993. *Surface processes and landforms.* New York: Macmillan.

Freeze, R. A. 1994. Henry Darcy and the fountains of Dijon. *Ground Water* 32 (1): 23–30.

Hubbert, M. K. 1956. Nuclear energy and the fossil fuels. Paper presented at the spring meeting of the Southern District Division of Production, American Petroleum Institute, San Antonio, TX.

National Research Council (NRC). 2012. *A framework for K–12 science education: Practices, crosscutting concepts, and core ideas.* Washington, DC: National Academies Press.

National Weather Service, Environmental Modeling Center. 2016. The Global Forecast System (GFS)–Global Spectral Model (GSM). *www.emc.ncep.noaa.gov/GFS/doc.php.*

NGSS Lead States. 2013. *Next Generation Science Standards: For states, by states.* Washington, DC: National Academies Press. *www.nextgenscience.org/next-generation-science-standards.*

EXAMPLE ACTIVITY DESIGN

THE IMPACT-CRATERING EXPERIMENT

IDENTIFYING AND TESTING A MATHEMATICAL MODEL

In this example activity design (EAD), we extend the cratering activity we began in the experiments of Chapter 2, "The Controlled Experiment," and the graphing analysis of Chapter 4, "Analyzing and Interpreting Data, Part 1." We walk through how to coach students in testing a selection of simple mathematical models, with the goal of identifying the mathematical model that best fits their data. As with several of the math-heavy activities in this chapter, this lab might work best as part of an extended "special project" lasting several weeks, through which students gain the expertise needed to understand this lab.

TEACHER PREPARATION AND PLANNING

Look over your graphs from the Chapter 4 EAD (based on data from the Chapter 2 EAD). We suggest that students convert the drop height values to velocities because velocity can be more directly related to kinetic energy, a value that is useful in mathematically modeling crater size. Before turning the following tasks over to your students, work through your own analysis of which models provide the best fit to the data so you have a feel for the statistical uncertainty involved in the analysis and what obstacles your students will encounter. Think about how much guidance your students will need versus how much you want them to figure out on their own. In the next section, we suggest limiting the exercise to a consideration of four predefined mathematical models, although you could choose more or fewer, or you and your students could create a different set of mathematical models to test against their results.

EXAMPLE PROMPTS AND LIMITING OPTIONS

Getting Data Into a Form to Compare to Models

Review the graphs of the impact-cratering experiments with your students. If they haven't already done so, have them convert drop height to velocity. Students can use the following expression to make the conversion:

$$velocity = \sqrt{2 \times 9.8 \text{ m/s}^2 \times drop\ height\ in\ meters}$$

Have students plot two graphs: (1) velocity versus impact crater diameter for multiple experiments done with a single mass but different drop heights and (2) crater diameter versus mass for experiments done at a single velocity but with different masses.

Testing Models Against the Velocity Data

Offer students a variety of different possible mathematical models for the relationship between impactor velocity and crater diameter. We suggest four example models in this section; in each model, a = diameter of the impactor, k = constant, and v = velocity. Which model matches the students' experimental results? Can they eliminate some of the models by inspection of the equation? For example, one of the models implies a linear relationship between crater size and impactor velocity. Is that consistent with their results?

Model 1: *crater diameter* $= a + kv^{1/2}$

Model 2: *crater diameter* $= a + kv^{0.67}$

Model 3: *crater diameter* $= a + kv$

Model 4: *crater diameter* $= a + kv^2$

Each of these models implies a relationship between crater size and velocity that can be plotted on a graph. Have students simulate data trends for each of the models to get a feel for the relationship each implies between impactor velocity and crater diameter. For example, have students choose values for velocity and calculate expected crater diameter for each chosen value (picking arbitrary constant numbers for a and k, such as 2.5 and 2.0). Students might do this either with a computer spreadsheet or by hand calculations plotted on a graph. We show example simulated data trends in Figure 7.4.

Figure 7.4

Simulated data trends for the four example models

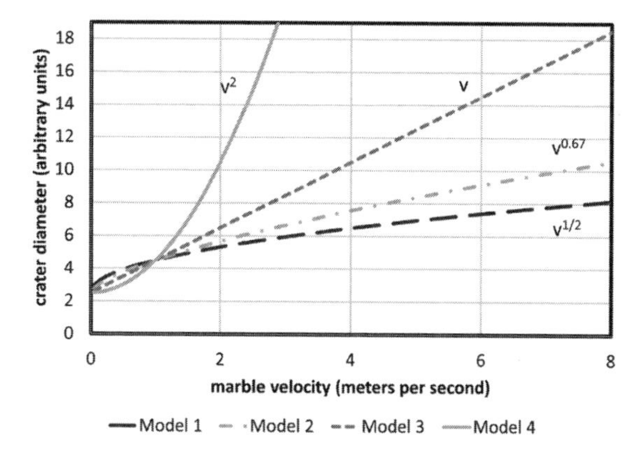

Students can compare their simulated graph to the graph of their data also plotted against velocity. Can students eliminate one or more of the models based on this comparison? For example, notice that the slope of model 4 increases at increasing velocity. Is this consistent with their data? The slopes of models 1 and 2 decrease with velocity. Is this consistent with their data?

To further test each model, students need to plot impact data gathered for a single mass but different velocities, against each model parameter. For example, they can plot crater diameter versus the parameter $v^{1/2}$, crater diameter versus the parameter $v^{2/3}$, and so on. From this, they can see which model provides the best fit to their data. This is done for our example data in Figure 7.5. In the case of the example data, model 2 provides the best fit by a slight margin over model one. Students can consider whether model one provides a statistically significant better fit or whether they are statistically equivalent. This will require thinking about the uncertainty in their measurements. Even model 2 is not a perfect fit, which means that students can consider whether model 2, even though it provides the best fit, is also incorrect.

Figure 7.5

Example of crater diameter data plotted against each model's velocity function with the intercept fixed as described in the text

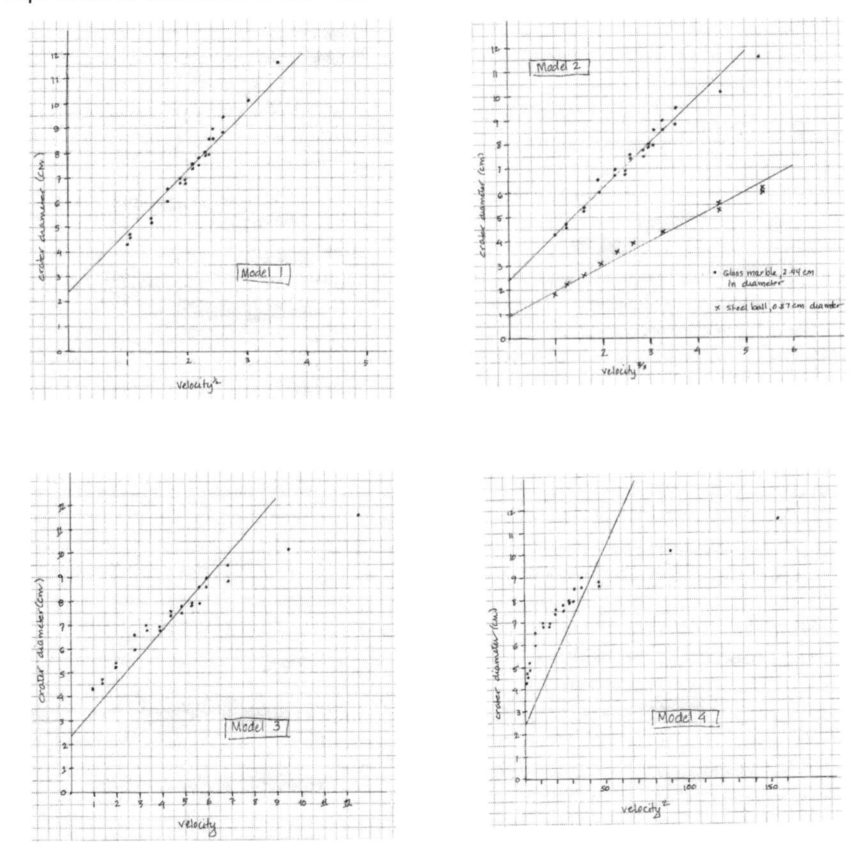

The theoretical model for real (hypersonic) impact craters is *crater diameter* = $k*v^{2/3}$. This model is similar to model 2 except that the intercept value *a* is zero. You might challenge your students to show that their results are either (1) consistent with this model

(i.e., the model for natural impact craters fits their impact data within uncertainty) or (2) inconsistent with this model (i.e., the sand and marble impacts differ from hypersonic impacts). What characteristics of the data indicate they are either consistent or inconsistent with the model?

Testing Models Against the Mass Data

Students can do a similar analysis of experimental results in which they hold velocity (drop height) constant but use variable masses. For this case, crater diameter will depend on mass *(m)* and a new constant *k*, but not on velocity. We suggest four example models:

Model 1: *crater diameter* = $km^{1/4}$
Model 2: *crater diameter* = $km^{1/3}$
Model 3: *crater diameter* = $km^{1/2}$
Model 4: *crater diameter* = km

As with the relationships between crater size and velocity, prompt your students to consider whether they can eliminate some of the models based on their experimental results (taking into account experimental uncertainty). Do their results fail to distinguish between one or more models (again taking into account uncertainty)?

Connecting to a Theoretical Model

So far, we've tested how well the experimental impact data fit a variety of different mathematical expressions without consideration of any relationships that might be expected from other areas of science. However, we can consider that it makes theoretical sense for the volume of material moved by an impact to be related to the energy of the impactor. We know that kinetic energy $(KE) = \frac{1}{2} mv^2$, where m = mass and v = velocity. Since volume of displaced material is proportional to crater diameter cubed, we can infer an expected theoretical relationship *crater diameter* = $k(KE^{1/3})$. Relating the crater size to energy not only provides a theoretical foundation for an otherwise purely empirical relationship, but it provides a strong cause-and-effect link between drop height and crater size.

There are two ways that you might introduce this theoretical expression into your activity with students:

1. Ask your students to consider whether their results are consistent with this theoretical model that relates crater size to energy of the impactor. Notice that the four models we suggest in this activity each correspond to a linear relationship between crater diameter and one of the following: $KE^{1/4}$ (model 1 for both mass and velocity), $KE^{1/3}$ (model 2), $KE^{1/2}$ (model 3), and KE (model 4).

Figure 7.6

Model trend line showing the relationship between kinetic energy (mass and velocity) and crater diameter

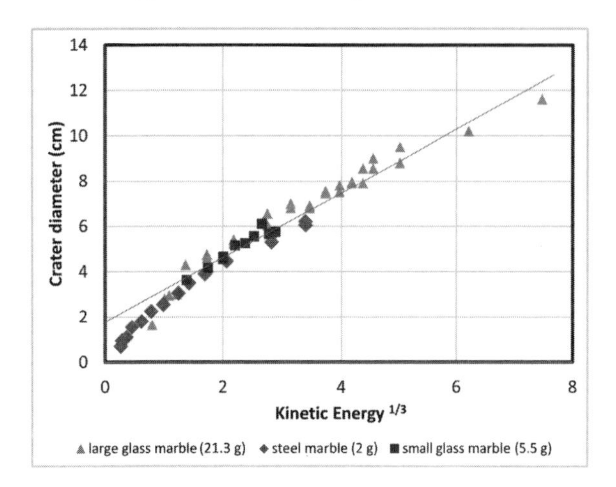

2. Have your students consider that the kinetic energy of the impacting marble = $\frac{1}{2} mv^2$, and encourage them to take the two models that they chose as "best fit" in the exercises above and come up with a single composite model that includes terms for both velocity and mass. Then, they can consider how this model relates to kinetic energy. They might come up with something such as *crater diameter* $= a + k\,(m^{1/3})(v^{2/3}) = a + k\,(KE^{1/3})$. We illustrate this relationship for our example data in Figure 7.6, showing that data from many different drop heights and masses can be brought into a single trend line by this model.

EXAMPLE INTERACTION

Student question: How am I supposed to compare my data to a model?

Teacher prompt: Does your plot of diameter versus velocity have the same shape as any of the model curves? *(Comparing data to the graph with all four models plotted)*

Student observation: Well, kind of. I'm not sure. My graph isn't on the same page as the model curves so I can't compare them.

Teacher prompt: Well, just consider the three basic shapes of the four curves: curve up, straight, or curve down. What does your plot of diameter versus velocity look like?

Student observation: My curve kind of goes up and then flattens out.

Teacher prompt: So that pattern is most like …

Student observation: I suppose most like the curve for models 1 or 2. Model 3 is a straight line, and 4 curves up. Right?

Teacher prompt: So you suspect that mathematical model 1 or 2 will best fit your observations. You can test this.

Student comment: I have no idea how to do that.

Teacher prompt: All four of the models represent different proportional relationships. For example, you can see if your diameters are proportional to the square root of the velocity, or to velocity to the two-third power. You already think that diameter won't be proportional to the velocity or velocity squared, but you can check those too.

Student question: What do I do first?

Teacher prompt: Think of it this way. If two variables are proportionally related to each other, what would the plot of those variables look like?

Student explanation: Well—they'd make a straight slanted line.

Teacher explanation: So suppose your crater diameter depends on the square root of the velocity of the impactor. You'll need to plot the measured diameter versus the square root of each of your velocities. If those data cluster along a line of best fit, you know you have a proportional relationship.

Student question: So I would do the same thing with the other three models, right?

Teacher answer: Yes.

SUMMARY CHECKLIST FOR TEACHER AS PRACTITIONER OF SCIENCE

- Have students think about the mathematical models and what they mean before calculating simulations of the models.
- Give students a chance to compute the shape of the curves for the models and compare with the shape of the curves defined by their data.
- Plot data against the variables of the different models to see which provides the best fit. Consider whether the best fit actually explains the variation in the data, or whether there might be a different model yet that provides a better fit (taking into account the variation in multiple measurements as a way to estimate uncertainty).
- Encourage students to think about the effects of velocity on crater size and the effects of mass on crater size and come up with a composite mathematical model.

TEACHER REFLECTION

What did you (the teacher) learn about mathematical modeling or about engaging students with this practice? As you develop future mathematical modeling experiences for your students, think about working with your colleagues in math and physical sciences. You might make discoveries together that none of you could have made separately, and the interdisciplinary connections lend a real-world flavor to the classroom.

You shouldn't be afraid to try something new, even though new aspects of even well-practiced classroom activities almost always lead to results that you don't understand at first. Although we have been doing crater impact experiments in the classroom for more than 20 years, neither of us had previously had students plot their results versus velocity to the 0.67 power or consider the relationship with kinetic energy. It was in adding this mathematical rigor that we discovered the unexpected (by us) non-zero intercept equal to the impactor size. In retrospect, this makes sense: If the velocity of the impactor is high enough to penetrate the sand, then the crater diameter can't be less than the diameter of the impactor—thus the value a in the equations corresponds to the value that the crater size approaches as the velocity gets small. However, we had always assumed that the theoretical model for natural impact craters would work for small-scale lab experiments. Figuring out why our mathematical model was not working for our experiments, and then modifying the model to work, proved quite challenging. It's this kind of unexpected challenge that you engage in with your students that creates a truly collegial environment of discovery in the classroom.

DIGGING DEEPER

Mathematical Modeling Related to Flooding

Getting students to engage with a science investigation often begins with giving students ownership of the investigation. Ownership often develops during special projects that go beyond the normal limits of the curriculum. In this section, we offer some activities for a special project following a unit in which students examine a variety of hydrology problems, such as considering how much water is generated by an inch of rainfall in a drainage basin and how they can account for that water through storage, interception, and runoff. An example unit on modeling runoff to which I (Mary) contributed is "Where Did the Water Go? Watershed Study," available at *http://nextgenscience.org/sites/default/files/MS-ESS_Watershed_Study_version2.pdf.*

The first activity in this section is a set of four challenges in which students apply their understanding of runoff to real problems in geo-engineering of water flow through their city. The second activity allows students to develop a computer algorithm for predicting runoff and flood potential.

MATH-READING CHALLENGES: STREAM VELOCITY AND DISCHARGE

Variables Affecting Velocity

Modeling is not only a practice of science but is an important crosscutting concept in the *NGSS;* for grades 6–8, Appendix G of the *NGSS* says that students can use "models to represent systems and their interactions—such as inputs, processes and outputs—and energy, matter, and information flows within systems."

Mathematical models cut across many disciplines and provide a particularly strong way to predict the behavior of systems given a variety of input parameters. For example, stream discharge depends on the velocity of a stream, and velocity depends on slope, shape of the channel, and how much friction the channel provides. The Manning equation provides an empirical mathematical model for stream velocity (v) in that it makes testable predictions about the effect of multiple variables (Easterbrook 1993):

$$v = (1.49/n)\, R^{2/3} S^{1/2},$$

where n = coefficient of roughness, R = cross-sectional area of the channel divided by its wetted perimeter, and S = slope. If we imagine a stream with a perfectly rectangular cross-section, then the cross-sectional area is the width times the depth (e.g., in square feet). The wetted perimeter is the depth times 2 plus the width (e.g., in feet).

As in the Reynolds number activity, students can work in teams to interpret what the mathematical model predicts about changes in velocity for given changes in the parameters of the system. Here are some challenges to help students think about the connection between the equation and real-world phenomena:

Challenge 1: Suppose that civil engineers plan to straighten a meandering stream where it passes through a medium-size city. Straightening the stream channel decreases the stream length, but the overall drop in elevation of the stream stays the same. Thus, the slope increases. If the slope doubles, how do you expect the velocity of the stream to change? Will it also double? More than double? Less than double? *(It will increase by the square root of slope, so less than double.)*[C20]

Challenge 2: Suppose that geological engineers are designing a channel around a city to divert floodwater from the downtown area. They want to maximize the velocity so that as much floodwater as possible gets through in the least time possible. Suppose that they are considering several designs for the diversion channel all of which have the same cross-sectional area, but different slopes, widths, and depths. In which of the cases shown below will stream velocity be highest?

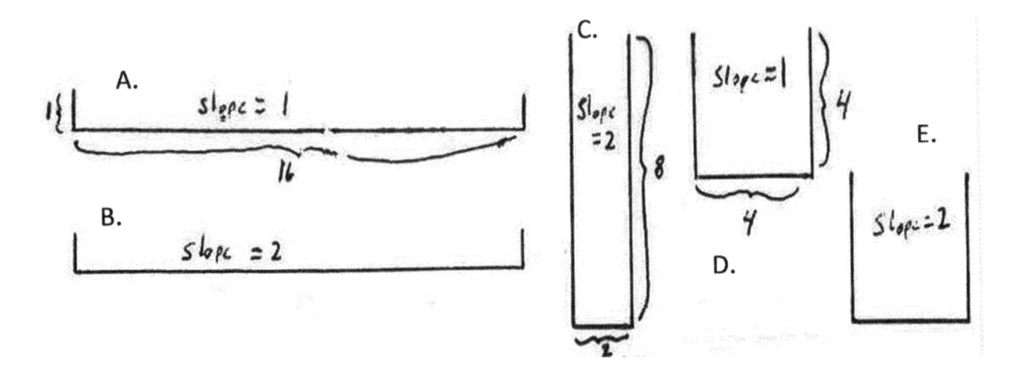

(Although stream cross-sectional area is constant, the wetted perimeter changes, causing changes in R *[cross-sectional area divided by wetted perimeter]. Options A, B, and C have larger wetted perimeters than D and E and therefore lower* R *values and lower velocities. The least wetted perimeter, greatest* R*, and greatest slope as shown in option E yields the highest predicted stream velocity.)*

Variables Affecting Discharge

The velocity calculated from the Manning equation can be combined with an expression for stream discharge (the volume of water passing a location in a unit of time, e.g., cubic feet per second) such as the following: *discharge* = *velocity* × *depth* × *width.* The combination of the two equations models how the discharge of a stream will vary with the shape of the stream channel, depth, width, and slope.

Challenge 3: If we double the slope of the stream but keep the depth and width the same (as in challenge 1 where the civil engineers straighten a meandering stream channel), how will the discharge change? Will it also double? More than double? Less than double? *(It will increase by the square root of slope, so less than double.)*

Challenge 4: Suppose that we keep the slope constant and the width of the stream constant, but we double its depth (such as might happen when water rises in a rectangular concrete channel under a bridge), will the discharge double? Halve? Less than double? More than double? *(This will increase both depth of water in the discharge equation and the value of* R *in the Manning equation. The value of* R *changes because, although both area and wetted perimeter increase when the depth increases, the area increases more than the wetted perimeter. So the discharge will more than double by an amount depending on the width. You can encourage students to plug in some example numbers to see this.)*

A COMPUTATIONAL SIMULATION FOR RAINFALL RUNOFF

This activity might be given as a special project following a unit in which students examine runoff more generally, as discussed at the start of the "Digging Deeper" section (p. 143).

The mathematical expressions used in this simulation are from a model developed by the Soil Conservation Service (now the National Resources Conservation Service). This empirical model, sometimes called the runoff curve number model, is readily available in a variety of books and online sources and has been widely used since the 1970s to predict how soil properties and human activity impact rainfall runoff. With this simulation, students can investigate the effects of four input parameters: (1) ground type (woodland, farmland, or urban), (2) rainfall amount, (3) rainfall rate, and (4) slope. The outputs of the simulation are total runoff, peak runoff rate, and time to peak runoff.

If algorithmic thinking is a key learning objective, you might give your students a portion of the published model from the original scientific paper (USDA Soil Conservation Service 1973) and have them develop a flow chart algorithm like that shown in Figure 7.7 (p. 146). If you have less time, you might start your students off with the flow chart in Figure 7.7.

A first step in this activity might be for you and your students to choose values for the input variables and crunch through a few calculations with a calculator to get an understanding of the algorithm and a feel for the meaning of the variables. This experience helps develop a sense of how the input and output variables are empirically related and

Figure 7.7

Example algorithm for a computer simulation activity simplified from the empirical rainfall runoff model published in 1973 by the Soil Conservation Service

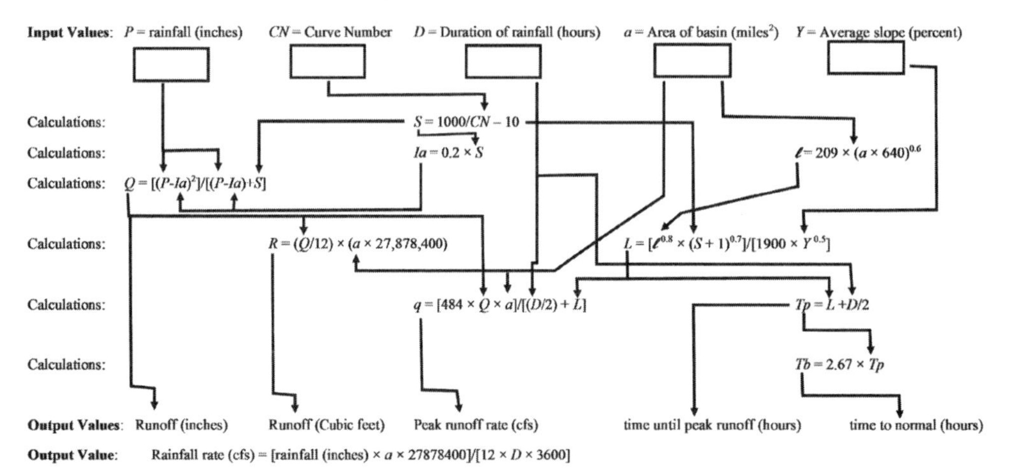

Curve number = *CN* = a measure of how much water runs off versus how much soaks in, evaporates, transpires through plants, accumulates in pools, etc. This number depends on the type of soil, the condition of the soil, and the type of cover over and in the soil.

Selected *CN* Values (for good hydrologic conditions):	Sandy Soil	Clay-rich Soil
Straight row small grain crops	65	88
Woodlands	25	77
Urban: Commercial and business	89	95

This model considers only steady-state rain that is the same over the drainage basin. The model could be further simplified by choosing constant values for some of the input variables. Students can construct similar algorithms from the original paper, make hand calculations from this simulation model, or use the model to construct a computer simulation in a spreadsheet (an example spreadsheet simulation using MS Excel can be obtained from the authors on request or downloaded at *http://web.mnstate.edu/colson/D2L-unlinked/Simple_runoff_algorithm.xlsx*.)

Note: Ia = water storage, infiltration into surface; *l* = length of main stream to farthest divide; *L* = lag time; *q* = peak runoff rate in cubic feet per second; *Q* = runoff in inches; *R* = total volume of runoff; *S* = retention of water (doesn't run off); *Tb* = time to base; *Tp* = time to peak.

Source: Additional curve number (CN) values come from *Hydrology Training Series, Module 104 Runoff Curve Number Calculations, Study Guide,* published by the Soil Conservation Service in 1989 (see also *www.nrcs.usda.gov/Internet/FSE_DOCUMENTS/stelprdb1082992.pdf*).

the sequence in which the calculations need to be done to work. Students can recognize the relative importance of different input values—for example, the slope only affects the speed at which the water runoff reaches its crest and recedes, whereas the curve number influences all of the output values.

The tedious number-crunching has the additional benefit of highlighting the value of computer-assisted computations. It might even motivate students to develop a short spreadsheet program to run the simulations. After doing one set of calculations by hand,

I (Mary) certainly did not want to redo calculations to fix mistakes or try out new values for the input variables.

The formulas can be typed into spreadsheet cells that refer to the input values and to each other as appropriate to implement the algorithm in Figure 7.7. If students are familiar with using spreadsheet formulas, the simulation can be created in a couple of hours or less. If they aren't familiar with spreadsheet formulas, then you will need to decide whether learning to program is worth the extra class time. If not, you may just have them do the simulations by hand, although that misses the fun moment when the computer simulation *they made* gives almost instantaneous feedback for different input parameters.

Once your students get their simulation written into the spreadsheet, they can compare output values against their hand calculations. They might also test their simulation program by considering whether the results make physical sense (e.g., total runoff can't be more than initial rainfall). They can check the model outputs against boundary conditions (e.g., runoff should equal total rainfall for the case where all the water runs off—curve number = 100). One thing to be aware of is that the math in this published model gives nonsensical results beyond the boundary condition where the runoff goes to zero.

Give students time to play around with their result, trying out different values for the curve numbers, changing the amount of rainfall, duration of rainfall, landscape use, or slope, to see the effect on volume and timing of flood crest. We show output for a couple of simulation cases in Figure 7.8. If students are sufficiently familiar with spreadsheets, they can include a graph in their spreadsheet program with values linked to the simulation output values, such that the results of each new simulation pop up on their graph.

Figure 7.8

Simulation using the model illustrated in Figure 7.7 for the case in which 2 inches of rain falls in 2 hours over a drainage basin 4 square miles in size with an average slope of 3%

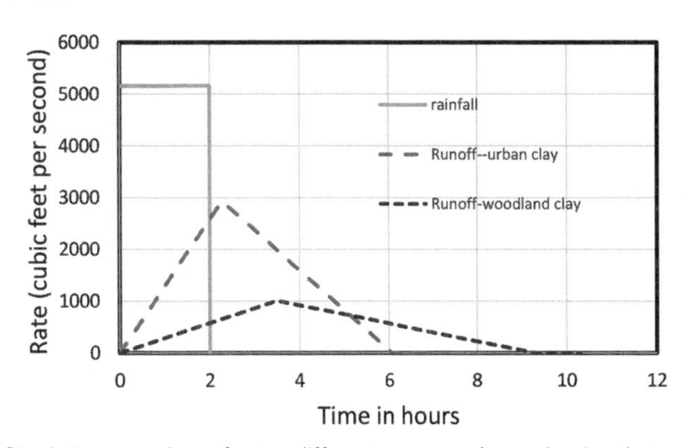

Simulations are shown for two different curve numbers, showing that both the amount of runoff and how quickly the runoff reaches a crest are influenced by human activities.

This simulation gives students an opportunity to consider how to construct a series of calculations in sequence to yield a desired output. Students can think about how the simulation reflects underlying real processes in nature and how input variables affect several seemingly unrelated outcomes.

REFERENCES

Easterbrook, D. J. 1993. *Surface processes and landforms.* New York: Macmillan.

U.S. Department of Agriculture (USDA), Soil Conservation Service. 1973. *A method for estimating volume and rate of runoff in small watersheds.* SCS-TP-149. Washington, DC: USDA Soil Conservation Service.

ARGUING FROM EVIDENCE, PART 1

FOCUS ON EVIDENCE

Evidence includes all the things that we observe. If we can see it, hear it, feel it, taste it, smell it, detect it, or measure it—that's evidence.

> *Evidence is at the heart of scientific practice. Proficiency in science entails generating and evaluating evidence as part of building and refining models and explanations of the natural world.* (Michaels, Shouse, and Schweingruber 2008, p. 19)

Despite the concrete nature of evidence, students struggle to distinguish evidence from its interpretation. When asked to identify the evidence for a theory, students routinely explain the theory instead. In fact, even among scientists, our identification of what counts as evidence is tied up in our interpretations and theoretical models, causing recognition of what counts as evidence to be less straightforward than we might think.

EVIDENCE AND PLATE TECTONICS

Plate tectonics is one of the great explanatory theories of geology, yet scientists resisted the evidence for it for more than 50 years because they couldn't reconcile the evidence with their theoretical models of how the Earth works. In his 1915 book *The Origins of Continents and Oceans,* Alfred Wegener presented extensive evidence that the continents have moved through geological time. Even before Wegener, people had noticed that continental shelves on either side of the Atlantic "match up." Wegener added a body of additional evidence, including (a) correlations in fossil types on adjacent continents, indicating that those areas, now separated by sea, must have been connected in the past; (b) the presence of fossil tropical plants, such as cycads, in regions that today are polar; and (c) the presence of glacial striations and deposits of glacial till that only make sense if there were continent where there is now coastal sea. Even today, these observations are cited as evidence for plate tectonics in textbooks for college and secondary earth science.

The problem is that this evidence didn't actually convince people at the time. Rollin Chamberlin, professor of geology at the University of Chicago, wrote "Wegener's hypothesis in general is of the footloose type, in that it takes considerable liberty with our globe, and is less bound by restrictions or tied down by awkward, ugly facts than most of its rival theories" (quoted at *www.ucmp.berkeley.edu/history/wegener.html;* see also Powell 2001, p. 92).

Ouch.

So, what were those "ugly facts" that Chamberlin thought did not properly bind Wegener's hypothesis? Not the observational evidence. They agreed on that. The "ugly facts" were theoretical models for how the Earth works and how continents might move.

Scientists understood that the Earth's crust and mantle were solid. It would take an unimaginable force to plow the continents through solid rock. Wegener's model for how to generate that force didn't make sense, and nobody had a better idea for how it might happen. Without a means to move the continents, scientists concluded that the continents could not have moved regardless of the observational evidence for it. It made more sense to interpret the evidence differently; for example, maybe land bridges once spanned the Atlantic Ocean between regions where the like fossils were found. Maybe Earth's rotational pole wandered, causing changes in climate as continents found themselves at different latitudes.

The evidence for plate tectonics continued to mount, thanks especially to the Cold War between the United States and the former Soviet Union. With submarines patrolling the seas, ready to launch a deterrent nuclear strike at a moment's notice, mapping the sea became a priority. The mountain ranges, volcanoes, and deep trenches at the bottom of the sea fell into patterns that paralleled the current boundaries of continents. Ages of rock and patterns of the topography suggested that the seafloor was moving as well as the continents, diverging at mid-ocean ridges and plunging back into the mantle at deep trenches in the seafloor. New measurements of the magnetic orientation of minerals in continental rocks showed that wandering poles could not account for climate change, because the values measured from each continent put the pole at different locations, a physical impossibility. New measurements of rock ages showed that ocean basins were younger than continents, consistent with the opening of oceans by seafloor spreading.

As observational evidence that continents moved became undeniable, it was clear that some force had to move them. By the 1960s, scientists realized that although the mantle is solid, the upper part of it is soft enough to ooze, providing a substrate for movement of tectonic plates. In the article "Continental Drift," published in *Scientific American* in 1963, J. T. Wilson noted that, although convection in Earth's mantle is too slow and deep to observe, the high heat flow at mid-ocean ridges and the low heat flow at trenches are consistent with movement of heat by underlying convection cells. New theoretical modeling of convection in Earth's mantle provided a mechanism to move the continents on the plastic substrate of the oozing asthenosphere.

On Testing Theories: Plate Tectonics

In earth science, evidence rarely consists of a single observation; rather, many observations gathered over a long period of time underpin geological theories such as plate tectonics, as explained by Patrick Hurley (1968) during the time when plate tectonics was becoming accepted:

> *The slow acceptance of what is actually a very old idea provides a good example of the intensive scrutiny to which scientific theories are subjected, particularly in the earth sciences, where the evidence is often conflicting and where experimental demonstrations are usually not possible. (p. 52)*
>
> *Any hypothesis must be tested on all points of observational fact. The balance of evidence must be strongly in its favor before it is even tentatively accepted, and it must always be able to meet the challenge of new observations and experiments. (p. 58)*

Plate tectonics was accepted once the observational evidence surmounted preconceived notions and once new theoretical models described a mechanism for plate motion.

Theoretical models for forces and motions in Earth's interior influenced the acceptance of plate tectonic theory, but in the end the observational evidence prevailed. Before the acceptance of plate tectonic theory, scientists agreed on the observational evidence, but they didn't agree on the "facts." Some of those "facts" came from theoretical models of the Earth that later proved to be false. In arguing from evidence, it's important for both scientists and students to recognize which of our "facts" are based on observation—evidence—and which of them are based on our mental models, which might be wrong and need to be changed.

Given that it took scientists 50 years to sort out evidence and theory, how can we offer classroom students an authentic experience with identifying—and arguing from—evidence within a much shorter time horizon?

Observation Trumps Theory

Another example of a mismatch between observational data and theoretical preconceptions came with the first departure of a spacecraft from Earth into interstellar space. *Voyager 1* left our solar system's heliosphere in late August 2012. However, the changes observed by the spacecraft were not what we expected:

> *A long-running debate on whether or not Voyager 1 was in interstellar space became a dispute between observations versus theory and modeling. (Krimigis and Decker 2015, p. 288)*

In the end, theory changed to account for new observation.

In the following sections, we explore examples of common difficulties that students have in separating evidence from assumptions, claims, explanations, theories, hypotheses, and models.[C21] These difficulties, we suggest, arise in part from a classroom culture that has long focused on knowing the conclusions of science rather than participating in how we know what we know.

Even when we think of "doing science," we tend to think of students designing and running classroom experiments or graphing and interpreting their data. "Doing" doesn't call to mind *talking* about one's new ideas and supporting those budding ideas with evidence and reasoning. Argumentation, perhaps ignored in the classroom as too time-consuming or nonproductive, is a necessary step in grounding scientific theory in the observational world. In this chapter, we offer some solutions to help students learn to recognize evidence.[C22] In the next chapter, we'll explore how reasoning connects evidence to explanation.

DIFFICULTIES STUDENTS HAVE IN IDENTIFYING EVIDENCE

Theories Are Based on Observation, Not the Other Way Around

Scientific theories are not evidence, but rather are based on evidence. But that's not how students see it. For example, one college student wrote in a term paper, "Several theories prove that the dinosaurs were driven to extinction by a meteorite impact."

Really? No observation needed? It's as if the theory itself is the evidence.

In an introductory college geology class, I (Russ) presented several lines of evidence for plate tectonics, including the idea that rock gets progressively older away from a mid-ocean ridge, supporting the interpretation that new rock forms there as the plates diverge. On an exam, one student wrote, "The new rock is pushed upward and the old rock is pushed outward supporting that the rock in the center is younger." In this case, the student reversed the relationship between observational evidence (rock ages) and theoretical model (movement of crust), as though we wouldn't believe the observation unless the theoretical model supported it. This is not uncommon.

On the same exam, another student wrote, "Evidence is Pangea or the belief that all the continents used to fit together." This statement makes the belief in the model the primary evidence for it.

In a report in a different class, another student was more direct: "Some people don't understand that (an observation) can't be true if it goes against scientific theory." Yikes. It would indeed be impossible to argue from evidence in a scientific way if you thought that the theory was the primary line of evidence and that the behavior of the natural world (the observations) had to conform to that theory! In science, it is we and our natural laws that are constrained by nature, not the other way around.

This idea is pointed out in a funny way in an old Bugs Bunny cartoon ("High Diving Hare," directed by I. Freleng, Vitaphone Corporation, 1948). Yosemite Sam is trying to force Bugs off a high diving board into a basin of water, but no matter what he tries, Bugs manages some new trick to avoid diving. Finally, Yosemite ties Bugs up on the end of the diving board and cuts the board off with a saw. He laughs. "This time, you're a-divin'!" However, instead of the diving board falling, the platform and ladder to which it was attached collapse to the ground, leaving Bugs and the diving board hovering in midair. Bugs looks at the audience and comments, "I know this defies the law of gravity. But, uh, you see, I never studied law!"

We find this funny because we know that Bugs's failure to study the law of gravity does not mean he isn't subject to it. In science, observation directs our theories, and Bugs's having a different model of reality—because he hasn't studied the law of gravity—doesn't mean that the observation will change to match his theory. If our theory doesn't match observation, then we have to change the theory, not the observation.

Explaining the Model Is Not the Same as Citing Evidence

High school graduates, including college students, often mistakenly think that the description of a theory or model is in fact evidence for the theory.[C23] This suggests that we are not giving students sufficient opportunity in the classroom to make this distinction and to articulate it to their peers.[C24]

I (Russ) asked my college elementary education majors the following question on a 2014 exam: "List and explain one line of evidence supporting the theory of plate tectonics." The most common answer was a description of the theory. One student wrote "When two plates split in the ocean, they create underwater mountains and basaltic volcanoes. They form where the faults of the plates appear. The magma comes from partial melting under the faults."

This tendency to describe models instead of identifying evidence continues among college science majors with three to four years of college science courses under their belts. This suggests that not only are students missing this key point in high school science courses, but they aren't getting it in college courses either. For example, in an upper-level Planetary Science course for secondary science teaching majors, I asked, "List and explain one observation supporting the theory or interpretation that the Earth differentiated to form a core." These two student responses were typical:

- A homogeneous Earth had gravitational differentiation where the dense metals (Fe) sunk to the core of the Earth creating a core.
- Earth started out as homogeneous and as Fe and Si started to partition out the silica would start to go to the crust and the Fe and siderophile [iron-loving] elements would go with Fe to the core. This is because Fe is more dense than Si.

These answers do a reasonable job of explaining key ideas of the theory but fail to list a single observation.

This answer gets a bit closer to addressing the evidence: "There are higher amounts of nickel and gold in the core than in the crust." This student had the right idea—evidence should be observational--but nickel and gold concentrations in Earth's core can't be measured with modern technology and so in reality this information was an untested prediction of the model rather than observational evidence.

About 20% of students were able to answer my question correctly, for example: "The Earth's mantle and crust are severely depleted in Fe and the other siderophiles relative to undifferentiated chondrites." Scientists have indeed measured the concentration of iron (Fe) in both the Earth's crust and in chondrites (a type of meteorite thought to be close to Earth's bulk composition). The observation that these concentrations differ is consistent with the model that the Earth's core formed by differentiation from a chondrite-like starting composition.

These experiences with college students' efforts to identify evidence suggest that separating evidence from theory is difficult, perhaps one of the most difficult of the practices of science. These students had not learned to recognize evidence in high school science or in multiple college science courses. Only 20% "got it" in my class where I specifically addressed the difference between evidence and description of a model repeatedly and in a variety of ways. That's not a particularly good percentage, and it suggests that something was missing from all the students' previous secondary and college science courses, including mine. This experience spurred me to change my approach—I inferred that the students needed more time to discuss the relationship between evidence and model and to practice arguing from evidence with other students.

In writing this chapter, I (Mary) also realized that I was not building in enough time for argumentation among my students, nor was I consistent in prompting students to clarify the difference between their evidence and their explanations. Classroom argumentation is hard, and it takes a lot of class time to help students challenge each other's ideas in productive ways. Some of the instructional complexity involved in developing this scientific practice in students can be seen in the guidelines of *A Framework for K–12 Science Education: Practices, Crosscutting Concepts, and Core Ideas* (NRC 2012, p. 73):

> *Young students can begin by constructing an argument for their own interpretation of the phenomena they observe and of any data they collect. They need instructional support to go beyond simply making claims—that is, to include reasons or references to evidence and to begin to distinguish evidence from opinion. As they grow in their ability to construct scientific arguments, students can draw on a wider range of reasons or evidence, so that their arguments become more sophisticated. In addition, they should be expected to discern what aspects of the evidence are potentially significant for supporting or refuting a particular argument.*

Other Difficulties in Identifying Evidence

Students (and scientists, and teachers) can overgeneralize, thinking that an observation supports their model when in fact the observation supports a whole swarm of models indiscriminately. For the question "List and explain one line of evidence supporting the theory of plate tectonics" mentioned earlier, one student responded "Evidence of plate tectonics is that there is seismic, volcanic, and geothermal activity."

So, is this evidence for plate tectonics? Not really. People have known about seismic, volcanic, and geothermal activity for centuries, yet it convinced no one that Earth's crust was broken into plates that moved around. Also, seismic, volcanic, and geothermal activity exists on other planets that don't have plate tectonics, indicating that those processes don't uniquely prove plate tectonics. These observations might be evidence for a simpler idea, that the Earth is geologically active, but not for plate tectonics specifically.

Students can be too vague in identifying evidence, thinking a broad observation without details supports their model when the observation is too general to support any particular model. For example, for the question above, one student responded, "The age of the Earth tells us there is plate tectonics." Presumably the student was thinking about the age structure of the ocean crust, which gets older outward from the mid-ocean ridges, or perhaps the idea that the ocean crust is consistently younger than the continents. However, that isn't specified.

Students (and scientists, and teachers) can fail to account for the intrinsic limitations of observations, overvaluing the strength of the evidence. I (Mary) once had my eighth-grade students looking for animal tracks at a local park. They reported back that there were no animal tracks in the park.

No tracks at all? How could that be true? It turned out that students had only looked in the grassy areas of the park. They hadn't wanted to explore down by the river where the trail was muddy and the mosquitoes thick (and where the animals left tracks in the mud).

Arguing from evidence involves evaluating the quality and strength of the evidence, sometimes in cases where problems with the evidence are not obvious.

POSSIBLE SOLUTIONS TO THE DIFFICULTIES IN IDENTIFYING EVIDENCE

Start With the Evidence, Not the Theory

The sequence in which we teach material can affect how students perceive the connection between evidence and theory. For example, many college and secondary earth science textbooks begin with a review of the theory of plate tectonics, with the belief that this theory provides a scaffolding for other earth science ideas. In this way, the theory of plate tectonics becomes a teaching strategy—providing an organizing framework into which students can plug ideas and observations.

The problem with this approach is that it presents the theory before the evidence, the opposite sequence to the practice of science. This sequence undermines any effort to emphasize how scientific models are developed from observation and it circumvents the "steps" that researchers go through in proceeding from the confusing fog of observations, through multiple working hypotheses, to a tidy theoretical model. It encourages students to see the theories as the primary facts, accepted on faith, that subsequent observations must fit.

There is an expectation among many people that students need to know the theory of evolution, the laws of motion, the theory of plate tectonics, the model for the atom, and so on as the primary content goal of science education. These explanatory ideas are indeed important, but we argue that they should not take precedence over the practices of science and should not precede an understanding of the relationship between evidence and theory. Students should first have the opportunity to grapple with some of the observations and experimental measurements on which those theoretical models are based.

In teaching, we tend to emphasize broader principles and theories because they help us understand and summarize the almost limitless number of facts that we can gain from observation, which are too large in number and sometimes too trivial in nature to commit to memory. If you design airplanes, do you want to memorize the magnitude of forces that act on an airplane under all possible conditions of velocity, wind, and air character? No. Instead, you want to understand a few principles that allow you to predict those forces for any particular set of conditions. Although the broad principles are clearly important to our understanding of nature, we propose that it's equally important to distinguish these interpretations of observations from the observations themselves, the evidence on which the principles are based.

I (Russ) had the good fortune to learn plate tectonics from one of the "old guard" faculty at the University of Kansas, Dr. William Merrill. Dr. Merrill entered geology before the era of plate tectonics. In the introductory geology class I took from him, he did not address the theory of plate tectonics until near the end of the term—after we had explored sequences of ocean crust found in the Ural Mountains far from any modern sea. After we had considered the bull's-eye pattern of rock ages formed on continents where younger rocks surround the older like rings around a target. After we had carefully built up the foundational concepts for subsidence and erosion, compression and mountain building. I remember being particularly befuddled by eugeosynclines and miogeosynclines, which are thick sequences of rock adjacent to the continents, representing profoundly different depositional environments. Deep water versus shallow water; volcanics versus limestone. Whoa. What did it all mean?

Then came plate tectonics, and one by one all of those strange observations and patterns fell into a picture that made perfect sense. I remember my exhilaration at seeing how all the pieces fit together. Black shale, pelagic fossils, and volcanics—yes—deep trenches adjacent to convergent boundaries. Limestone, shale, sandstone—yes—wedge of accumulated sediment on continental shelves adjacent to the continents.

Would it have been simpler had Dr. Merrill started with the plate tectonic theory and then hung all the observations—the evidence—on that framework? Of course. Would Dr. Merrill's lectures have been easier to follow and the concepts more clear? Of course. Would I have gotten a better understanding of geology and plate tectonics? No.

My excited understanding of plate tectonics came because we started with the evidence, and when all the pieces of evidence came together, I understood the nature of science and how complex theories such as plate tectonics come to be accepted. I would not have gotten that epiphany had we started with the theory fait accompli and then traveled backward to examine the evidence for it. I would not have had the chance to grapple with the confusing mountain of data from which theories emerge.

Knowing theories is not where science happens. Science happens where the fog of observation turns slowly into theoretical (or conceptual) understanding. Arguing from evidence is about giving students the chance to take that journey. There isn't time in a classroom for students to grapple with all of the evidence or figure out all of the theories. But some of them? Yes.

"What Is the Evidence?" Challenges

In Chapters 4 and 7, we presented "challenges" for helping students understand graphs, models, and equations. Here, we offer some example science reasoning challenges for understanding evidence and the relationship between evidence and model. As in previous chapters, we offer these examples as a prompt for you, the teacher, to create challenges tailored to your own curriculum.

One approach to creating science reasoning challenges is to consider whether some new observation, if made, would support a particular model, theory, or hypothesis. For example, consider the following challenge:

> *Going outside one day, you notice that the clouds in a jet plane trail occur in narrow bands—like ripples—perpendicular to the direction the jet plane is traveling. You conclude that the shock wave from the jet plane exhaust causes the rippled clouds. For each of the following observations, indicate whether that observation, if made, would support the conclusion, be inconsistent with the conclusion, or be immaterial to the conclusion.*
>
> 1. *In an experiment with shock waves like those behind a jet, you observe that the banded cloud formation forms both with and without the shock wave. (inconsistent with conclusion)*
> 2. *After watching clouds over an entire year, you notice that you never again see clouds like those you saw behind the jet except once when the only other jet of the year passes over. (supports the conclusion)*
> 3. *You go to the library to research weather and find that over the past three years, rain has been slightly more common in areas with more jet planes passing over. (immaterial to conclusion)*

Another approach to science reasoning challenges is to have students identify evidence and conclusion from their textbook or other source of information.[C25] The following passage of text (in bulleted list format) comes from an article about the Mars Curiosity rover that was posted on NASA's website in 2012. The passage includes four key conclusions with four key supporting observations (evidence). You can challenge your students to identify each conclusion and for each conclusion identify the supporting line of evidence. For a more difficult challenge, you might have your students read the entire original article and extract for themselves major conclusions and the evidence to support them.

- "From the size of gravels it carried, we can interpret the water was moving about 3 feet per second, with a depth somewhere between ankle and hip deep," said Curiosity science co-investigator William Dietrich of the University of California, Berkeley.

- The rounded shape of some stones in the conglomerate indicates long-distance transport from above the rim, where a channel named Peace Vallis feeds into the alluvial fan.
- The abundance of channels in the fan between the rim and conglomerate suggests flows continued or repeated over a long time, not just once or for a few years.
- "The shapes tell you they were transported and the sizes tell you they couldn't be transported by wind. They were transported by water flow," said Curiosity science co-investigator Rebecca Williams of the Planetary Science Institute in Tucson, Arizona.

You might ask students some questions to prompt a deeper understanding of the difference between the evidence for stream channels on Mars and the conclusions about the presence and nature of the stream channels. For example, what aspect of the shapes of the particles tells us that they were transported a long distance? *(the rounding of the individual particles)* Why does rounding tell us that the particles were transported a long distance? *(Long transport is needed to knock off all the sharp edges.)* How does size tell us that the particles couldn't be transported by wind and that the water was moving 3 feet per second? *(Wind can't lift big particles, and the big size of the particles tells us the water was moving fast.)*

The observations on Mars deal with the role of water in erosion and deposition and are most meaningful when integrated into a related unit of study. We offer an experimental activity that might also fit within such a unit in "Suggestions for Designing Activities: Experimental Sedimentary Petrology With Downspouts," in Chapter 10 (pp. 205–208).

In addition to giving students practice at recognizing evidence, science reasoning challenges like those presented here give teachers the opportunity to assess, in real time, student progress. These challenges are not intended as stand-alone activities but can be a supplement to working with evidence and conclusion in the context of a classroom investigation. It's easier for students to identify evidence and conclusion in cases where they are participants in the investigation, making that the place for them to start. However, much of science comes to us from other people, from scientists or from reporters who tell us about scientific discoveries. Recognizing what is evidence, assumption, and conclusion is important in understanding that communication.

Language-of-the-Argument Challenges

Science reasoning challenges can also serve the purpose of helping students separate and articulate the parts of a scientific argument, including evidence, conclusion, assumption, and other types of unsupported claims.[C26] We aren't big fans of learning vocabulary or terminology as a substitute for science, but it's difficult to understand a scientific argument without being able to distinguish—and name—these different types of statements.

I (Mary) have noticed that students can often tell theory from evidence when they've been specifically told which is which, but they aren't able to generalize that to new scientific problems. The following challenge provides an opportunity to practice identifying different components of an argument in which students have not been told the answer previously and for which they haven't developed any preconceived notions. Presenting a "made-up" abstract challenge might seem disconnected from science "content," but remember that your goal with this practice of science is not to teach a theory but rather to help students recognize and articulate the logical relationship between evidence and conclusion.

> *One day while in your time machine in the future, you meet a fellow time traveler watching a strange dinosaur-like creature drag the carcass of an elephant into a cave. When you ask her what she is doing, she replies, "I'm watching that creature take elephants into her nest to feed her young." Both of you wait until the creature leaves, then you go explore the cave. You find three elephant carcasses, and stuffed down the throat of each of them is a large, white egg. The other traveler states that this evidence supports her hypothesis.*
>
> *From the text above, identify the evidence, conclusion, an assumption, and two claims that do not bear on the conclusion. You might consider the following list of statements from the story.*
>
> *a. Each of the three elephants has an egg in its throat.*
> *b. The eggs were laid by the dinosaur-like creature.*
> *c. The dinosaur-like creature dug the cave.*
> *d. The dinosaur-like creature will soon be back with another elephant.*
> *e. The dinosaur-like creature is storing elephants to feed her offspring.*

Inspection of the text of the question reveals that "e" is the conclusion. It is a conclusion based on the observation that the creature is taking elephants into the cave and on the experience of knowing that living things have offspring which they feed. This conclusion is further supported by the observation that there are eggs in the elephants' throats, an observation which we might expect if we think the eggs will hatch and eat the elephants' carcasses from the inside out. However, there is no independent evidence that the eggs belong to the dinosaur-like creature. That is an assumption, a claim unsupported by observational evidence used as a starting point for an argument. Maybe the dinosaur-like creature found the elephants with the eggs already in their throat and is gathering the elephants to rescue them before the eggs hatch. (See Table 8.1 for an example chart separating observations and evidence from implications and assumptions in this challenge.)

Table 8.1

Example chart separating observations and evidence from implications and assumptions in the time traveler challenge

OBSERVATIONS AND EVIDENCE *What do we see? What was measured?*	IMPLICATIONS AND ASSUMPTIONS *What do the observations mean?*
• The creatures carried the elephant carcass into a cave. • Each of the three elephants has an egg in its throat.	• The cave is the creature's nest/home. • The creature is feeding her young. • The eggs were laid by the creature.

Note: Implications of evidence, as used in this table, can include claims, conclusions, explanations, and theories. As we use the word here, a *claim* isn't necessarily underpinned by evidence or reasoning—for example, it might be an assumption or an untested hypothesis—and it may or may not be testable. *Assumptions* are claims that are not directly supported by the observations. A *conclusion* is a claim that is based to some measure on observation and reasoning. A *hypothesis* is a testable claim, although it may not have been tested. Note that we don't include the important step of argumentation in this chart—we will examine arguing from evidence and the chain of reasoning between evidence and theory in Chapter 9.

This puzzle is modified from "Critical Thinking Questions for Examinations and Exercises," a test developed by Donovan and Allen and published in *Enhancing Critical Thinking in the Sciences* (Crow 1989). It works as a way to practice thinking about the different components of a scientific argument but works less well as a test of student understanding of those components. I (Russ) gave a simplified version of this question (asking only for the assumption) to more than 100 introductory college students and faculty in the mid-1990s. All of the nonscience students and faculty incorrectly identified "e" ("The dinosaur-like creature is storing elephants to feed her offspring") as the assumption, suggesting that the question did not function well to test critical thinking. The question worked far better as an exercise to illustrate the different types of claims in scientific arguments, especially the difference between assumption and conclusion—two types of claims that are sometimes used interchangeably outside of science.

By developing an imaginary science argument like this, you can think about how different claims and observations might contribute to an argument. Drawing from your students' discussions in the classroom, you can craft a meaningful opportunity for them to articulate the differences among claims and better distinguish evidence from claim.

Cultivate a Classroom Culture of Asking "What's the Evidence?"

Identifying evidence and connecting that evidence to a conclusion is a creative and nonlinear step of science, and thus one of the harder steps to learn or teach. One has to extract the evidence from a large body of observations, some of which may not be relevant or key. One has to visualize the connection between concrete observations and an abstract and emerging mental model for what caused those observations to occur. There is no fixed formula or predefined set of "steps" for doing this; each problem, each situation, will require somewhat different solutions. However, you and your students can always step back from a discussion, lesson, or lab, and ask "What's our evidence?"[C22]

As a way to encourage students to focus on the evidence, I (Russ) sometimes ask students to think of a prediction that a model makes and how an observation addresses that prediction. For example, consider the model of wandering poles proposed to explain changes in climate before the acceptance of plate tectonic theory. If the Earth's poles had wandered through time, then paleomagnetic evidence from all the continents would point to a single location of the poles. This prediction is not supported by observation: Rocks from different continents indicate that the poles had to be in many different locations at the same time—a nonsensical result. Thus, the evidence supports drifting continents instead. Thinking about predictions of a model directs students of any age toward observational evidence rather than falling back on explaining what the model says.

Charting the Components of Argument

Charting evidence and claims, as we proposed for the Mars stream channel activity or as shown more formally in Table 8.1, can help students recognize the difference between a theoretical model and what we see or measure. Another approach to linking evidence to model is offered by Doug Lombardi, Bret Sibley, and Kristoffer Carroll in their 2013 article "What's the Alternative? Using Model-Evidence Link Diagrams to Weigh Alternative Models in Argumentation" in *The Science Teacher.*

Sometimes what counts as evidence gets obscured by the different meanings of words flung about by scientists, educators, and the public. For example, the words *theory, hypothesis, model,* and *claim* are used differently by different groups of people. A construction such as Figure 8.1 can help students distinguish evidence from argument, argument from hypothesis, and hypothesis from theory—all of which are important aspects of learning how to argue from evidence.

FINAL THOUGHTS

New models and new understandings develop when we recognize that an incongruity exists between our observational evidence and our existing mental model. Arguing from evidence takes place at that very point of discovery where we change our understanding of the universe. It shouldn't be surprising that arguing from evidence is the hardest part of science.

Figure 8.1
Relationships among different aspects of the scientific enterprise

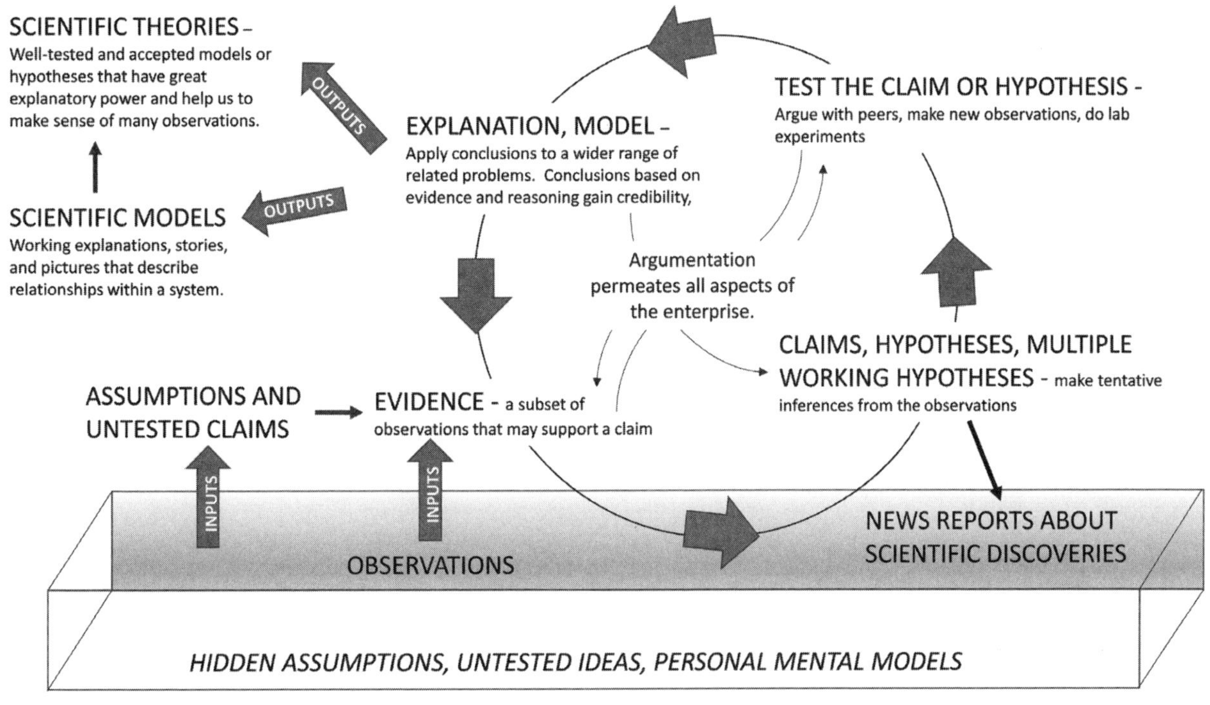

This illustration shows one way to look at these relationships. Initial observations and untested claims arise from the milieu of our untested ideas. Scientific research typically takes those observations (evidence) and proceeds through the cycle of testing, new observations, argument, and model development. Classroom students and the public often encounter science quite differently, exposed first to the claim from the scientists. An emphasis of the *Next Generation Science Standards* is that students, like the scientists, should proceed through the cycle of testing, new observations, argument, and model development.

For the teacher who wants to be a practitioner of science and do science with his or her students, this can be a daunting challenge. The interplay between observation, model, and revised model (see Figure 8.1) often only makes sense when you've been through the research process many times before. Even then, the next project may be a stumper.

Don't let that stop you.

Really? I should dive into things I don't fully understand and let my students see that I don't have all the answers?

You bet. Maybe you've had an opportunity to do real research at some point. If so, that's great. It's in doing research that we come to understand what constitutes evidence and the connection between evidence and conclusion. If you haven't been involved with a research project, then there's no time like the present to start. Make your classroom your laboratory and join your students in the great adventure of scientific discovery.

Connecting evidence to theory is something that has to happen in your students' own heads; it's not something that you, the teacher, can give them or make happen. You can, however, structure your class time to give those connections a chance to develop. When they do develop, it can be a true delight. After completing the plate tectonics investigations described in Chapter 5, I (Mary) commented to my students, "The investigations get more complex as you go along, but things start to make more sense." One of my students responded, with a fair bit of amazement, "Yeah! It's like you start to notice things. Like at this divergent boundary the rocks are really young, because magma is rising here."

That made my day.

REFERENCES

Donovan, M. P., and R. D. Allen. 1989. Critical thinking questions for examinations and exercises. In *Enhancing critical thinking in the sciences*, ed. L. W. Crow, 13–16. Washington, DC: Society for College Science Teachers.

Hurley, P. M. 1968. The confirmation of continental drift. *Scientific American* 218 (4): 52–64.

Krimigis, S. M., and R. B. Decker. 2015. The *Voyagers'* odyssey. *American Scientist* 103 (4): 284–291.

Lombardi, D., B. Sibley, and K. Carroll. 2013. What's the alternative? Using model-evidence link diagrams to weigh alternative models in argumentation. *The Science Teacher* 80 (5): 50–55.

Michaels, S., A. W. Shouse, and H. A. Schweingruber. 2008. *Ready, set, science! Putting research to work in K–8 classrooms.* Board on Science Education, Center for Education, Division of Behavioral and Social Sciences and Education. Washington, DC: National Academies Press.

National Aeronautics and Space Administration (NASA). 2012. NASA rover finds old streambed on Martian surface. *www.nasa.gov/home/hqnews/2012/sep/HQ_12-338_Mars_Water_Stream.html.*

National Research Council (NRC). 2012. *A framework for K–12 science education: Practices, crosscutting concepts, and core ideas.* Washington, DC: National Academies Press.

Powell, J. L. 2001. *Mysteries of terra firma: The age and evolution of the Earth.* New York: Free Press.

Wegener, A. 1915. *Die Entstehung der Kontinente und Ozeane* [The origins of continents and oceans]. Braunschweig, Germany: F. Vieweg.

Wilson, J. T. 1963. Continental drift. *Scientific American* 208 (4): 86–103.

EXAMPLE ACTIVITY DESIGN

WEGENER'S EVIDENCE FOR CONTINENTAL DRIFT

Arguing From Evidence

This example activity design engages students in both identifying evidence and arguing from evidence. Sources for the evidence include a reading assignment and previous classroom activities.

TEACHER PREPARATION AND PLANNING

Identify a reading that addresses the evidence cited by Wegener for continental drift. It's important that the reading focus primarily on the evidence, not on describing the theory. It's also important that the reading cite multiple lines of evidence because this "primes the pump" of information and makes students feel as if they have something to say in the discussion.

Students need to realize that the text itself is not their evidence. Therefore, it's better to do this activity after students have gained experience with experimental and field evidence in their own investigations, regardless of the specific topic. This personal work with making observations gives students a foundation for understanding the source of the evidence cited in their reading assignment.

EXAMPLE PROMPTS AND LIMITING OPTIONS

Readings

Have students read the chosen text. You might have them make a chart with "observational evidence" on one side and "implications of the evidence" on the other side, similar to Table 8.1 (p. 161). For example, if the evidence is "presence of *Mesosaurus* fossils found in small regions of Africa and South America," the implications might be either that the continents must have once been together or that there was some other way for these creatures to cross the Atlantic Ocean.

Encourage students to look for multiple lines of evidence, not just one or two. Also encourage students to be detailed and specific, not to just write down "fossils" as evidence for continental drift.

Discussion

Establish procedures for the discussion, so everyone gets a chance to contribute. Consider arranging your students in a circle so they can speak to each other. You might start the discussion with a seed question such as one of these:

- Why did Wegener think the continents had moved?
- Why didn't other people believe the continents had moved?
- Would you have been convinced that the continents moved if you had lived at the time of Wegener? Why?

Have prompts visible to all students, to help them know how to ask questions and engage with other students' ideas. Here are some examples:

- I think _______ because of _________.
- I didn't understand _______.
- Could you explain what you mean by ________?
- I agree with [idea one] but I don't understand (or agree with) [idea two].
- How do you know _________?
- What is the evidence for _________?
- Couldn't that observation also be explained by ___________?

Arguing from the evidence will be hard initially. Students will tend to tell stories from the models, not the evidence. Try to find a balance between letting students discuss as they choose and providing guidance. For a successful experience with arguing from evidence, there needs to be a point where the student questions his or her understanding based on new evidence or based on old evidence seen in a new light. Keep that goal in mind as you facilitate the discussion.

You might remind students to focus on the evidence if the discussion begins to wander into overly speculative territory. Listen for descriptions of the theory instead of descriptions of the evidence and ask a question to refocus the discussion where needed. If students are too quick to believe the plate tectonic model, you might ask how the evidence excludes other models. Can the students think of other models that fit some or all of the data? Why not believe that a land bridge crossed the Atlantic; is there evidence that doesn't support the existence of a land bridge?

EXAMPLE INTERACTION

Student 1 question: Couldn't the mesosaurs have just started out on both continents? Why does there have to have been a land bridge or the continents together, either one?

Student 2 question: Then why would the mesosaur fossils only be in a small area in Africa and a small area in South America?

Student 3 question: And why would those areas match up on opposite sides of the Atlantic?

Student 1 interpretation: Maybe there was a particular climate that they lived in and it was on opposite sides. That would explain why they were only in a small area and on matching sides.

Student 4 claim: We know that climate changes. Plate tectonic theory causes the climate to change.

Student 5 claim: Yeah, that's why there used to be ice ages in Africa.

Teacher question (seeing the discussion going toward telling stories based on theory): Is there any other observational evidence that the idea of mesosaurs starting out on both continents couldn't explain?

Student 1 argument from evidence: Well, the glaciers. If a glacier was coming out of the ocean, that doesn't make sense. Unless the continents were together. Even if there were glaciers on both continents separately, they would still move downhill, not uphill.

SUMMARY CHECKLIST FOR TEACHER AS PRACTITIONER OF SCIENCE

- Read a text focused on evidence and citing multiple lines of evidence.
- Have students identify, in writing, the evidence and the implications of each observation.
- Prompt discussion, giving guidance for how to participate.
- Encourage students to stay focused on the evidence but to consider alternative models and whether the evidence truly distinguishes between models.

TEACHER REFLECTION

How will you continue to cultivate a classroom culture that listens well and values the contributions of the other students? Depending on how you structured your discussion opportunities, different classroom management problems probably popped up. With small-group discussions, the hard part is providing enough structure so that every student stays on task. With large-group discussions, the hard part is keeping 30 students engaged in active listening and taking their turn to respond.

With either size of discussion group, having students write down a few of their ideas before the discussion helps everyone feel as if they have something to contribute. Sometimes, it's useful to pause in mid-discussion and have students write about what they think they know so far. This gives you a chance to see what all the students are thinking in real time. It also gives students a chance to consolidate some ideas in their own mind. On reflection, are there ways you could have done the practice of arguing from evidence more effectively?

Learning to identify evidence and then supporting one's claims or explanations is a skill that your students will need to practice. Lots. It's OK if your early tries at having students argue from evidence fall short of your expectations. What did you learn from the experience? What will you do differently the next time? Look for ways, during future lessons, to provide mini-opportunities for students to argue from evidence; it doesn't always have to be a full-blown lesson.

ARGUING FROM EVIDENCE, PART 2

CLAIM AND ARGUMENT

Theory. Science's most maligned and misaligned word. Beliefs about this word range from the idea that it refers to a vague notion ("It's just a theory") to the idea that it represents absolute truth ("An observation can't be true if it goes against theory"). In the minds of many people, theories even have the power to cause events. Consider these two statements from college term papers:

"Dinosaur extinction is attributed to two theories."

"Personally, I believe it is a combination of theories that killed the dinosaurs."

Wow—if a couple of theories can wipe out an entire group of creatures that ruled the Earth for 200 million years, just think what a couple of theory-making scientists can do! Or a classroom full of eager students! We feel powerful already!

THE HEART OF SCIENCE IS THE ARGUMENT, NOT THE THEORY

The following statement, from a college term paper, captures a common misunderstanding of theory among the general public—something to either accept or not, but not something that we can evaluate through reasoning: "In summary, there are many different theories and there is no way to know which of them is true." Really? Some arguments aren't stronger than others? Evidence doesn't support one over another? We can't test the theories with new observations?

Our students—and the public in general—often encounter scientific claims as facts to be accepted (or not) rather than ideas to be supported by observation and defended by argument. We teachers often emphasize learning the theories and de-emphasize arguing about them—perhaps because understanding the theories simplifies and organizes classroom instruction or perhaps because we don't want to call into question the "certainty" of the scientific enterprise. This means that students enter college very able to answer the question "In terms of the theory, explain ..." but less able to answer the question "Why should I believe this theory is correct?" Yet arguing from evidence is at the heart of the practice of science, and understanding scientific argument is a far more important goal of science education than merely knowing the "facts and theories."

Questioning the Theories

Hypothesis, theory, and *law* are often confusing terms for students. Patrick Schuette, a secondary life science and earth science teacher in Minnesota, had this to say:

> *Students across the country are asked to "define" scientific hypothesis, theory and law as part of state and national standards. Teachers often have students classify examples into one of the three definitions. They stress to students that theories are widely accepted fundamental explanations of natural phenomena. However, despite this memorizing, classifying and listening to explanations of terms, students often fail to understand that a theory is linked to underpinning evidence.*
>
> *Students generally can classify plate tectonics as a theory because they are told it is a theory. They accept it as theory. ... However, students should examine and question the evidence. ... As students question the evidence and look for alternatives they gain appreciation for what a theory is and can truly start to see how a hypothesis is different from a theory. (Schuette 2015)*

A Framework for K–12 Science Education: Practices, Crosscutting Concepts, and Core Ideas (NRC 2012, p. 73) poses the following challenge:

> *The study of science and engineering should produce a sense of the process of argument necessary for advancing and defending a new idea or an explanation of a phenomenon and the norms for conducting such arguments. In that spirit, students should argue for the explanations they construct, defend their interpretations of the associated data, and advocate for the designs they propose.*

The *Next Generation Science Standards* (*NGSS;* NGSS Lead States 2013) interprets this challenge, in part, with this element of science and engineering practice 7, Arguing From Evidence, for grades 6–8:

> *Construct, use, and/or present an oral and written argument supported by empirical evidence and scientific reasoning to support or refute an explanation or a model for a phenomenon or a solution to a problem.*

It would be convenient if arguing from evidence were achievable through some set of reproducible steps, like the steps of the scientific method that was taught for many years. The problem with the scientific method is that it never was the way that real science proceeds—many teachers now refer to it as the scientific "mythod." It's not that the scientific method had no foundation in the scientific enterprise, but what started out as a way to summarize a whole set of complex processes became simplistic steps memorized from the first chapter of the textbook.

We now use new terminology to try to capture the "steps" of scientific investigation and argument—for example, *claim, evidence,* and *reasoning.* Like the scientific method, these words attempt to "lasso the whirlwind," to simplify and codify the complex enterprise of scientific observation and argument. *Claim, evidence,* and *reasoning* are good reminders of what we need to include in our thinking, but they are not steps, and so, in using these ideas, we need to take care that we help students to think like scientists and not simply teach them the jargon of an educational scaffolding.[C27]

Jennifer Lepper, a teaching colleague of mine (Russ), did an experiment in teaching the scientific method with two sections of her earth science class for elementary education majors. She taught the nuanced process of scientific investigation in both classes. In one class she also taught the "steps" of the scientific method (ask question, construct hypothesis, test hypothesis, analyze result, report conclusions).

The class that only encountered the nuanced version was able to understand much of the nuance. However, the class that encountered the scientific method only had the scientific method and none of the nuance. The powerful imagery and simple outline of the scientific method overwhelmed any subtleties in the real process of science.

Narrative Evidence

We emphasize anecdotal illustrations from the classroom over statistical research in this book. This approach can be compared with the ethnographic study of human cultures. Anthropologists live with a group of people, become part of their society, and from the varied and nuanced observations of that experience, gain an understanding of the culture that is conveyed to other anthropologists through narrative. In contrast, sociologists often study groups of people through surveys and interviews, and then communicate with other sociologists through statistical analysis. Both approaches are useful, but they yield different kinds of insight.

Too much reliance on statistical data over holistic experience can lead to absurd results. Sharon Salzberg, a teacher of Buddhist meditation, was once asked by a neuroscientist, "Without further scientific validation, how will you know if meditation works?" (Salzberg 2015). Uh, maybe because I and others have experienced it working in the laboratory of our own bodies?

In the ethnographic tradition, we offer a narrative exploration of teaching focused on our experience as participants in discovery with students, which readers can compare to their own experiences. In this way, *Learning to Read the Earth and Sky* is a holistic conversation between teachers.

So, in teaching students to argue from evidence, with all its complexity, nuance, and variation, how do we lasso the whirlwind? It won't happen with one lesson on arguing from evidence, however well-constructed that lesson might be. Arguing from evidence is not an educational check box—yup, got that one covered. Arguing from evidence permeates every aspect of the scientific enterprise, as we showed in Figure 8.1 in Chapter 8 (p. 163). Likewise, arguing from evidence needs to permeate every activity in the science classroom, throughout all the years of learning science.

Compared with other goals of classroom learning such as doing experiments, applying models, and understanding core ideas, arguing from evidence is particularly time-consuming. It takes a lot of time for students to think deeply about their reasoning, share with others in a formal way, and give and get feedback. They aren't very good at it at first. What's more, transferring the skill to a new context with new "content" is hard for them. They need lots of opportunities to try it out in different contexts before they can learn to generalize to new situations.[C28]

The teacher has to choose a balance between how much time to devote to arguing from evidence and how much to devote to giving students evidence and understanding to argue from. However, based on our experience, we think that most of us have lots of room for growth in developing student abilities to argue from evidence. In the past, we (Mary and Russ) have focused on encouraging students to explain models and cite evidence to support their explanations—but we have allowed less time for argumentation. We have come to believe that it is in participating in argumentation that the real sense of the scientific process can crystallize for students.[C29]

FINDING A PLACE TO START

Students can distinguish evidence from theory, explain models, and generate hypotheses to answer new questions without ever asking "Wait, does this make sense?" When asked to address a new question, students will often refer, indiscriminately, to concepts learned in past lessons without arguing from evidence.

After I (Mary) had completed lessons on erosion, deposition, and plate tectonics, I asked my students to write answers to the question "Why are continents higher than the oceans?" Here are a few responses:

- "The continents are higher because people need to have a place to live. On continents, there is land, where people can build homes and be safe. Lower places on Earth, like the ocean floor, would be a very hard place to live."
- "Because, with all of the ocean water, it erodes and makes high and low areas. If the ocean went away, the low areas I think would grow back up with the high areas."
- "[Because] the continents collect rock [and] soil over time and build higher."

In studying plate tectonics, my students had learned about the correlation between seafloor topography and the temperature and density of rock, yet that correlation did not show up in any of the responses. My students weren't able to apply the previous concepts to a slightly different question. More important, the idea of making an argument based on evidence didn't occur to them. Students invoked what they knew about erosion and deposition, but without critical consideration of whether it made sense.

This would have been a good place to engage students in arguing from evidence: They had some observations and experience to bring to bear, they had identified a variety of models to test against those observations and experiences, and the driving question ("Why are continents higher than the oceans?") was a compelling one. I could have taken a few of the student responses and addressed them in class or group discussion. *Do* continents collect rock? Is that what makes them higher? How can we know?

I didn't engage students in arguing from evidence in this particular situation because I had not yet recognized the important distinction between arguing from evidence and other practices of science that I'd already addressed, such as explaining a model and interpreting observations. In seeing the student responses, I realized that unless students have the opportunity to argue from evidence, explaining models and interpreting evidence is not enough. It's in arguing from evidence that students learn to apply ideas to a broader context and to recognize the difference between making a claim and supporting a claim with evidence and reasoning.

An initial class discussion on a question can provide insight into student preconceptions and help them generate hypotheses to test against the evidence. The following introductory exchange took place in my eighth-grade earth science classroom when the class was looking at a photograph of Mauna Kea with snow at the top. I asked, "Hawaii is in the tropics, at 19° north latitude. How can there be snow there?"

Terry: There is snow there, because it's high up.

Mary (me): What does high up have to do with snow?

Terry: Well, it's colder up there.

Mary: But what direct experience do you have with the air being colder as you go up?

Hamza: I went up a mountain on a gondola and it was lots colder outside the gondola on the mountaintop.

Emily: Isn't it colder because the air is thinner and there is less oxygen?

Doug: Is it colder because it's closer to space?

Luis: If you get closer to the Sun wouldn't it be warmer?

Alyssa: Wouldn't it be colder because there's less oxygen by the clouds?

Leena: When you go up in an airplane it's really cold because of the air conditioning.

David: Aren't planes sealed off so the air doesn't suck us through?

Emily: Wouldn't the plane be colder in general because it's moving through the air?

Philip: When I was on an airplane it was –60° … I was watching the flight [tracker info] on the screen.

Mary: We've identified a problem—maybe the airplane is a sealed environment. Maybe we should shift gears and think about the experience people have had with driving in the mountains, rather than being in an airplane.

Christine: When I was in the mountains, the air was really thin so your ears pop. You don't notice the change in temperature, but it is harder to breathe.

Zoey: We were in the mountains and we hiked at a waterfall, and we couldn't breathe very well hiking up the steep path.

Jeremy: We got warmer higher up, because the Sun is closer; but it was still hard to breathe.

David: If it's closer to the Sun and warmer, we wouldn't expect snow high up on Mt. Everest.

Mei: We drove from Seattle to the mountains, still snow 'cause it was closer to the atmosphere, close to the end of the atmosphere and then as you cross through and get close to the Sun it gets warmer.

Mary: Even at sea level aren't you still close to the atmosphere?

David: Don't the greenhouse gases make a thin blanket covering to insulate us?

Jacob: But if it's thinner then there would be less material; wouldn't there be less material to trap heat?

Leena: That is a really good question. If it's harder to breathe, how come it's not harder to breathe in the plane when you are high up in the atmosphere? Does the plane bring its own oxygen?

David: It's air-conditioned and they put the oxygen in with it.

The conversation continued, focused on air-conditioning. Notice how quickly students abandoned observational evidence and gravitated toward invoking theoretical models. There were quite a number of theoretical models invoked by the students during this short exchange. Air is colder because of (1) the lack of oxygen, (2) the proximity to space, (3) the presence of atmosphere, (4) the thinness of the greenhouse gas blanket,

(5) the proximity to clouds, and (6) the movement of a plane through the air (wind chill factor). The air is warmer due to (1) proximity to the Sun, (2) disappearance of atmosphere, and (3) presence of a greenhouse gas blanket. One student reported a temperature change that appears to be based on an incorrect theoretical expectation rather than actual observation, bringing to our attention how difficult it is to sort out which "facts" are observational evidence and which "facts" are theoretical interpretations.

I tried a couple of times to prompt students to argue from evidence, with limited success. Even so, I was happy with the exchange. It's in trying out their ideas, thinking about evidence, and doing their best to argue from evidence that students come to understand the relationship between evidence and theory. No amount of having the evidence and theory explained to them can substitute for this experience.

These experiences with starter discussions brought to my attention how difficult it is to structure a class so that students marshal evidence and make arguments. It's easy to tell them the answer or to have an open "brainstorming" discussion, but encouraging them to question models, identify relevant observations, and reason through what the observations mean is hard. It's also hard to assess their efforts or provide guidance on how to do better. When students ask, "What do I need to say next time to get a better grade?" how do I respond?

There is no simple answer to give them, no set of steps, except, perhaps, to identify what they know and what they don't know, to ask questions, and to be ready to charge into the "fog."

ARGUING THROUGH THE FOG

Once we accept the idea that letting students argue through a confusing pile of information to reach conclusions on their own is sometimes of more educational value than clear and straightforward presentation of theories, it's tempting to throw students into an open-ended discussion of evidence and conclusion with a challenge such as "look at pictures and maps of the Grand Canyon and argue about how you think the Grand Canyon formed." However, this question is very broad and complex and there isn't a lot of evidence in pictures and maps that even scientists could bring to bear on this question, so the discussion is likely to devolve into idle speculation. In fact, charging into the arguing-from-evidence part of your unit without students having enough information to make reasonable arguments can reinforce wrong preconceptions and create muddled models of reality.

Often in an initial classroom discussion such as the discussion about snow on the top of Mauna Kea described in the previous section, it becomes clear that students don't have enough observations to answer the question. Students can't argue from evidence in the absence of knowledge of evidence. Rather than letting students continue to engage in idle speculation or hypothesis generation, the teacher can "prime the pump" with information.

This information needs to be more than a description of theoretical models or correction of students' wrong models—students need to learn about observations and measurements that people have made. Or, they can make their own new observations.

For example, what observations might inform student discussion of the question "Why is there snow on the mountaintops in Hawaii?" You might start by asking them which things they know and which they only speculate. Then, what new things do they need to learn to answer the question?

If there is less air higher up, what does that mean about the pressure? Pressure increases as one goes deeper into water. By analogy, how will pressure change deeper in air? What if you took a barometer up a mountain? How would it change? Has somebody done that? How does an altimeter let the airplane pilot know how high the plane is? Students and the teacher can get answers to these questions, gathering observational evidence that bears on the discussion.

If pressure decreases upward, how does that affect temperature? Has someone measured the effect of changing pressure on temperature? Students could play with balloons or tire pumps and see how the temperature changes as air decompresses out of the balloon or how the temperature changes in the tire pump when the air is compressed.

Through a combination of research into published observations made by scientists and doing their own experiments, students can gather enough information to feel confident in their contribution to arguing from evidence. After gathering information, doing experiments, and making observations in the field, students can go through their discussion again, gaining deeper understanding of the role of observation—as opposed to speculation—in addressing a question such as "Why is there snow on the mountaintops?" They will experience the initial fog of observation—sorting through multiple models and making unsure arguments—and then see those observations turn into understanding as the process of investigating and arguing proceeds.

The fog of argument—and not having enough information to dispel the fog—happens in real research as well as the classroom. Hap McSween, one of my (Russ's) dissertation committee members at the University of Tennessee, was one of the first people to study the mysterious meteorite Allan Hills 77005 (AH 77005; McSween, Taylor, and Stolper 1979). Why mysterious? It was a differentiated meteorite, meaning it had been through extensive chemical processing such as happens on planets. Its composition was similar to compositions derived from Earth's mantle, again suggesting a planet-size object of origin. It almost seemed that the meteorite must have come from Mars. Other meteorites that seemed to have the same origin as AH 77005 were found to have geologically young ages. This observation also suggested they came from Mars because, unlike most asteroids, Mars was big enough to have stayed hot inside so it could melt rock in geologically recent times.

Problem was, physicists calculated that no meteorite could possibly get knocked off the surface of Mars and find its way to Earth. Thus, existing theory suggested that the meteorites could not have come from Mars.

The chemical evidence continued to mount. The ratio of different isotopes of oxygen matched that of Mars and differed from other bodies, such as Earth, the Moon, and other meteorites. In 1983, Don Bogard and Pratt Johnson found that gases trapped in tiny bubbles in one of the meteorites had the same composition as gas sampled from the Martian atmosphere by the Viking landers. At this point, the physicists went back to their calculations and, with a few tweaks to the model, found a way to get those meteorites from Mars to Earth.

Evidence that this group of meteorites came from Mars has continued to accumulate—scientists never stop testing their theories. Most recently, measurements of isotopes of argon by the Curiosity rover showed that the argon isotopes of Mars match those in the meteorites (NASA Jet Propulsion Laboratory 2013).

In the end, observation trumped theory, but only after years of arguing through the fog.

Classroom Example 1

One fall, I (Mary) engaged my students in looking at photos of trees growing from cracks in rock. We'd been studying mechanical weathering and had read about root wedging and ice wedging. The photos showed saplings growing from a layer of siltstone in southeastern Minnesota (part of a colleague's EarthCache project; for more information about EarthCache, go to *www.earthcache.org*).

I talked through the photos to help them orient to the view and get a sense of scale and then asked the driving question "A tree growing out of a rock! How is this possible?" In response, students started generating explanations with little focus on what was observable: "The crack was already in the rock and a seed fell through the crack to the soil, sprouted and grew up in the existing crack." "The seed was under the rock and sprouted and pushed its way up through the rock and broke the rock apart." "The crack was already in the rock, but the tree widened the crack."

Students then asked me, "So what's the right answer?" The answer, I said, is in the observations, not in an answer key. We're going to try to figure it out!

I encouraged students to look more closely at the photos and they came up with several key observations:

- There is always a crack in the rock if a tree is coming out of the rock.
- One crack might have more than one sapling growing out of it.

- There are other cracks in the rock that don't have trees in them.
- Some cracks are vertical parallel to the tree trunk, and others are horizontal so the tree trunk goes perpendicular to the crack.

To expand our data set, we searched online for other pictures of trees growing from cracks in rocks. Students found illustrations of root wedging and ice wedging online and compiled example pictures in their notebooks. They identified key observations for each picture and then summarized key observations.

- There are lots of cracks in other places, too, where there aren't any trees.
- Cracks with trees in them are often wider than neighboring cracks.
- Sometimes the crack is widest right where the tree is.
- Sometimes it looks like the root, and not the trunk, fills the crack.
- Sometimes the tree isn't as wide as the crack that it's in.

I then prompted them to complete one of the following sentences about the Minnesota photograph:

- I think the cracks formed by frost wedging because …
- I think the cracks formed by root wedging because …

Some of the students' responses focused on climate and did not address observational evidence; for example: "I think the cracks formed by frost wedging because there is lots of cold weather in Minnesota so when water gets in and freezes it makes cracks."

Other responses cited the evidence but didn't explain the connection between observation and conclusion; for example: "I think the cracks formed by frost wedging because the tree doesn't take up the entire space of the crack."

Some cited multiple lines of evidence, and hinted at reasoning, although the reasoning wasn't always clear; for example: "I think the cracks formed by frost wedging because it's more of a smooth crack and doesn't have a root present in the crack to push it apart."

Only a few students included reasoning in their argument—most simply stated the evidence without explaining why they thought that evidence supported the claim. My takeaway was that my sentence prompts were too limiting. For example, my prompts led students away from more complex and realistic interpretations such as "the cracks may have predated the trees but been widened by them" or "the cracks might result from more than one process." More important, the prompts did not make a distinction between citing evidence and explaining why that evidence supports the claim. This led students to cite evidence but not explain it.

When I do something like this again, I'll have students first write a claim and then write a sentence structured like this: "I think this because I can see ______ in the photo and that means ______." This structure allows students to participate in all parts of argumentation—claim, evidence, and reasoning—and it sets out the evidence and explanation of evidence as separate parts of the argument.

Classroom Example 2

Classroom experimental activities are a great place to practice arguing through the fog because the classroom results may not match those "expected" from the activity guideline, the experimental results from different groups often disagree, and students engage well when their own experiments give them skin in the game (see Figure 9.1). For example, I (Mary) had my eighth-grade students measure the effect of different substrates (soil or water) on the rate of heating in sunlight. The results did not match the prediction in our textbook, nor did results from the different groups exactly match each other. We put our results up on the whiteboard and talked about what they meant.

Figure 9.1
Student discussion of results

Providing opportunities for students to discuss and debate the cause of differences in experimental results between groups is one way to facilitate arguing from evidence.

Students were quick to identify possible errors in their measurements: "I think I may have been measuring in Fahrenheit instead of Celsius—otherwise my temperatures don't make any sense." "I noticed that some other groups had their light a lot closer to the thermometer than we did." "We didn't have the light coming at exactly the same angle for our soil and water experiments and that might have affected what we got."

Interpretation of the effects of experimental design was a bit harder for students to wrap their head around, but they did well. "What about how deep in the water or soil we put the thermometer, won't that make a difference?" "Should we have used the same mass of water and soil?" "The other groups took their measurements after longer time periods than we did—how long is long enough?"

The hardest thing for students to recognize and argue was the idea that their experimental results might be telling us something real about the universe even if they didn't match the book. There was a tendency to believe that the book had to be right and to ask "What did we do wrong?"

"Your results are your results," I said. "What do *your results* mean?" The students gave a variety of responses, as reflected in the conversation that follows:

> "Maybe the kind of ground makes a difference. Soil can be either fluffy or hard."
>
> "But my group smashed our soil down and got about the same result as the other groups that didn't smash theirs down."
>
> "Maybe it depends on the angle of sunlight. Over the course of a day, it comes in at lots of different angles."
>
> "But nobody's results are even close to the book's value, and different ones of us did different angles."
>
> "Maybe our measurements weren't accurate enough."
>
> "The book says that land heats up five times faster than water, but all of us got a result less than that, about two times faster. If the difference was due to some random effect, then some of us should have got it different in the other direction."
>
> "Maybe it's because all the water heats up at once because the light goes into the water and because the water stirs the heat in. In the soil, we only heated the top part so the heat was more concentrated."
>
> "But wouldn't that make us wrong in the other direction? Wouldn't the land heat up more than five times faster?"
>
> "But what if it also depends on how deep we had our thermometers? If the heat doesn't mix in, then the ground might heat up slower just a little bit deeper."
>
> "Maybe so," I said. "*How can we test that?*"

Classroom science, like real science, yields experimental and measurement problems to figure out and seemingly contradictory results to interpret. These challenges invite questions from students similar to those asked by scientists and engage students in problem solving and arguing from evidence. This is one reason teachers can't offer a problem-free, "teacher-proofed," "student-proofed" activity and hope that students learn science from it. If the activity yields the "right answer" with no problems to solve, then that activity has very little resemblance to real science and provides no opportunity to argue through the fog.

WRAPPING UP THE ARGUMENT

Arguing through the fog is inductive and requires significant creativity to identify the evidence and possible alternative explanations. In contrast, wrapping up an argument is a linear, deductive part of science. For example, using evidence for how we know that Earth's outer core is liquid (which we explore further in Chapter 12, "Stories of Places We Can't See but Can Hear"), students might make the following three (simplified) points in logical sequence: (1) laboratory experiments show that S waves can't travel through liquids but do travel through solids; (2) seismographic observations show that S waves from earthquakes don't travel (directly) through the Earth's core; and (3) therefore, Earth's core must be liquid.

It isn't practical to create "steps" for arguing though the fog because each situation will be different. However, we can identify common steps for the wrap-up argument, such as the following: (1) state evidence, (2) give reasoning that connects the evidence to the conclusions, and (3) summarize the conclusion.

As we write this, the first exciting pictures of the close encounter with Pluto are coming back from the New Horizons spacecraft. A July 14, 2015, report by Joel Achenbach in the *Washington Post* includes the following statement: "Linear features on the surface 'could be scarps, or faults,' said New Horizons team member and planetary scientist Cathy Olkin."

How might we summarize the argument for this statement, using the three-step format above? One solution might be the following:

On Earth, we notice that tectonic features are often linear or slightly curved—not winding, or lobate, or circular. This makes sense because, just like your hands can't make complicated turns with sharp angles while sliding past each other, rocks moving past each other will also tend to break along a linear trend. Given that the features seen on Pluto are linear, it makes sense that they might be tectonic features, like faults.

Having students write out simplified, three-point arguments of this sort can be a useful class exercise. Students might start from news articles, classroom experiments, or other information. For example, consider the argument in the "Arguing Through the Fog" section that some meteorites originated on Mars. You might encourage students to write a summary in their own words. They might work toward something such as the following:

The differentiated character of AH 77005 suggested that it came from a planet-size body like Mars. However, impact modelers were not able to identify a way to knock a rock off of Mars so that it could reach Earth. Once scientists learned that gas bubbles in the meteorite match the composition of the Martian atmosphere, a consensus grew among scientists that the rock came from Mars.

Wrap-up arguments are most appropriate for well-established ideas of science in which conclusions and key lines of evidence are widely accepted. For example, some aspects of climate change are still in the "foggy," making-sense stage of understanding among scientists, while other aspects are well established. Encouraging students to write wrap-up arguments on the well-established parts can both give them practice in arguing from evidence and help them think about the difference between conclusions that scientists agree on and tentative ideas that scientists continue to grapple with.

Suppose you've been through a unit on greenhouse warming and have examined how gases such as carbon dioxide (CO_2) can cause warming of the atmosphere. In summarizing the arguments, students might work toward something such as the following.

In 1859, John Tyndall made measurements showing that CO_2 absorbs infrared radiation but not visible light. This means that CO_2 in the atmosphere does not change the amount of visible light reaching the Earth's surface from the Sun but it does absorb infrared radiation emitted from the Earth's surface. Thus, increasing CO_2 in the atmosphere slows the loss of energy to space and contributes to an increase in Earth's average temperature.

This type of exercise requires that students identify the key observations, conclusions, and arguments from a selected text or other classroom lesson. Summarizing well-established arguments in their own words gives students practice in the logical-reasoning part of arguing from evidence. This kind of practice removes the cognitive load of figuring out their conclusions and identifying evidence concurrently with making the arguments and allows students to focus only on articulating the reasoning that connects evidence to conclusion. It gives them practice making sense of the observations and making that understanding their own.

THE TYRANNY OF THE ANSWER KEY

In science, being right or having the right theory is not the point. The point is, does our theory match all the observations and do we have sound reasons for believing it?

> **The Latest in Science Education**
>
> "There is a great danger in the present day lest science-teaching should degenerate into the accumulation of disconnected facts and unexplained formulae, which burden the memory without cultivating the understanding." (J. D. Everett, in the 1873 translation of *Elementary Treatise on Natural Philosophy* by A. Privat Deschanel; quoted in Willingham 2010)

Being right, but unable to argue from evidence, is of little significance in the history of science. For example, consider the claim of Greek astronomer Aristarchus in the third century BC that the Sun, not the Earth, is the center of our solar system (Ferguson 1999). We now know that this view was correct, but it was not accepted for 2,000 years. There is a tendency to think that Aristarchus was prescient, or that his contemporaries were either stupid or unwilling to accept new ideas. But the truth is that Aristarchus had no observational evidence from which to argue his case. In fact, the evidence available at the time supported an Earth-centered view.

Scholars of the day understood the idea of parallax—that distant objects appear to shift when viewed from a different perspective. Hold out your thumb and look at it with first one eye closed and then the other. Your thumb appears to move even though in reality it is only your perspective that moves. Contemporaries of Aristarchus realized that if the Earth were moving through space, then the stars should appear to shift in the sky due to parallax, and *they did not*. Thus, the evidence supported the idea that the Earth was stationary.

Aristarchus had an explanation for that: If the stars were unimaginably far away, then the parallax shift would be too small to see, a conjecture that we now know was true. But claiming that the stars shift by an amount we can't see is not an observation, and so Aristarchus had no evidence to back his argument.

Aristarchus was right—the Sun, not the Earth, is the center of our solar system. But his claim was not science because it was not based on observation and could not be argued from evidence. What he could argue from evidence was this: If the Earth orbits the Sun (an assumption for which there was no evidence), then the stars must be extremely far away.

Even middle and high school students know that science should be about more than knowing the right answer. In a December 17, 2013, NPR education piece with reporter Eric Westervelt, a 15-year-old high school student said, "Right now, [in] chemistry it seems like—Why are we learning this? It would be a better learning experience if we were almost forced to really understand instead of just memorizing things. … And [went] more in-depth in knowing why the answer is that answer."

Despite this insight into the learning process, many students and teachers develop a drive to get the right answer—the answer they know they are "supposed" to get, or the answer from the answer key, or from the book. Getting the "right answer" reinforces concepts learned in class, but it inhibits the creative analysis of experimental results that is an essential part of scientific exploration.

Students tend to explain their experiments in terms of what they were supposed to get, not in terms of their actual results, eliminating the ambiguity and concept construction that goes with real science. Often, their conclusion is either "I got the right answer—ta-da!" or "Oh, no, I must have done something wrong!" Incoming college students, when prompted to report experimental results, typically write either "Results proved the hypothesis" or "Results wrong due to human error," reflecting the answer key–driven thinking they've encountered in high school science.

Instead of fretting over getting the right answer, students should try to explain how they know their answer is right or wrong. Or perhaps they might explore why their results could both be "wrong" and yet give true insight into something real about the universe. Sometimes they might have actually made a mistake in the experiments, but if so, what? And why did that give a "wrong" result?

Some of the best questions in inquiry curricula come with the phrase "answers will vary" on the answer key. One (wrong) interpretation of this is that any answer is just fine. No, some answers are better than others. Some answers might even be wrong. Even right answers need justification. But in science, every answer, right or wrong, is a prompt for the next question: Who says that? What is their evidence? How can we test it?

In the classroom, looking to "get the right answer" can lead to bad experimental science. For several years, I (Russ) had elementary education majors do a three-week investigative activity in which they chose a question to investigate, developed and ran experiments, and reported the results. A popular experimental activity was to measure the absorption of infrared radiation by CO_2 and thus prove global warming—lots of materials for teachers were available online. Students acquired heat lamps (producing mostly infrared light) and dry ice (CO_2). They put the dry ice in a Mason jar and allowed it to evaporate, thus displacing the air in the jar with CO_2 gas. They then put a thermometer in the Mason jar, closed it up, and put it under the heat lamp. They placed another thermometer equally far from the lamp, but not inside the jar. Sure enough, the temperature rose faster in the Mason jar full of CO_2. Ta-da! Global warming proven.

I asked them, "How do you know that it was the CO_2 that caused the jar to warm up faster?"

They did another experiment using a Mason jar filled with plain air. Again, the temperature rose much faster inside the glass jar. There was no distinguishable difference between CO_2 and plain air.

Yes, my students had gotten the "right" answer, but for the wrong reasons. Glass absorbs virtually all of the infrared light, so the gas inside the jar had no significant effect on the overall absorption. Thus the glass jar heats up faster regardless of what's inside it, and this experiment tells us nothing about the role of CO_2 in Earth's greenhouse effect. In his experiments measuring the amount of infrared energy absorbed by different gases, John Tyndall contained the CO_2 in a long brass tube sealed on both ends with salt—salt absorbs neither visible light nor infrared.

Science exploration, and arguing from evidence, is not about getting the right answer from an answer key. Instead, science exploration is about making observations and connecting those observations to conclusions through reasoning. It's important that students test their results against observation and reasoning and not just check with an answer key to see if they "got the right answer." A vital part of developing an I-don't-need-the-answer-key mind-set is that students see their teacher figuring things out, too, not just relying on an answer key.

REFERENCES

Achenbach, J. *Washington Post*. 2015. After a Wait, Spacecraft Confirms That It Survived Its Close Pass of Pluto. July 14. *www.washingtonpost.com/national/health-science/new-horizons-finally-makes-it-to-pluto-sees-craters-and-great-mounds/2015/07/14/9bcb0f04-2a1f-11e5-bd33-395c05608059_story.html.*

Bogard, D. D., and P. Johnson. 1983. Martian gases in an Antarctic meteorite? *Science* 221 (4611): 651–654.

Ferguson, K. 1999. *Measuring the universe: Our historic quest to chart the horizons of space and time.* New York: Walker Books.

McSween, H. Y. Jr., L. A. Taylor, and E. M. Stolper. 1979. Allan Hills 77005: A new meteorite type found in Antarctica. *Science* 204 (4398): 1201–1203.

National Aeronautics and Space Administration (NASA) Jet Propulsion Laboratory. 2013. NASA rover confirms Mars origins of some meteorites. *http://mars.jpl.nasa.gov/msl/news/whatsnew/index.cfm?FuseAction=ShowNews&NewsID=1525.*

National Research Council (NRC). 2012. *A framework for K–12 science education: Practices, crosscutting concepts, and core ideas.* Washington, DC: National Academies Press.

NGSS Lead States. 2013. *Next Generation Science Standards: For states, by states.* Washington, DC: National Academies Press. *www.nextgenscience.org/next-generation-science-standards.*

Salzberg, S. 2015. The challenge of seeing meditation only through a scientific lens. On Being. *www.onbeing.org/blog/the-challenges-of-seeing-meditation-only-through-a-scientific-lens/7492.*

Schuette, P. 2015. Hypothesis, theory, and law: Discovery through questioning. *Issues in Earth Science,* issue 4. *http://earthscienceissues.net/Essays/Hypothesis-Theory-Law.pdf.*

Westervelt, E. 2013. To make science real, kids want more fun. *www.npr.org/2013/12/17/251675532/to-make-science-real-kids-want-more-fun-and-fewer-facts.*

Willingham, D. T. 2010. *Why don't students like school? A cognitive scientist answers questions about how the mind works and what it means for the classroom.* San Francisco: Jossey-Bass.

EXAMPLE ACTIVITY DESIGN

THE COLUMNS OF POZZUOLI

ARGUING FROM EVIDENCE

Arguing from evidence in the classroom can quickly spiral into idle chatter unless prompts and guidance are provided throughout the discussion. The less structured the activity, the tougher it can be to keep the discussion on track. Here, we offer a structured activity specifically targeting arguing from evidence.

TEACHER PREPARATION AND PLANNING

This example activity design (EAD) is based on a reading that introduces students to evidence, arguments, and counterarguments that address a particular claim. We provide an example text below, but you may want to use a different text that corresponds more closely to your curricular sequence. In that case, you will need to either find or write an appropriate text. This text should identify several observational points that address the claim, including observations that do not seem completely consistent with the claim so that students can not only identify arguments for the claim but also think about alternative explanations for the observations. The following example text is written at about a 10th-grade reading level, although younger students can chew through it given more time.

EXAMPLE READING

Mediterranean clams and limestone columns built by the Roman Empire in Pozzuoli, Italy, played an unexpected role in the 19th-century's emerging field of geology. Observations of the columns supported the idea that large landscape features—such as mountains, canyons, and plains—could result from slow, everyday processes operating over long periods of time. This Italian site played such an important role in Charles Lyell's principle of uniformitarianism that a picture of the columns found its way to the frontispiece of his influential book *Principles of Geology,* first published in 1830 by John Murray (see Figure 9.2, p. 188).

The key mystery of the marble columns was the origin of holes drilled into the rock. The pattern of the holes made a dark wide ring on each of the 40-foot-tall columns. Each ring extended from about 12 feet above the base to about 21 feet above the base, as seen in Figure 9.2. The character of these holes matches the borings of the marine

Figure 9.2
Illustration of the columns of Pozzuoli

Source: Reprinted from frontispiece of *Principles of Geology*, Vol. 1, by Charles Lyell (London: John Murray, 1832). Available at *https://en.wikipedia.org/wiki/Macellum_of_Pozzuoli#/media/File:Charles_Lyell_-_Pillars_of_Pozzuoli.jpg.*

clam *Lithodomus,* and some of the holes still have shells in them. Today, this clam lives in seawater (not freshwater) and bores into stones, piers, and boat moorings below low tide. The mystery was, How did the borings of a marine clam get into stone columns that were nearly 600 feet from the shoreline?

Charles Lyell believed that this feature showed that sea level had changed through time, not suddenly by some catastrophic event but gradually. Clearly, the columns were built above sea level and are above sea level today, but in between the sea must have partly covered them.

The fact that there were no borings in the lower part of the columns puzzled people. If the middle of the columns was below sea level, then wouldn't the bottom of the columns also be below sea level and also bored by the clams?

When the columns were excavated in the 1700s, information about the sediment that had been around them was lost. However, Lyell and others proposed that the lowest part of the columns had been covered by sediment or volcanic ash at the time of the sea-level rise. Thus, the bottom of the columns did not get bored by the clams because this variety of clam did not live within the sediment.

Other people had different ideas about what the holes in the columns meant. Some people argued that convulsions of the Earth significant enough to change sea level would have toppled the columns. In response, Lyell noted that other buildings were submerged beneath the sea without being destroyed. He gave the example of houses of the fort of Sindree on the Indus River delta, which were covered with seawater when the ground of the delta subsided (lowered) in 1819.

Other people proposed that the sediment that covered the bottoms of the columns may have dammed up the outlet to the sea, causing rainwater to pond around the columns. If so, they argued, then the observations could be explained without a gradual sea-level change.

EXAMPLE PROMPTS AND LIMITING OPTIONS

Encourage students, in groups, to identify in writing each observation and each claim in the text. For example, they might note the following observations:

- The columns are still standing.
- The columns don't have borings at the top or bottom.
- The columns do have borings in the middle (actually, what is observed are *holes* in the stone columns—that they are *borings* is an interpretation—you might talk about this distinction with your students).
- These boring are above today's sea level.
- The borings match borings that are made by a species of modern clam.
- Shells are found in some of the holes.
- This modern clam does not make borings above low tide.
- This modern clam does not live in freshwater.
- These modern clams live only in the water column above the mud line, not within the sediment.
- The houses of the fort of Sindree were not destroyed when they sank beneath the sea.

They might note the following claims:

- The columns were built above sea level.
- The sea rose and covered part of the columns.
- Mud covered the lowest part of the columns.
- The sea then lowered again, exposing the lower part of the columns.
- The mud that once covered the lowest part of the columns has washed away or has been removed by excavation.
- Sediment dammed up the area around the columns and caused rainwater to pool.
- Any event energetic enough to cause the sea level to change would have toppled the columns.

- These observations played an important role in developing the principle of uniformitarianism.

Encourage students to challenge Lyell's claim or to think of alternative explanations for some of the observations, making reference to the observations.

- For example, they might ask why there are borings in the middle of the columns, but not at the bottom. How can the middle of the columns be below water, and get bored, while the bottoms of the columns aren't bored?
- What about ponding of water around the columns? That doesn't require slow changes in sea level. Does that interpretation also fit the observations? Why or why not? (The rainwater would be fresh, not marine, which might be an important factor in the discussion.)
- What about the possibility that the columns were not built on land but were originally built under seawater? Or what if the columns were "reused" support columns for a pier and already had the borings in them when they were placed at their present location?
- Other ideas?

You might challenge your students to identify whether particular aspects of the arguments that seem like observations really are. For example, are the following observations? *(None are.)*

- The lowest part of the columns was covered by sediment or volcanic ash at the time of the sea-level rise.
- The columns were built above sea level.
- The columns are still underwater today.
- The temple was originally built underwater.
- The columns were submerged under the sea after construction.

CLASS/GROUP DISCUSSION

1. Have students, in groups or as a class, identify the key arguments offered in the text (not the same as claim or evidence) and rephrase each of the arguments in their own words.
2. Have students address each of the claims one by one, including any claims

they have identified that challenge Lyell's ideas, and explain whether each observation is supportive of that claim and why: How strongly does each observation and argument support or refute the claim? What is your reasoning?

3. Have a final open discussion: What you think the evidence means? Identify problems, alternative interpretations, argue how the alternative explanations can (or can't) be excluded.

WRITING

Have students write a summary three-point wrap-up argument, as described in the text of this chapter. Here is an example:

> *Lyell noticed that marble columns at Pozzuoli had holes, most likely drilled by rock-boring clams at a level well above modern sea level. Since this variety of clam lives only in an environment constantly covered by seawater, he concluded that the level of the sea relative to land had changed through time at that location, first being lower when the columns were built, then being higher when the clams drilled their holes, and then being lower again in the modern world.*

EXAMPLE INTERACTION

Teacher prompt: What about the other ideas for the borings—the ones that don't require a change in sea level? Can we rule them out, or are they still possibilities?

Student 1 observation: The idea that rainwater may have ponded around the columns doesn't work because the clams don't live in rainwater, only seawater.

Student 2 argument from evidence: Plus, nobody says anything about the dam. Wouldn't the dam still be there, at least part of it?

Student 3 interpretation: Maybe it was removed during the excavation. The text says information about the sediments was lost then.

Student 1 question: It seems like we can't be sure that the columns didn't already have borings in them when they were put where they are now.

Student 4 argument from evidence: If the columns were supports for an earlier pier, and got bored then, why wouldn't there be borings all the way to the top, not just in the middle?

Student 5 interpretation: Maybe the bottom was in mud.

Student 4 question: But what about the top? Supports for piers don't poke up into the air.

Student 5 reasoning: Maybe this one did.

Student 6 argument from evidence: If it was part of a pier before the Romans came and put it where it is now, it wouldn't be a Roman pier, right? Are the columns of a different architecture than the Romans used?

Student 7 question: We can find out the architecture from the picture of the columns, right? Can we look it up?

SUMMARY CHECKLIST FOR TEACHER AS PRACTITIONER OF SCIENCE

- Choose or write a text that presents a variety of claims within the context of the observational evidence.
- Have students identify the observations and claims.
- Have students come up with alternative interpretations that might fit at least some of the observations.
- Identify, discuss, and argue the different claims in terms of supporting observations.
- Summarize the key observations, arguments, and conclusion in a few sentences.

TEACHER REFLECTION

Arguing from evidence should be integrated into most classroom investigative activities. Students can be much more thoroughly involved in arguing from evidence that they have generated themselves. In that more "free-for-all" environment, the type of conclusions that students might draw and the number of observations and arguments they might make, are less controlled than in an activity such as the Pozzuoli EAD.

Take your experience with the more controlled arguing-from-evidence activity presented here and think about how it can inform your role when students argue from evidence in a less structured investigation:

- What challenges did students face in recognizing the differences among observation, argument, and conclusion?
- What prompts helped students think about alternative interpretations?
- What prompts helped them to test those alternative interpretations against the observations?

- What prompts helped them know how to explain and describe their reasoning as they connected observation to conclusion?
- Was there any time that you felt you offered too much or too little prompting?

PART II

THE LANGUAGE OF THE EARTH

Throughout Earth's history and continuing today, movement of matter and energy has changed landscapes, cycled water, and recombined the elements into a wide variety of rocks and natural resources.

PART II

> *Rather than misrepresenting science as a dry catechism of certainties, we should emphasize the excitement of the quest for answers to our questions. The joy is in the chase, for, as Robert Louis Stevenson said, "It is better to travel hopefully than to arrive."*
>
> —Robert H. Dott, Jr., "What Is Unique About Geological Reasoning?"
> *GSA Today,* October 1998

The important ideas in earth science focus less on what we know than on how we know it. For example, the principle of superposition is less a universal law for how a planet's rock and sediment cycle and change than it is a starting place and methodology for how to read the story written in that rock and sediment. The concepts of geochemical differentiation are less a conclusion of a study than they are a place to begin one. In some sense, how we read the story of the earth and sky is a more fundamental and important idea in earth science than any of the individual stories that we have read, perhaps giving us a response to the following challenge posed by Robert Dott, Jr., in "What Is Unique About Geological Reasoning?"

> *[M]any authors have argued that (geology) has unique modes of reasoning and unique laws of its own. … Each generation of geologists has worried about what, if anything other than the geo-, is unique about our science. Deep time, the fossil record, uniformitarianism, the method of multiple working hypotheses, and historical science are among the special claims. We need to look further for differences, for geology certainly is much more than simply applied physics and chemistry.*

In this section, we consider five key ways that we can read the earth. We read the earth in the rocks (Chapter 10), in the layers of rock (Chapter 11), with seismic waves (Chapter 12), with light (Chapter 13), and in the atoms (Chapter 14). These approaches to inquiry are different from the other sciences, and yet coincide with the key idea of all science—and with the *Next Generation Science Standards*—that science is the process of figuring things out, not the conclusions that are figured out in the end.

10

STORIES IN ROCKS

A friend who homeschooled told us a story of encountering historical science on a walk with her children one spring day. Snow still covered the ground but the March sun was growing warmer. She and her kids found a leaf inset into the snow, and they puzzled over how the leaf got sunk into the snow that way. If the leaf had fallen on the snow, it should be lying on the surface. If the leaf had fallen first and then the snow, the leaf should be covered up. So, how did the leaf get *sunk in?* Without a time machine, it seemed like an impossible question to answer. But, with a little bit of prior experience (dark things heat up faster in the sun) and a little bit of experimental science (trying to poke the leaf into the snow with a finger), they soon realized that the leaf was not poked into the snow but rather melted into the snow. It was a puzzle in historical science, telling a story of how things came to be the way they are.

ARGUING FROM EVIDENCE IN HISTORICAL SCIENCE

Historical geology differs fundamentally from other sciences. The physical scientist does an experiment and observes the result. But the historical scientist sees the result and tries to figure out how it got that way. The historical scientist deals with events that cannot be repeated or observed.

I (Russ) often illustrate the difference by referring to a beaker of water. Physical scientists might heat the water and observe that it boils away. But suppose you had an empty beaker. Was it always empty, or did it once contain water that boiled away? Students can usually come up with ways to answer the question—is there residue in the beaker? Can we boil away a beaker full of water and see if the residue formed by that process matches what's in the empty beaker?

Rocks are the residue of past geological processes. That residue—the rock—holds the clues to how the rock formed long ago.

Just as our friend had to understand the process that set the leaf into the snow in order to tell the story of what happened, geologists need to understand the processes that made rocks in order to read the story written in the earth. Those processes reflect the laws of nature. The physical scientist assumes that the laws of nature at work today will continue to be at work in the future, allowing science to be predictive. The historical scientist assumes that the laws of nature at work today have been at work in the past, allowing us to infer past events. Thus, when we see sandstone, we think of beaches or deserts because that's where modern processes cause sand to accumulate over time. When we see ripples in the sandstone, we think of winds or currents.

Geology found its origins in telling stories of the past. In the 1800s, the new science of geology captured the imagination of the scientific world with its answers to seemingly unanswerable questions. Has the Earth always existed, or did it have a beginning? If it had a beginning, was it long ago or recent? Has the Earth changed through time or stayed the same? If it changed, did it change slowly by gradual processes or abruptly in great catastrophes?

Much of what we understand about Earth history and the lineage of life is based on the practice of historical science. Most key concepts of earth science, including overarching theories such as plate tectonics, cannot be understood as more than "theories from the scientists" unless students understand how stories are recorded in rock and how they can read those stories on their own.

Today, we believe we've answered many of the big historical science questions of the 1800s, but historical science remains an important source of evidence for predicting the future of our planet. For example, it provides our main observational evidence for understanding climate change. In predicting how climate might change in the future, we first have to understand how it has changed in the past. How fast did it change? What caused it to change?

People often think about mathematical modeling when they think about climate forecasting, but those models need to be calibrated by and tested against the evidence of past climate change. That means that traditional geology, examining the story of Earth's past, remains an important part of modern science.

Even so, in modern secondary curricula, historical science often gets overshadowed by experimental aspects of the physical and life sciences. The practices of science unique to historical science—by which we infer temporal relationships from the rock record, compare modern and ancient processes, and use multiple working hypotheses—are not specifically called out in the *Next Generation Science Standards (NGSS)* science and engineering practices (NGSS Lead States 2013). Historical science shows up in the Earth and Space Sciences disciplinary core ideas (DCIs), but it is somewhat hidden within a structure that emphasizes Earth's place in the universe and the modern cycle of earth processes. For example, only two subheadings of the DCIs clearly invoke historical science: ESS1.C: The History of Planet Earth and ESS3.D: Global Climate Change. These fall under the DCIs Earth's Place in the Universe and Earth and Human Activity, respectively. Sadly, there is no mention of historical science in the grade band endpoints for Global Climate Change or in the grade band endpoints for any of the subcategories of the DCI Earth's Systems, where historical science has played a key role in understanding the cycles and systems of the Earth.

The depth of misunderstanding of historical science is illustrated by a quote from one of my (Russ's) geoscience majors. In an oral report on Devils Tower in Wyoming—a feature with a notoriously controversial history of formation, this student said, "Because the Tower was formed in the past, we can't know how it formed, we only have theories." Ouch.

READING THE STORIES

Rocks tell the story of Earth's past, and we believe that lessons on rocks should include a focus on reading those stories. Unfortunately, many students encounter rocks as an exercise in classification and identification. This approach misses the opportunity to understand how historical science, like all science, is based on observation and arguing from evidence.[C30] Instead of naming rocks, how can we create rock lessons in which students argue from evidence to tell stories of Earth's past?

Like any scientific investigation, reading stories in rocks can get complicated. However, each type of rock tells a few key stories that can be taught at any age level.

Learning Progressions

The disciplinary core ideas described in *A Framework for K–12 Science Education: Practices, Crosscutting Concepts, and Core Ideas* (*Framework*; NRC 2012) are broken into "grade band endpoints," which serve as markers for how student understanding might mature over time. For ESS1.C: The History of Planet Earth, K–2 students look at patterns of change on Earth's surface; students in grades 3–5 think about landforms and rock formations changing over time; students in grades 6–8 consider relative ages and the geological time scale; and students in grades 9–12 think about absolute ages for the geological time scale and evidence for Earth's early history.

We want to emphasize that these learning progressions are tentative, an educational experiment if you will. The *Framework* authors write, "All in all, the endpoints provide a set of initial hypotheses about the progression of learning that can inform standards and serve as a basis for additional research" (p. 33).

We point this out with the personal experience that all of the grade band endpoints for K–12 remain challenging for college students. Conversely, many aspects of understanding the evidence for Earth's early history, listed as a 9–12 grade band endpoint, can be readily grasped by first graders—such as simple stories told by rocks.

For example, *igneous* rocks tell the story of how fast or slowly molten rock cooled. Fast cooling happens at the surface of the earth where magma emerges from a volcano, producing rock with crystals that are too small to see with the naked eye, rock such as basalt or rhyolite. Slow cooling happens deeper in the crust and produces rock with the larger, sparkly crystals, such as granite.

Metamorphic rocks—with their foliated, layer-like textures that form when a rock is squeezed in one direction more than another and with their unique set of minerals that are only stable at high pressure and temperature—tell the story of deep burial such as in the roots of a mountain range. When you find a metamorphic rock at the surface of the earth, you know it formed deep and has been brought to the surface, perhaps by erosion.

Sedimentary rocks such as conglomerate, sandstone, or shale tell the story of the energy of the environment of deposition. In a swift stream or energetic beach, small particles get washed away, leaving pebbles behind that become a conglomerate. A deep lake, with little water movement, allows tiny particles to settle, leaving clay and silt that become a shale. Intermediate energy, such as energy along a typical beach or in a windswept desert, may leave sand that becomes a sandstone.

Safety Note

Caution students that rocks can have sharp edges and cut skin. Handle them with care!

Pick a few rock types to examine (perhaps based on what rocks your students might see on field trips in your region) and learn their stories.[C31] Maybe you can put the stories in context with a driving question; for example: "Did this rock come from a volcano or did it cool slowly inside the Earth?" "Was there once a mountain range here, and how do we know?" "What do the rocks *of our particular region* tell us about the geological past of *our home area*?"

I (Russ) have worked with both elementary and college students to address the driving question "Did the dinosaurs live in a quiet swamp or along swift rivers up in the hills?" Students can base their answers on the particle size of sediment in the rocks that the dinosaur bones were found in. When I visit elementary classrooms, I often talk about the stories that rocks tell. One favorite is to talk about dinosaurs—not dinosaur biology, but rather what the environment of deposition tells us about where the creatures lived and died. I bring in *Edmontosaurus* bones that I found in northwestern South Dakota, along with samples of the rock in which the bones were found (see Figure 10.1). Students consider the sediment grains that make up the rock and what the sediment size tells us about the environment—was it a fast river? A wave-swept beach? A swamp with still water? Students draw pictures, based on evidence and reasoning, of what the place was like where the *Edmontosaurus* died. In the classroom, students look at real rock samples. I have used this activity with students from first grade through college, adjusting how much guidance I offer and limiting the complexity of observations and conclusions depending on my audience.

Figure 10.1
Rocks and images for the *Edmontosaurus* activity

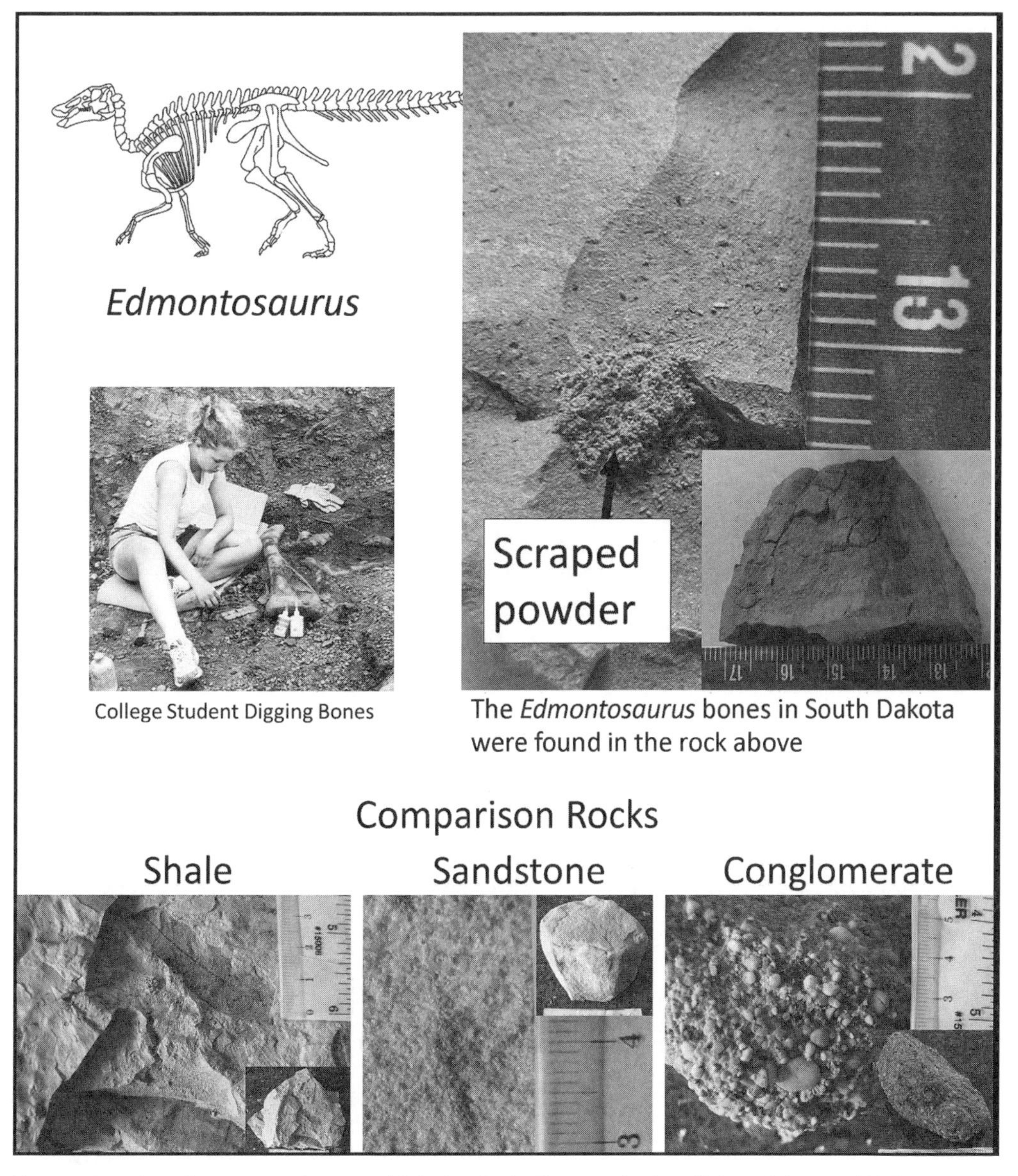

This activity is based on a paper that we published in *Rocky Mountain Geology* (Colson, Colson, and Nellermoe 2004), which concluded that the hadrosaurs lived in a low-energy swamp by the sea.

Source: Hadrosaur skeletal image courtesy of Gregory M. Erickson, PhD, Florida State University. All other images courtesy of the authors.

Students often struggle to understand the distinction between particle size and size of the rock sample. Younger students in particular need a lot of help to observe the grains or crystals that make up the rock. You also may need to point out that the size of the rock sample tells you about weathering processes—or people with hammers—but not about environment of deposition.

There are many stories told by sediment size in sedimentary rocks that you can read with your students. Drawing pictures of environments of deposition at different times in the past based on the size of sediment particles in the rock, as in the *Edmontosaurus* activity, is a great way to engage students in thinking about the stories rocks tell. Here is another example of such a challenge, using a photo of a rock from the Gale Crater on Mars (see Figure 10.2):

> *Suppose that you want to take your time machine and travel to Mars's past where you plan to build your home. What was Mars like in the distant past? As part of your planning, you want to draw a picture of the scenery at different times and places so you can decide where you want to live. One place that you're thinking about is Gale Crater during the Noachian-Hesperian time period (yes, Mars has its own time periods). You know that the Curiosity rover explored rocks from that era at Gale Crater, and we have a photo of one of those rocks. Can you draw a picture, based on evidence, of what that place might have looked like then? Was this spot a desert with large, windblown dunes? A deep, quiet lake? A swift, mountain river? A sluggish river meandering across a broad plain?*

In this case, the pebble-size particles in the rock tells us that significant energy was needed to move those particles and wash away the smaller particles—a swift, mountain river flowing down from the high rim of Gale Crater into a lake that once filled the crater. Rounding of the grains tells us that the sediments in the rock were tumbled in a stream and not brought in suddenly by volcanic or impact activity.

This activity needn't end with telling a story with a picture. Students might talk about their different pictures, explaining why they drew the picture that they drew and engaging in arguing from evidence. See also Chapter 6, where we consider a model of Gale Crater that is based in part on the evidence in this rock, and consider Chapter 8, where we distinguish between evidence and conclusion in a news article that is based in part on this and similar rocks.

You can work in more complex stories as appropriate for your region and age of students. Sedimentary evaporites might tell a story of an arid, restricted lagoon or lake. Glacial till—a poorly sorted mixture of mud, sand, and gravel—might tell of deposition from a now-melted glacier.

Doing some lab work to understand the process that forms each rock brings historical science into the realm of experimental science. For example, you might make measurements to show how particle size depends on velocity of a stream or how igneous crystal size depends on cooling rate. In the two "Suggestions for Designing an Activity" sections at the end of this chapter, we offer some seed ideas for experimental activities with stream velocity and crystal growth.

Figure 10.2
Rock from the Gale Crater on Mars

Source: Image courtesy of NASA via the Curiosity rover, 2012, NASA/JPL-Caltech/MSSS and PSI.

FINAL THOUGHTS

Cooking food is a favorite activity when teaching rocks. An internet search on the phrase "fudge to teach about rocks" (without the quotation marks) pulls up dozens of edible lessons. However, these lessons, at best, provide only a symbolic reference to rocks and geological processes.[C32] Cooking all the fudge in the world is not going to teach students how to read the cooling history of an igneous rock based on the size of its crystals. Building sedimentary rock layers with bread, peanut butter, and M&M's "gravel" doesn't give students the opportunity to experiment with natural processes that segregate sediment into layers.

Not all classroom lessons need to engage students in the process of scientific investigation. For example, edible lessons can create memorable social experiences that help students remember a learned concept. Other non-experimental activities, such as the beans and humidity exercise in Chapter 2, help students understand a difficult-to-visualize concept. These types of activities help convey ideas that have been figured out by scientific methods, but they do not in themselves engage students in the practice of science. For example, no amount of exploration with beans and paper cups would have ever convinced even one scientist that water forms by condensation from vapor, but a lesson with beans and cups can help students understand the idea of saturation and condensation. We conclude that these kinds of non-experimental activities can be useful, but should not be confused with activities that engage students in doing investigative science.

REFERENCES

Colson, M. C., R. O. Colson, and R. Nellermoe. 2004. Stratigraphy and depositional environments of the upper Fox Hills and lower Hell Creek formations at the Concordia hadrosaur site in northwestern South Dakota. *Rocky Mountain Geology* 39 (2): 93–111.

National Research Council (NRC). 2012. *A framework for K–12 science education: Practices, crosscutting concepts, and core ideas.* Washington, DC: National Academies Press.

NGSS Lead States. 2013. *Next Generation Science Standards: For states, by states.* Washington, DC: National Academies Press. *www.nextgenscience.org/next-generation-science-standards.*

SUGGESTIONS FOR DESIGNING ACTIVITIES

EXPERIMENTAL SEDIMENTARY PETROLOGY WITH DOWNSPOUTS

The Hjulström diagram is a key graph in sedimentary geology (Hjulström 1935). This graph shows how the deposition or erosion of sediment depends on both the sediment size and the velocity of the water in which the sediment might be deposited or eroded. Filip Hjulström developed the graph as part of his doctoral study of the Fyris River in Sweden in 1935. The activity in this section offers an experimental way to reproduce key elements of the graph (see Figure 10.3).

Figure 10.3

A simplified Hjulström diagram and the results of a group of fourth, fifth, and sixth graders doing an activity similar to the one described in the "Lab Assignment" section but using only two sizes of particle

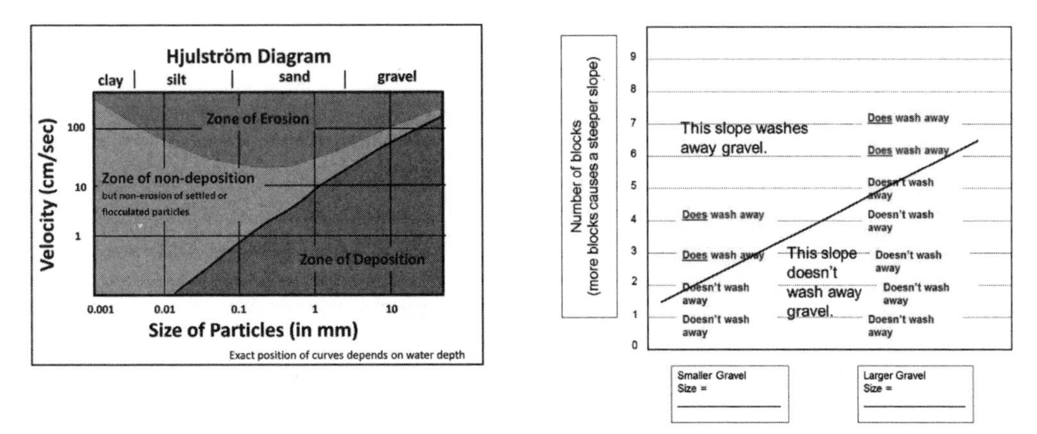

Several years ago, I (Russ) gave teachers in a workshop the following prompt: "Design an experiment to measure how the slope of a river influences whether sediment is deposited or eroded." My goal was to offer a lesson in experimental design. The teachers built elaborate simulations of rivers—sticks for logs, pebbles for boulders, and meanders defined by modeling clay. Participants could watch water go down the river, watch sand grains rolling along in the current, and get some sense of change (although the paths of their rivers were constrained by the modeling clay and couldn't change much). Unfortunately, none of the designs included any means to measure the dependence of one variable on another. Thus, there was no controlled experiment, nor any real way to address the original experimental design challenge. The teachers built dioramas, not experiments.

The following activity offers an experimental way to address the task above but defines it more carefully than I did for my teachers long ago, offering ways to prompt and limit the scope of the activity while allowing for creative student contributions. Students think about what to measure and how they can eliminate the influence of all but one of the variables. Students can try out their experimental ideas, the teacher can offer prompts on experimental design (students will recognize many problems on their own), and then students can take another crack at solving experimental challenges with a second round of experiments. The activity bears not only on deposition and erosion in modern rivers, but on interpreting the story that sedimentary rocks tell us about past environments.

Safety Notes

1. Have students wear indirectly vented chemical-splash goggles and aprons during the whole activity, including setup, hands-on piece, and takedown.
2. Immediately wipe up any spilled water to avoid a slip and fall hazard.

LAB ASSIGNMENT

Keep the assignment general; don't overspecify what students should do. Emphasize the importance of controlling variables. An assignment for students might look like this:

Your materials include the provided downspout to act as a stream channel, the provided sediment particles, and whatever other materials you think you need and can scrounge: for example, water buckets, marking pens, rulers, materials for propping up the downspout, etc. (*Note to teacher:* For younger students, you may want to be the one who scrounges, but students should be encouraged to request supplies.)

Your goal is to design an experiment that eliminates all variables from the system except for the two you are interested in—stream slope (our proxy for velocity) and particle size. You want to construct a graph that shows how erosion or deposition (non-erosion for the purposes of this experiment) depends on sediment size and slope of the stream.

There are many experimental problems you need to overcome. The main point of this lab is for you to try to anticipate those problems and either avoid or overcome them. One suggestion is the following: Rather than do experiments with all sediment sizes included simultaneously, use only one sediment size at a time. Otherwise, on the timescales that you have to do the experiments, the bigger particles will shelter the smaller particles from the stream and provide a result that doesn't completely reflect the ability of the stream to transport the smaller particles. The provided particle sizes are about 3 cm, 2 cm, and 1 cm gravel.

Plan to communicate your results to your classmates with a presentation or report.

QUESTIONS TO ASK YOUR STUDENTS TO HELP THEM IDENTIFY EXPERIMENTAL CHALLENGES

Did you measure the slope at which particles of a given size began to move? Or perhaps where they were washed entirely off the downspout? However you defined *erosion,* did you keep that definition the same for all of your experiments? Did you include an explanation of your definition in your report?

One way to find the slope at which the particles begin to erode is to gradually increase the slope in a series of increments, such as by adding small blocks to one end of the downspout. Did you report the method that you used? Also, how did you measure the slope of the downspout? Knowing the slope requires that you know both the height of one end and the length of the downspout (2 inches makes a big difference in slope for something that's only 2 inches long, but not so much for something that's 2 miles long). You could also measure the slope with a protractor; although, given the small size of classroom protractors and the cumbersome nature of trying to measure the orientation of the downspout with it, it might be less accurate for most students. Did you record your slope and your method of measuring it?

Did you keep all the other variables constant? For example, was the water deep enough to cover all of the sediment particles? Obviously, if you had enough water to cover the small particles but not enough to cover the big ones, that might have a bigger effect than the slope! Also, the depth of the water affects the velocity. Keeping the water depth constant for all experiments, at a depth that will cover the biggest particles, eliminates this error from your experiments. Did you come up with some means to control this variable?

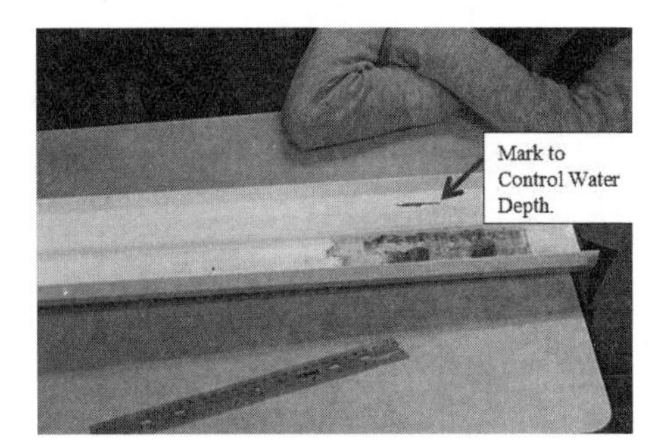

A student's experimental control is shown.

One way to control this variable is to make a mark on the side of the downspout and then maintain the flow of water at that depth by controlled pouring of the water.

A related problem arises if you pour the water into the downspout from a significant height. If you create a waterfall in your experiment, the height of the waterfall will have a much greater effect than the relatively tiny changes in the slope of the downspout!

The bottom of the downspout might be very slippery in some places and rough in others. Did you control for the effect of this variable?

Did you do multiple experiments at each slope to make sure your results were reproducible? Repetition is a key element of science investigation.

SUMMARY CHECKLIST FOR TEACHER AS PRACTITIONER OF SCIENCE

- Make sure students have access to materials they will need, such as a downspout, buckets, and water. Leave some materials for them to use or not, as they see fit.
- Decide how much guidance you want to provide, given your students' abilities and time constraints.
- Rather than provide lots of upfront directions and experimental setup, consider letting your students try out their own experimental designs. You can provide guidance and prompts along the way, as well as give time for students to critique each other's designs. Who knows, maybe someone will invent an interesting experiment that challenges you!
- Circulate as your students do their experiments, offering guiding questions and observations where appropriate.
- Provide opportunities for you and your student to talk about what the results mean in terms of both modern rivers and interpreting past depositional environments.

REFERENCE

Hjulström, F. 1935. Studies of the morphological activity of rivers as illustrated by the River Fyris. *Bulletin of the Geological Institution of the University of Upsala* 25: 221–527.

EXPERIMENTAL IGNEOUS PETROLOGY WITH THYMOL

One of the original clues to reading the story told by igneous rocks was discovered in the late 1700s at a glass factory in Leith, Scotland. Geologist James Hall noticed, after a furnace accident, that the molten glass crystallized into an opaque white mass. Curious about this result, he did experiments to measure the effect of cooling rate on melted natural rocks and found that the faster the molten rock cooled in his small graphite crucibles, the smaller the crystals were, until eventually, at fast enough cooling rates, no crystals grew at all, making a glass (Geike 1962). He applied these observations to igneous rocks, inferring that coarse-grained rocks such as granite cooled slowly, fine-grained rocks such as rhyolite cooled quickly, and glassy rocks like obsidian cooled the fastest of all.

The melting temperature of molten rock, on the order of 1200°C, is too high to make Hall's work a good activity for the classroom. However, with appropriate safety precautions, the organic compound thymol, with a melting temperature of 48°C, can be used to give students an opportunity to experiment with the effect of cooling rate on crystal size. Like all good experimental activities, the results are not always straightforward, giving plenty of opportunity for students to problem solve and interpret.

SAFETY

Safety is an important issue in this lab, as it is in real research. Although safety is not listed as one of the *NGSS* science and engineering practices, it is one of the practices of real science and so taking safety precautions is an important experience for students of science.

Learning about safety measures is only meaningful if there are real safety concerns. Thus, restricting your classroom labs to those with no safety risk cannot give students experience with thinking about safety, even if you use safety equipment. We've seen pictures of students engaged in making layers of sediment with M&M's where everyone is wearing safety goggles. Yikes! Not only does this practice not give students any real understanding of safety, it can actually undermine safety by giving students the false impression that wearing goggles, or taking other safety measures, is simply a game that we play to follow lab rules. Taking safety measures where there are real safety risks provides a very different message, much more like the safety considerations during real research.

Although a safety exercise at the start of a term is one way to make sure that everyone has thought about safety, a lab such as this one probably deserves a refresher to reinforce ideas of safety and also make sure that safety is at the front of students' thinking.

I (Mary) have students sign a "safety contract" (an acknowledgment form) before starting this lab; that contract appears at the end of this section (p. 212). Also check out the National Science Teachers Association document *Safety in the Science Classroom, Laboratory, or Field Sites (www.nsta.org/docs/SafetyInTheScienceClassroomLabAndField.pdf).*

EXPERIMENTAL CONSIDERATIONS

- Find thymol activities online or in a book and think about how each might work for you. Try out one or two of them. In an online search while writing this activity suggestion, we found five thymol activities. However, to ensure you find something, we've posted our thymol activities online, one designed for college students (Russ's activity) and one for eighth graders (Mary's activity). These activities can be found at *http://web.mnstate.edu/colson/D2L-unlinked/Experimental_Igneous_Petrology_with_Thymol_Link.htm.*
- Is the method of temperature control safe for your students? Will it keep the thymol from getting too hot and stinking up your room? I (Mary) use small conduction hot plates, not coil-style hot plates.
- In general, a metal container will provide faster thermal equilibration and has the potential to provide better temperature and cooling-rate control. We've found that a 2-tablespoon metal coffee scoop works well. Think about what you want to use and why.
- Does the experiment provide for a uniform depth of thymol or does the depth vary? This can affect results.
- Decide how much freedom you want to give your students to explore. Then give them the guidance you choose, and do the experiments.
- Identify differences between student results and differences with what you expect. Can your students explain the results? Can you? Are there follow-up experiments that might help explain the results?
- Discuss the results and what they mean.
- Tie the experiments to both the role of experimentation in historical science and how historical science allows us to read the stories in the earth.

EXPERIMENTAL CHALLENGES THAT CAN LAUNCH TRUE INVESTIGATIONS

Depending on the way the experiments are done, it's possible to supercool the thymol—that is, cool it below its freezing temperature. For example, this might happen if you heat it well above the melting temperature or hold it above the melting temperature for a period of time (so that there are no crystals left). In this case, the slowest cooled experiment might yield the smallest crystals because the supercooled liquid crystallizes very fast once crystallization starts. Trying to figure this out, and under what conditions it occurs, can be a challenging classroom activity. Not getting the "right" result can be a good thing, a starting point for true investigation. Based on the results of your experiments, you might discuss how supercooling would affect the texture of the mineral grains in a natural igneous rock (the fast crystallization after supercooling often produces more featherlike dendritic crystals rather than the squarish, euhedral crystals of normal slow cooling).

REFERENCE

Geike, A. 1962. *The founders of geology.* New York: Dover.

Mary's Safety Contract for the Thymol Lab

1. The compound we will be using is called *thymol*. It is extracted from plants like thyme and bee balm. It is used in small amounts in Listerine. Thymol is not dangerous if used properly. It smells a little like IcyHot.
2. DO NOT PUT YOUR HEAD OVER THE METAL SCOOP AND INHALE DEEPLY. This will irritate your nose.
3. Thymol has a low melting temperature, about 48°C. Preheat your hot plate. Monitor the hot plate temperature with a thermometer until it rises to 48°C (about the point your hand would be uncomfortably warm). Then, turn the hot plate down to a lower heat setting to maintain that temperature.
4. Doing the above will keep the temperature of your hot plate as low as possible but still melt the compound. You should not see vapors or "smoke" rising from your watch glass. If you do, the hot plate is too hot.
5. The metal cup should never get too hot to handle with your bare hands.
6. Do not put your fingers in your mouth, nose, or eyes during lab. Wash your hands with soap and water at the end of lab today.
7. You must have your indirectly vented chemical-splash goggles, aprons, and nonlatex gloves on at all times during the whole activity, including setup, hands-on piece, and takedown. Thymol would irritate the tissues of your eyes and possibly your hands as well.
8. Keep the hot plate toward the center of the lab station. Don't put pens and pencils against the hot plate.
9. Make sure the electrical cords don't dangle off the edge of the table. Plug the hot plate in above the worktop, not below the "bridge" that connects the lab table to the countertop.

I have read and agree to follow all the safety rules set forth above. I will cooperate to the fullest extent with my teacher to ensure my safety and the safety of others in the classroom. I understand that not doing so may result in my being removed from the lab.

Signature ___________________________ Date ________

STORIES IN ROCK LAYERS

Whereas rocks tell the story of what a place was like at one time in the past, layers of rock tell the story of how that place changed through time. They're like a flip book of planet Earth's past in which snapshots change from page to page, providing a motion-picture view of change through time. During the 1800s, the academic world was swept by the realization that everything changes: life changes, earth changes, even the stars change. The big debate then, between catastrophists (those who thought planet Earth changed in a series of abrupt events) and uniformitarianists (those who thought planet Earth changed through the gradual operation of continuous processes), was not about whether Earth changed, or even how long the change took, but *how* it changed and *what caused* that change, all of which are stories revealed in the layers of rock.

STRATIGRAPHY: EARTH'S STORYBOOK

Most of us know that the history of our world is written in the layers of rock as stories are written on the pages of a book. When work piles up on our desk, we (Mary and Russ) often joke with friends that some piece of pressing responsibility is "back in the Mesozoic and we'll need to dig it out." And our friends understand what we mean. Stratigraphy has become part of our cultural language, showing up in a 1990s *Shoe* comic strip (see Figure 11.1, p. 214). We know at some level that strata make up the pages of Earth's storybook. Yet few of us can read those pages or even have a firm idea of the principles by which those stories are read.[C33] Students construct paper timelines to illustrate the immensity of Earth's past, but stratigraphy—the practice of science behind that timeline—is often not taught.

This is unfortunate, considering that stratigraphy not only is one of the key principles of geology—and unique among the sciences—but also provides a great opportunity to practice logical, scientific thinking and to argue from evidence. The simple reasoning involved in understanding crosscutting relationships and the principles of stratigraphy provide the foundation for how we know that the Earth is old, how we know when the dinosaurs walked the Earth and when they went extinct, and how we read the stories of ancient oceans that crept across the land—over and over—only to retreat.

Stratigraphy appears in both middle and high school performance expectations in the *Next Generation Science Standards* (*NGSS;* NGSS Lead States 2013). For example,

Figure 11.1

The timeliness of strata, illustrated by the *Shoe* comic strip from December 4, 1994

Source: © 2015 King Features Syndicate, Inc., World Rights Reserved.

performance expectation MS-ESS1-4 is "Construct a scientific explanation based on evidence from rock strata for how the geologic time scale is used to organize Earth's 4.6-billion-year-old history."

Stories of changes in Earth's past inform modern studies of how variations in climate might affect life on Earth and how human activity might disrupt the global ecosystem. One related *NGSS* performance expectation is HS-ESS2-2: "Analyze geoscience data to make the claim that one change to Earth's surface can create feedbacks that cause changes to other Earth systems."

Given its foundational importance in geology, why is it that so few of us understand the principles of stratigraphy and how we read the stories of the earth? Is it too hard?

We don't think so. In the 1860s and 1870s, John Wesley Powell read the stories of immense time and change in the rocks of the Grand Canyon. With no formal education in geology and little prior knowledge of the stratigraphy of the region, he made key contributions to the field of geology. With perseverance and a passion to understand, he read one of the most fundamental geological stories of the American West.

If Powell can do it, then teachers with degrees in earth science, and their students, can follow in those footsteps.

GAZING INTO THE ABYSS OF TIME

The abyss of deep time yawns indifferently at humanity, and perhaps we sometimes fear to peer into that vast indifference.[C34] It's more comfortable to stick with the snapshot stories told by single rocks and to make paper models of the immensity of time. Even so, the vastness of time, like the vastness of space, can provide an almost spiritual experience of equal importance to factual understanding—recognizing that we are only a very small part of something very big. Reflecting on his own journey through the immensity of time, Powell (1875) wrote:

> *We have looked back unnumbered centuries into the past, and seen the time when the schists in the depths of the Grand Cañon were first formed as sedimentary beds beneath the sea … and the geologist, in the light of the past history of the earth, makes prophecy of a time when this desolate land of Titanic rocks shall become a valley of many valleys, and yet again the sea will invade the land. (p. 214)*

Mystery, coupled with a sense of becoming, lends an exhilaration to the immense tapestry of time that spreads behind us. This emotion of discovery, while more difficult to communicate than scientific descriptions of nature, is an equally important part of our construction of an understanding of existence. This emotion of discovery is why dinosaurs are so exciting to children and why the sense of immense time and space is so fascinating and frightening to adults. These ideas capture a sense of wonder, awe, and mystery, which, while not completely scientific, cannot be amputated from science without leaving that science seriously disabled.

In the early 2000s, I (Russ) visited the Smithsonian National Museum of Natural History for the first time. My time was short, so I zipped through the museum and absorbed impressions of great creatures, change, and images of place and time. Rough bones too large for any creature living today materialized in the bustling, yet quiet, passageways. I saw bark and leaves of plants, similar to ones I know and yet fundamentally different. Creatures too weird for fiction, disregarding my sense of disbelief, stood before me in unquestionable reality.

I was astonished at the immensity of time needed to harbor so many cohorts of creatures that had never met each other. My epiphany arose not because I had never before examined such strange things, but because I had never examined so many at once. I saw not just the details of an immense past but a glimpse of the entire tapestry of life and earth and time.

I realized that the age of planet Earth can't be grasped as only a measured number.[C35] In fact, that's not even how Earth's age was first discovered. The discovery of deep time began, not with a measured date, or with one observation, or experiment, or story, but with fieldwork and many, many stories of the past. Stories such as those read by Powell at the Grand Canyon.

POWELL'S PASSION FOR READING STORIES IN ROCK

Powell had no degree in geology, but he did have a fierce determination to understand, a trait that we teachers can also practice and share with our students. This determination can be seen in several entries from his journal of the 1869 expedition, as recounted in a technical report published in 1875 and a book for the general public published in 1895 (the latter book was reprinted in 1961).

Often, he would leave his men with the boats at the river and climb up the cliffs to study the rocks and landscapes. On August 18, the fifth day in the canyon, he wrote:

> *The day is employed in making portages and we advance but two miles on our journey. Still it rains. ... While the men are at work making portages I climb up the granite to its summit and go away back over the rust-colored sandstones and greenish-yellow shales to the foot of the marble wall. I climb so high that the men and boats are lost in the black depths below and the dashing river is a rippling brook, and still there is more canyon above than below. All about me are interesting geologic records. The book is open and I can read as I run. (Powell 1895/1961, p. 263)*

His fierce determination to learn is better understood when put in context of the situation of the expedition. Powell and his nine companions were already low on supplies when they entered the canyon on August 13. Powell wrote then:

> *We are now ready to start on our way down the Great Unknown. Our boats, tied to a common stake, chafe each other as they are tossed by the fretful river. They ride high and buoyant, for their loads are lighter than we could desire. We have but a month's rations remaining. The flour has been resifted through the mosquito-net sieve; the spoiled bacon has been dried and the worst of it boiled; the few pounds of dried apples have been spread in the sun and reshrunken to their normal bulk. The sugar has all melted and gone on its way down the river. But we have a large sack of coffee. (Powell 1895/1961, p. 247)*

The concern for supplies became so sharp that, by August 27, three members of his team decided to leave the river and attempt to hike to a Mormon settlement to the north. Powell was sorely tempted to join them. But his passion to learn, and to finish, were too strong:

> *Then we cross the river and go into camp for the night on some rocks in the mouth of the little side canyon.*

> *After supper Captain Howland asks to have a talk with me. We walk up the little creek a short distance and I soon find that his object is to remonstrate against my determination to proceed. He thinks that we had better abandon the river here …*
>
> *All night long I pace up and down a little path, on a few yards of sand beach along by the river. Is it wise to go on? I go to the boats again to look at our rations. I feel satisfied that we can get over the danger immediately before us; what there may be below I know not. …*
>
> *But for years I have been contemplating this trip. To leave the exploration unfinished, to say that there is a part of the canyon which I cannot explore, having already nearly accomplished it, is more than I am willing to acknowledge, and I determine to go on. (Powell 1895/1961, p. 278)*

He sometimes put his drive to understand even ahead of his own safety. On one occasion, Powell became stranded on a small ledge and had to be rescued, as described in his journal entry for August 27, 1869:

> *High above the river we can walk along on the top of the granite, which is broken off at the edge and set with crags and pinnacles, so that it is very difficult to get a view of the river at all. In my eagerness to reach a point where I can see the roaring fall below, I go too far on the wall, and can neither advance nor retreat. I stand with one foot on a little projecting rock and cling with my hand fixed in a little crevice. Finding I am caught here, suspended 400 feet above the river, into which I must fall if my footing fails, I call for help. The men come and pass me a line, but I cannot let go of the rock long enough to take hold of it. Then they bring two or three of the largest oars. All this takes time which seems very precious to me; but at last they arrive. The blade of one of the oars is pushed into a little crevice in the rock beyond me in such a manner that they can hold me pressed against the wall. Then another is fixed in such a way that I can step on it; and thus I am extricated. (Powell 1875, p. 97)*

This event happened on the same day that Powell decided to finish the trip rather than leave the river and walk out. And, despite the risks and challenges, he drove on to explore and understand.

SIMPLE GRAMMAR, COMPLEX STORY

Powell was a self-taught geologist, and he pursued the scientific enterprise with a fierce determination that led to a successful outcome. His exploration of the Grand Canyon can inspire the teacher who hopes to become a practitioner of science and explore the earth with his or her students.

Powell's exploration also gives us data to work with. Although we can bring single rocks into the classroom, stratigraphy is much too big for classroom examination. So, it's helpful to let fieldworkers, such as Powell, bring the strata into the classroom for us. Given accounts and descriptions of the Grand Canyon, students can construct an understanding of how the layers of rock reveal the long story of events that Earth has experienced, just as Powell did on his journey down the Colorado River in the 1800s, and as geologists continue to do today.

To read this long story in the rocks at the Grand Canyon and follow in the footsteps of Powell, students need to understand a few key principles of geology, such as the concept of crosscutting relationships and the principle of superposition. I (Russ) often illustrate these concepts with simple classroom demonstrations. For example, I take a broken pencil to class, show it to my students, and ask, "So, which came first, somebody made this pencil, or somebody broke it?" I'll get a couple of faint laughs and someone will say, "It had to be made first."

Yes, something has to exist before subsequent processes can affect it. This is the concept of crosscutting relationships. But I warn them, "When we start applying this to rocks, it can get complicated!"

Then, I ask them to close their eyes, and I stack up three different textbooks on the table, dropping each book from a few inches off the table and letting it lie where it falls. They open their eyes and I ask, "So, which book did I drop first?"

Well, the one on the bottom.

Yes, and sediments act the same way. Gravity pulls them down to the surface of the earth (with a little *e*, meaning the underlying landscape, rock, and sediment). They can't float around up in the air waiting for later layers of sediment to slip in underneath them. What's more, the sediments will tend to be deposited in flat layers because gravity won't let them poke up into the air at arbitrary angles.

From these two absurdly simple principles—the grammar of the earth—we read the complex story of planet Earth's past.

CLASSROOM LESSONS AND CHALLENGES FROM THE GRAND CANYON

In his exploration of the Grand Canyon, Powell identified three great sets of rocks in the Grand Canyon, separated from each other by unconformities, which he illustrated as shown in Figure 11.2. In geology, distinctive sets of rocks are referred to as *units*. We discuss these units, which Powell identified as A, B, and C, in the following subsections and suggest some activities related to these units.

Powell's Unit A

Let's start with Powell's observations and interpretations of the lowest and oldest set of rocks, unit A. This passage is from his technical report of 1875, *Exploration of the Colorado River of the West and Its Tributaries:*

> *We find these lower rocks to be composed chiefly of metamorphosed sandstones and shales, which have been folded so many times, squeezed, and heated, that their original structure, as sandstones and shales, is greatly obscured, or entirely destroyed, so that they are called metamorphic crystalline schists. Dame Nature kneaded this batch of dough very thoroughly. (p. 213)*

Several classroom-size scientific questions can be extracted from this passage. For example, if the rocks were once sandstone and shale, in what kind of place might those sedimentary rocks have formed? Was it a canyon like today? Mountains?

Figure 11.2

Illustration of the three great sets of rocks and two great unconformities in the Grand Canyon from John Wesley Powell's 1875 report

There is an error in this illustration, which we discuss in the text. Can you spot the error—a feature that could not possibly form as drawn?

No, neither canyon nor mountains. Those are erosional environments—not depositional ones. To get sediments that became sandstone and shale, there must have once been a depositional environment where the Grand Canyon is today.

As Powell concluded in his 1875 report, "We have looked back unnumbered centuries into the past, and see the time when the schists in the depths of the Grand Cañon were first formed as sedimentary beds beneath the sea" (p. 214).

Another question might be, Which process came first, the sedimentary beds being deposited or their becoming folded and metamorphosed? Just as the pencil can't be broken until it's made, we can't fold rocks until they exist. The folding had to come later than the deposition on a sea floor.

Folding a rock isn't easy. Have you ever tried to fold one? Even if you're strong enough, the rock will break long before it folds. Folding a rock requires that it be very hot and under high pressure—like the way that taffy bends when it's hot but breaks with a snap when cold.

High temperature and pressure occur at great depth within the Earth's crust, such as within the roots of a mountain range. So, which came first, the presence of a sea where sediment accumulated or the uplift of a range of mountains whose roots were metamorphosed by heat and pressure?

Stories are about sequences of events, and telling the story of planet Earth involves putting events into proper sequence. Already, based on Powell's observations, we're starting to put together a sequence of events for Earth's past in the western United States. There was an ancient sea. Sediments of the sea piled up in deep layers and were uplifted into a mountain range.

In his 1875 report, Powell further described the lowest unit of rocks (unit A) as follows: "After these beds were deposited, after they were folded … they were fractured, and through the fissures came floods of molten granite, which now stands in dikes, or lies in beds" (p. 213). Powell did not specifically describe the evidence for his conclusion in this passage, but we can infer his observation based on our understanding of crosscutting relationships. You might have students draw pictures of what Powell must have seen to form these conclusions. Or you might present a thought puzzle for them to consider, such as that shown in Figure 11.3, in which several possible observations are shown and students can figure out which one matches Powell's conclusions.

If the granite had been there at the time of the folding, it would have been folded (see Figure 11.3C). Powell didn't see the granite folded like the schist, and so he concluded that the granite came after the folding, not before.

Neither did he find chunks of weathered granite in the schist (see Figure 11.3B). If chunks of granite were in the schist, it would mean that the granite had to already be there at the time the schist was formed. Yet, he concludes that the granite came after the schist.

He didn't find the schist cutting across the granite (see Figure 11.3D). Like the broken pencil, we can't cut across something until it's there first. If he had found this relationship, he would have thought that the granite must be older than the schist, not younger. Instead, he saw the granite cutting across the schist in relatively unfolded layers and dikes, as in Figure 11.3A.

But wait, if the granite is younger than the schist, why is part of it below the schist in Figure 11.3A? Because the molten rock was injected into the schist. Sometimes it leaked to the top, sometimes it stayed on the bottom, and sometimes it cut through the middle

Figure 11.3

Science reasoning challenge for Powell's unit A rocks in the Grand Canyon

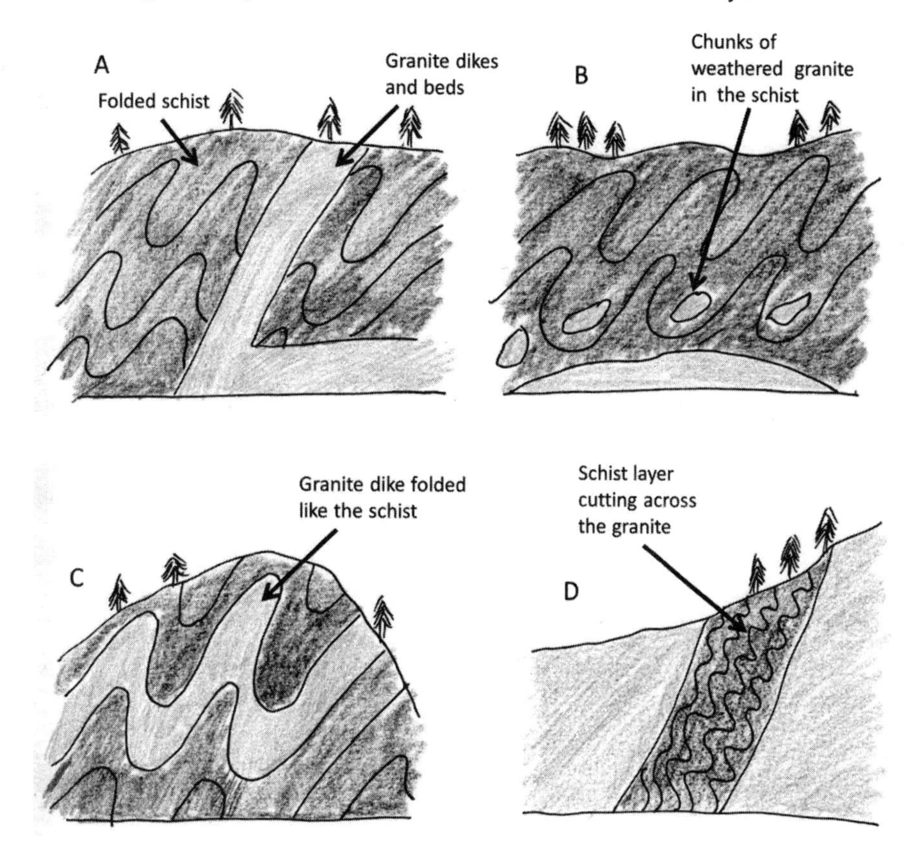

Which of the cross-sectional illustrations shows what Powell must have seen to allow him to figure out that the granite came after the formation and folding of the schist?

in dikes. If you look carefully, you can see both dikes and horizontal layers of granite in Powell's drawing in Figure 11.2.

A final question might be, Was the mountain range still there—stacked high above and pressing down on the schist—when the granite formed? The name "granite" tells us about this rock—that it is an igneous rock with large crystals. As we reminded ourselves in Chapter 10, large igneous crystals signify slow cooling. If it cooled slowly, did it cool near the surface or deep within the crust? The coarse crystals are evidence that the molten granite cooled slowly—meaning that the mountains above the schist and granite had not yet eroded away.

After students have worked through a number of sequencing puzzles from Powell's text, you might encourage them to put together an evidence-and-interpretation chart for unit A like those we introduced in Chapter 8. Table 11.1 shows an example evidence-and-interpretation chart for unit A as well as for Powell's units B and C, which we consider in the subsections that follow.

Table 11.1

Example evidence-and-interpretation chart using John Wesley Powell's observations

From two simple ideas, superposition and crosscutting relationships, Powell inferred three incursions of the sea and two mountain-building events for the Grand Canyon region, as well as the geologically recent uplift of the Colorado Plateau and erosion of the canyon.

Unit	Observations	Powell's interpretations/conclusions *The implications of the evidence*
Powell's unit A	Crystalline schist—made of metamorphosed sandstone and shale	The layers of sandstone and shale were deposited in an ancient sea, followed by burial and mountain building that metamorphosed those layers.
	Granite dikes in the schist	Magma cooled at depth, not at or near Earth's surface—again suggesting the presence of mountains.
	Unfolded granite dikes in the schist	Intrusion of magma occurred after the folding of the schist.
Erosional boundary	Unconformity between unit A and unit B cutting across granite dikes and schist	Mountains above the metamorphic and igneous rocks of unit A were eroded away after the formation of the dikes and schist.
Powell's unit B	Layers of "hard vitreous sandstone" (Powell called this quartzite)	Sandy sediments were deposited in another sea.
	Lava dikes and sills cutting across the schist, the erosion surface between A and B, and the quartzites	Volcanic activity came after the seas and mountain-building events of unit A and after the deposition of the quartzite of unit B.
	Non-horizontal quartzite layers	A second tectonic event—perhaps a mountain-building event—tilted the rocks after the deposition and burial of the layers of sandstone (quartzite).
Erosional boundary	Unconformity between unit B and unit C	The tilted layers of unit B were uplifted and eroded.
Powell's unit C	"Carboniferous" sedimentary layers	A third sea invaded the region. (Unknown to Powell, there were several incursions of the sea during this time.)
	Horizontal "Carboniferous" rocks	There have been no more tilting events since the Carboniferous* time.
Grand Canyon	Deep canyon with the Colorado River cutting across units A, B, and C	The "Carboniferous" and other rocks were uplifted and the river eroded down through them.

* Quotation marks are used around *Carboniferous* elsewhere in this table to indicate that the rocks of Powell's unit C are no longer considered to be entirely of Carboniferous age as Powell believed, but instead span the entire Paleozoic Era. Quotation marks are not used here because we continue to believe today that no major tilting has occurred since Carboniferous time.

Powell's Unit B

Let's turn our attention now to the middle unit of rocks (labeled B in Figure 11.2), set apart from unit A by an unconformity. Powell concluded that the granite was intruded before the erosion of the mountains, and before the later deposition of sediments in a *new* sea:

> *After they were folded ... they were fractured, and through the fissures came floods of molten granite, which now stands in dikes, or lies in beds, and the metamorphosed sandstones and shales, and the beds of granite, present evidences of erosion subsequent to the periods just mentioned, yet antecedent to the deposition of the non-conformable sandstones. (1875, p. 213)*

Which of the illustrations in Figure 11.4 shows what Powell must have seen to conclude that the erosion had *not yet occurred* at the time the granite was intruded?

Figure 11.4

First science reasoning challenge for Powell's unit B rocks in the Grand Canyon

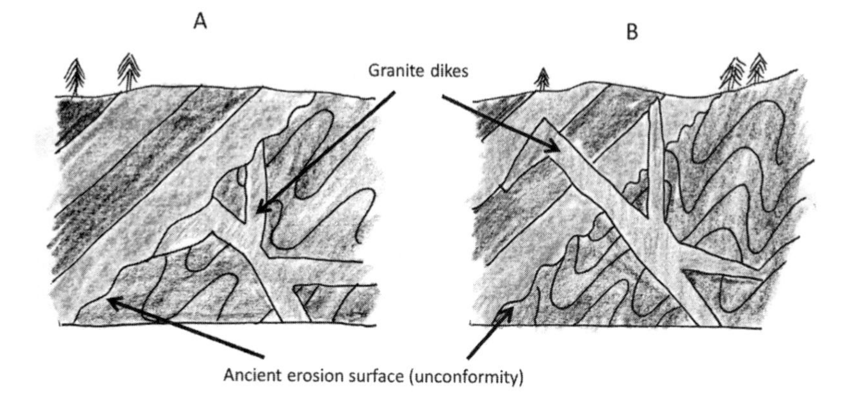

Which of the cross-sectional illustrations shows what Powell must have seen to allow him to figure out the sequence of events among the granite, erosion, and deposition of the unit B rocks?

He did not see the granite cutting across the erosion boundary and into the later rocks (Figure 11.4B). Like the breaking of a pencil, the granite could not cut across the erosion boundary or the later rocks if they did not yet exist. Instead, he saw the erosion boundary cutting across the granite, as shown in Figure 11.4A.

We previously concluded that for large crystals to grow, the granite had to be deeply buried when it formed. Thus, like the schist, the presence of the coarse-grained granite provides evidence for the former presence of a range of mountains. Yet, we've just reasoned that the rocks immediately above the granite today *were not there* when the granite

Figure 11.5
Second science reasoning challenge for Powell's unit B rocks in the Grand Canyon

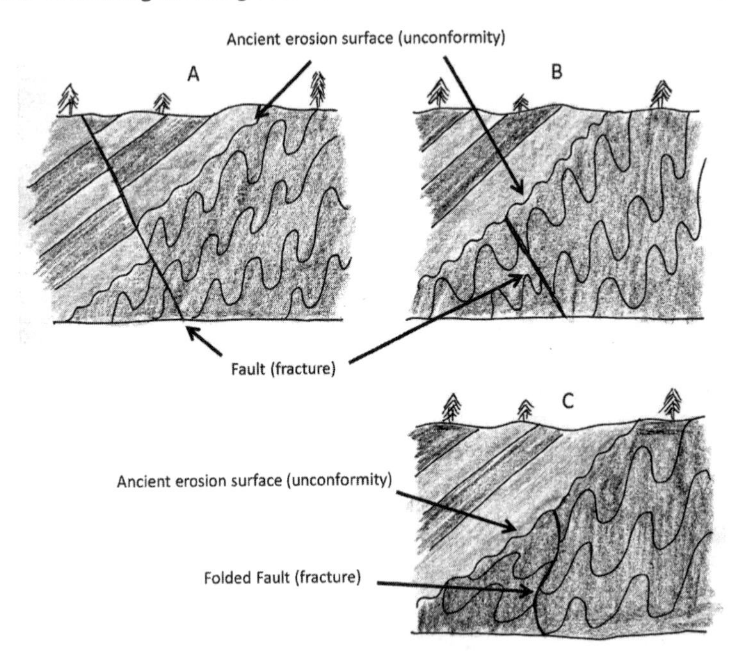

Which of the cross-sectional illustrations shows what Powell must have seen to allow him to figure out the sequence of events among the faulting (fractures), erosion, and deposition of the unit B rocks?

formed. Where did all the rocks and sediment go that were once part of the mountains that caused the folding and metamorphism of the schist? From our inferred sequence of events, we can conclude that any mountain of granite and other rocks that once existed above the erosion boundary eroded away long ago and washed to the sea, to be replaced by the rocks that we see there today.

We can do a similar analysis of the fractures (faults). Like the granite, Powell inferred that these fractures preceded the erosion of the ancient mountains and the deposition of new sediments that became unit B. Which of the illustrations in Figure 11.5 conceptualizes what he saw? Again, you could have your students make their own drawings.

Since the fracturing that he's talking about in this passage ("After they were folded …") happened before the erosion and deposition of new rocks, the faults will not cut across the erosion surface (Figure 11.5A). Since it happened after the folding, it will not be folded (Figure 11.5C). The fractures cut across unit A, but they are crosscut by the erosion surface at the base of unit B, as shown in Figure 11.5B.

Continuing with the description of the unit of rocks labeled B by Powell, he wrote that above the crystalline schists "we can see beds of hard, vitreous sandstone of many colors, but chiefly dark red. ... This group of rocks adds but little more than five hundred feet to the height of the walls and yet the beds are 10,000 feet in thickness. How can this be?" Powell explains further in his popular report of 1895:

> *Then over the black gneiss [the crystalline schist] are found 800 feet of quartzites [the vitreous sandstones], usually in very thin beds of many colors, but exceedingly hard, and ringing under the hammer like phonolite. These beds are dipping and unconformable with the rocks above; while they make but 800 feet of the wall or less, they have a geological thickness of 12,000 feet. Set up a row of books aslant; it is 10 inches from the shelf to the top of the line of books, but there may be 3 feet of the books measured directly through the leaves. So these quartzites are aslant, and though of great geologic thickness, they make but 800 feet of the wall. (p. 379)*

Remembering the principle that rock layers are deposited horizontally, ask yourself, Which came first, the sediments being deposited to form the layers, or the rock layers being tilted?

The rocks couldn't be tilted until they first existed. Tilting implies a tectonic event, such as mountain building. The tilting had to come after the rocks were deposited. Was this the same tectonic event that built mountains and produced the schist and the granite?

No. These quartzites were deposited on the old erosion surface that formed as those former mountains eroded away. That erosion cut across the schist and therefore came after its formation. We can conclude that the tilting resulted from a new uplift event.

In his 1875 report, Powell wrote "The beds, themselves, are records of the invasion of the sea" (p. 212). Which came first, the advance of this sea or the mountain-building event that formed the schist? Solving this puzzle involves considering multiple cross-cutting relationships at the same time. It might be helpful for students to summarize some of these relationships as done in Table 11.1. The advance of the sea comes after the first mountains that made the schist, and long after the sea that deposited the sediments that became the schist. We know this because the erosion surface crosscuts the schist and granite, meaning that the erosion surface must be younger. The stacking of rocks stratigraphically above the erosion surface means they had to come after the formation of the erosion surface.

But the sea came before the uplift that tilted the rocks. We know this because the rocks had to first form on the floor of the new sea before they could become tilted.

Referring to the unit of rocks labeled B, including the vitreous sandstone, Powell wrote: "This region of country was fissured, and the rocks displaced so as to form faults, and through the fissures, floods of lava were poured, which, on cooling, formed beds of

Figure 11.6

Example illustration of dikes and lava flow in Powell's unit B rocks

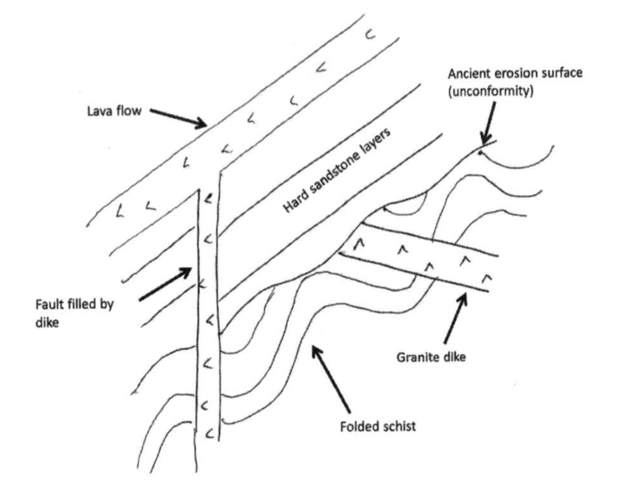

What did Powell see to allow him to figure out the sequence of events related to dikes and lava flows in unit B? In this example illustration (cross-sectional view), the lava flow lies stratigraphically above the sandstone. Notice that *stratigraphically higher* does not mean the same thing as *topographically higher.* In fact, because the rocks are tilted, some parts of the older sandstone layers are actually higher than parts of the younger lava flow.

trap, or greenstone" (1875, p. 213). If the fissures of lava cut across the sandstone, which had to come first, the lava or the sandstone?

You might encourage students to draw an illustration of this different set of faults and dikes and which rocks they must cut across. Although not shown in Powell's picture of unit B rocks, the lava flows lie stratigraphically above the sandstone. An example illustration is shown in Figure 11.6.

Powell's Unit C

Powell's third set of rocks, the C layer in his picture in Figure 11.2, are horizontal and not metamorphosed. They lie on top of yet another erosional surface. Powell wrote:

> *I have already mentioned that the summit of the plateau is also the summit of rocks of Carboniferous Age [Paleozoic by the modern time scale]. These beds are about three thousand five hundred feet in thickness, and beneath them we have a thousand feet of conformable rocks of undetermined age. This gives us 4500 feet, from the summit of the plateau down to the non-conformable beds. (1875, pp. 211–212)*

Measuring the Thickness of Rock

The method that Powell used to measure the thicknesses of the rock layers has an interesting connection to activities we've talked about in previous chapters. He used a barometer. As elevation increases, there is less air above, and so pressure is lower. From the difference in pressure between the top and bottom of a layer, vertical thickness can be calculated. However, pressure also changes with passing of high- and low-pressure weather systems, so it was necessary for two people to make pressure measurements at the top and bottom of each layer at exactly the same time. Here is a description of the procedure from Powell's July 10, 1869, journal entry:

Howland and I determine to climb out, and start up a lateral canyon, taking a barometer with us for the purpose of measuring the thickness of the strata over which we pass. The readings of the barometer below are recorded every half hour and our observations must be simultaneous. Where the beds which we desire to measure are very thick, we must climb with the utmost speed to reach their summits in time; where the beds are thinner, we must wait for the moment to arrive; and so, by hard and easy stages we make our way to the top of the canyon wall and reach the plateau above about two o'clock. (Powell 1895, p. 191)

Considering that the erosional surface cuts across the tilted layers of the B unit, did the deposition of the rocks of B come before or after the erosional surface on which the C rocks lie? We can't erode rocks away until they first exist, so the erosion had to come later.

Which came first, the rocks of unit C, or the tilting of the rocks of unit B? The rocks of set C were not yet present at the time of the tilting, otherwise they would be tilted. Therefore the mountains with their volcanoes were long gone at the time unit C was formed.

Which came first, the deposition of the rocks of unit C, or the erosion of the Grand Canyon? The rocks had to exist before the canyon could be eroded in them. Thus, the canyon itself is the youngster in this story, a relative newcomer in a long, long story.

In his 1875 report, Powell identified the present landscape as the youngest of three great emergences from the sea, during which rivers carved valleys into the rock:

> *Three times has this great region been left high and dry by the ever shifting seas; three times have the rocks been fractured and faulted; three times have floods of lava been poured up through the crevices, and three times have the clouds gathered over the rocks, and carved out valleys with their storms. The first time was after the deposition of the schists; the second was after the deposition of the red sandstones; the third time is the present time. (p. 213)*

A final classroom challenge based on Powell's expedition and report comes from a mistake in his illustration shown in Figure 11.2 (p. 219). Can you spot it?

Think about crosscutting relationships and sequences of events. The sequence of events implied by Powell's illustration is actually impossible, and the illustration does not correctly show what we see in the rocks of the Grand Canyon.

Look at the unconformity at the top of unit B. Which had to come first, the erosion event or the rocks of layer B? Since the erosion cuts across the layers, we know that the layers had to already be there when the erosion happened.

Now consider the unconformity at the bottom of unit B. See the problem? This erosion also cuts across the layers of unit B, meaning that this erosion must also be younger than the rocks of unit B. But for this to be true, the underlying schist would also have to be younger than unit B, which would mean that unit B should have also been folded and highly metamorphosed like the schist. And yet it was not. Unit B must be both older and younger than the underlying schist, which is impossible.

We provide a modern cross-section illustration of the rocks at the Grand Canyon on page 229 of the "Suggestions for Designing an Activity" section at the end of this chapter.

FINAL THOUGHTS

There is no place where an alert person cannot glimpse deep time, whether it be in glacial debris that speaks of changing past climates, sequences of rock layers that speak of changing past environments, ancient fossils that speak of extinct creatures, or eroding hills that speak of change in progress today. You, too, can discover planet Earth's deep past.

As you consider the work of Powell and whether you can reproduce in the classroom his classic discoveries—or whether it's too hard—remember that he did these explorations with only a single good arm, driven by an unquenchable desire to discover and understand. In the preface to his 1895 report, about a quarter-century after his first expedition, he wrote:

> *Many years have passed since the exploration, and those who were boys with me in the enterprise are—ah, most of them are dead, and the living are gray with age. Their bronzed, hardy, brave faces come before me as they appeared in the vigor of life; their lithe but powerful forms seem to move around me and the memory of the men and their heroic deeds, the men and their generous acts, overwhelms me with a joy that seems almost a grief, for it starts a fountain of tears. I was a maimed man; my right arm was gone; and these brave men, these good men, never forgot it. In every danger my safety was their first care, and in every waking hour some kind service was rendered me, and they transfigured my misfortune into a boon. (p. v)*

REFERENCES

NGSS Lead States. 2013. *Next Generation Science Standards: For states, by states.* Washington, DC: National Academies Press. *www.nextgenscience.org/next-generation-science-standards.*

Powell, J. W. 1875. *Exploration of the Colorado River of the West and its tributaries.* Washington, DC: U.S. Government Printing Office. *http://pubs.usgs.gov/unnumbered/70039238/report.pdf.*

Powell, J. W. 1895/1961. *Canyons of the Colorado / The exploration of the Colorado River and its canyons.* Meadville, PA: Flood & Vincent/New York: Dover.

SUGGESTIONS FOR DESIGNING AN ACTIVITY

READING STORIES OF DEEP TIME

Stratigraphy is big—too big to bring into the classroom and usually too big to see all at once in the field. Therefore, bringing stratigraphy into the classroom often means bringing in data collected by others. These data might be narrative, as with the text from John Wesley Powell, or graphical, such as the cross-section of the Grand Canyon illustrated in Figure 11.7. Working with both narrative and cross-sectional data gives students a chance to make connections between these modes of scientific communication. Ideally, students can make field observations of their own before beginning this activity. For example, seeing faults, dikes, or layers of rock in the field helps students understand the meaning of the cross-sections and the narrative accounts, even if the features they see in the field aren't at the same location as the classroom activity.

This activity might be used as a capstone activity after engaging students in classroom lessons like those suggested earlier in this chapter. Or it might be offered as a stand-alone exercise to give students a chance to make connections between observational data, as portrayed in the cross-section, and the interpretation of that data, as explained in Powell's 1875 report.

Figure 11.7

Modern cross-section illustration of the rocks at the Grand Canyon showing the relationship to Powell's units A, B, and C

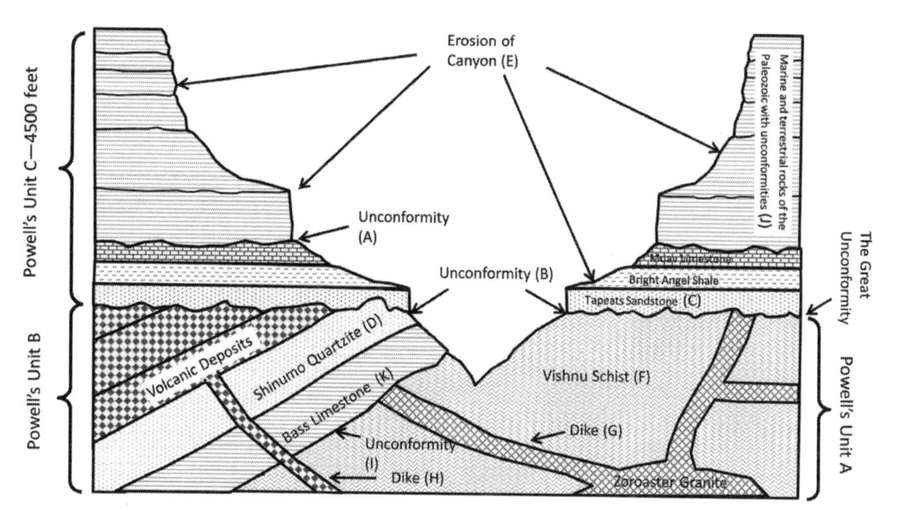

AN EXERCISE ON "THE PRESENT IS THE KEY TO THE PAST"

Rather than have students start by telling the whole story from start to finish, you might engage them with a small part of the big picture. For example, take a look at the sequence of rocks above the Great Unconformity: Tapeats Sandstone, Bright Angel Shale, and Muav Limestone. Which of these rocks formed first? Which last?

Powell proposed that these rocks were marine, but was that ocean advancing over the region at this time (transgressing), or was the ocean retreating (regressing)?

We learned in Chapter 10 that sandy sediment (sandstone) indicates a higher-energy environment of deposition than does muddy sediment (shale). Often, higher-energy environments occur near the shore where the waves wash, and lower-energy environments occur farther out to sea where the water grows deep and still. Even farther offshore, sand and mud from the land become rare, and more of the sediment is formed of living creatures that make sediment from components dissolved in the water. These carbonate sediments can form limestone.

We see this pattern along many (but not all) modern shorelines. For example, Figure 11.8 shows the pattern of sediment off the East Coast of the United States.

Given the pattern of sediment along the East Coast, and considering where different types of sediment might be deposited, what story can we infer from the sequence of Tapeats Sandstone, Bright Angel Shale, and Muav Limestone at the Grand Canyon? Was the ocean advancing over the land or was it retreating from the land? Which came first, the sandstone or the limestone, and what does this tell us about whether the sea was advancing or retreating in the region of the Grand Canyon?

This sequence of rocks is thought to record the advance of the Sauk Sea across North America. Eventually the Sauk Sea covered almost the entire continent except perhaps for a narrow strip of land in

Figure 11.8
Pattern of sediment off the East Coast of the United States

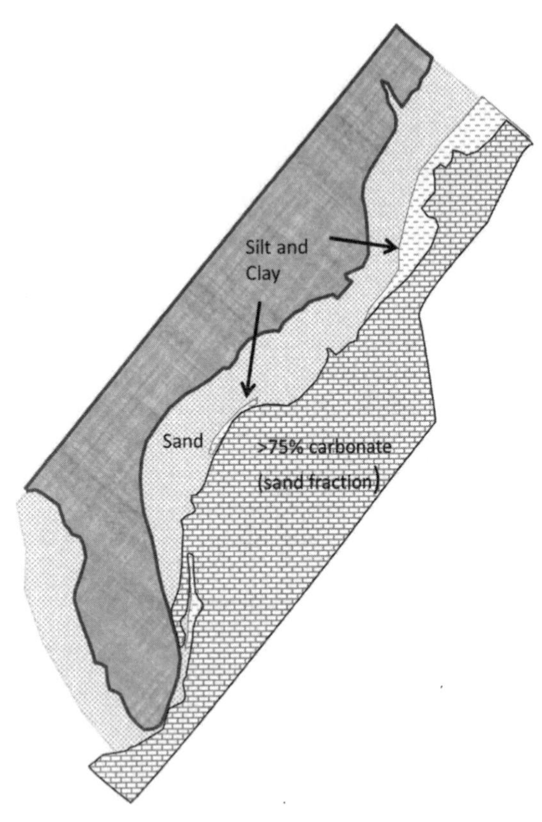

Source: Based on data from Milliman 1972.

Minnesota and central Canada. The sandstones of the Wisconsin Dells are believed to have formed along this same shoreline as the sea advanced from the other side of the continent.

"The present is the key to the past" is a foundational idea in geology—we study modern geological processes and environments to understand those of the past. However, we have to remember that conditions may have been different in the past. For example, in the Cambrian Period when the Tapeats Sandstone, Bright Angel Shale, and Muav Limestone were deposited, there were no land plants, so erosion and supply of sand and mud were likely quite different. Climate differences can also affect deposition of limestone, since limestone is less stable in colder water. This effect can be seen to some extent in the deposition of carbonate sediments off the modern East Coast of the United States in that carbonate sediments are much more abundant in the warmer, subtropical waters of Florida.

ARGUING FROM EVIDENCE: SEQUENCING FEATURES IN ROCK BASED ON CROSSCUTTING RELATIONSHIPS

The letters *A* through *K* indicate features in the rock at the Grand Canyon that Powell saw and that can be seen today. They can be put in chronological sequence using crosscutting relationships and the principle of superposition. This sequencing provides students with an exercise in logical thinking, especially if you encourage students to articulate their reasoning for their order of choices. This sequencing also provides the basis to begin constructing an explanation for the formation of the Grand Canyon. (The correct sequence, from first to last, is F, G, I, K, D, H, B, C, A, J, E.)

ARGUING FROM EVIDENCE: TRANSLATING ROCKS AND FEATURES INTO A STORY LINE OF EVENTS

Putting the features in rock into chronological order as done by using crosscutting relationships and superposition is not quite the same thing as inferring events from those features and putting the *events* into chronological order. The following sequence of events can be inferred for the Grand Canyon region through interpretation of the rocks and the principles of stratigraphy. The first event is listed at the bottom, as is traditional in stratigraphy, and the most recent event is listed at the top. The letters refer to units in the Grand Canyon cross-section shown in Figure 11.7.

- Uplift of the Colorado Plateau and erosion of the Grand Canyon (E)
- Deposition and erosion of sediments as seas advance and retreat through the Paleozoic Era (J)

- Deposition of sediments in a sea (C)
- Erosion of highland areas and old mountains (B)
- Tilting of rock layers, perhaps by uplift of mountains
- Eruption of volcanoes
- Deposition of sediments in a sea (K)
- Erosion of a mountain range
- Intrusion of igneous rock deep within the crust
- Metamorphism of sediments in the roots of a mountain range
- Uplift of a range of mountains
- Deposition of sediments in a sea (F)

You might give students the above list of events (scrambled) to put in order. Or you might have them infer events themselves. In our experience, giving a list of events, or at least how many events are expected, makes assessing the exercise a bit more doable. However, if the goal is a classroom discussion where students argue from evidence about their inferred stories, then that may not be necessary.

Sequencing the observations is an easier exercise than sequencing the events inferred from the observations. Middle school students can order the features and articulate the evidence for their chosen order. Interpretation of how rocks formed based on the rock clues and sequencing the inferred events may be too complex for some middle school students.

OBTAINING, EVALUATING, AND COMMUNICATING INFORMATION: UNDERSTANDING POWELL'S TEXT

You might have students read a portion of Powell's 1875 technical report to develop skills in thoughtful reading of an historical, nonfiction text. As students read, they can correlate his statements with one or more lines of evidence cited in the text of the report. Students might separate evidence from conclusion (interpretation), in a fashion similar to that shown in Table 11.1 (p. 222).

Alternatively, you might have students associate conclusions from Powell's report with evidence seen in the Figure 11.7 cross-section. In the excerpt from Powell's report at the end of this section, sections of text are specified by arrows and circles. Students can place the appropriate letter from Figure 11.7 in the small circles to identify which rock or features Powell is referring to, or to identify the primary evidence on which he bases a claim.

SUMMARY CHECKLIST FOR TEACHER AS PRACTITIONER OF SCIENCE

- Identify stratigraphic and crosscutting data for a real region of the Earth, such as that shown in the cross-section of the Grand Canyon (Figure 11.7). Combining the graphical data with a narrative interpretation of the region provides a good opportunity for students to make the connection between what they read and the features they see.
- Give students a chance to see a few of these features in the field (or at least photos from the field). This helps keep the graphical data in the cross-section from being too abstract. Figure 11.9 provides a large-scale photo of the Great Unconformity.

Figure 11.9

The Great Uncomformity: An angular uncomformity between Powell's unit B rocks and the overlying Tapeats Sandstone

Source: Courtesy of Doug Dolde, *https://en.wikipedia.org/wiki/Grand_Canyon_Supergroup#/media/File:View_from_Lipan_Point.jpg.*

- Give students a chance to interpret a small part of the big picture before tackling the entire big picture—such as the inference of an advancing ocean in the activity earlier in this section.
- Give students a chance to put features into sequence.
- Give students a chance to make the connection between the sequence of features in the rock and the sequence of events implied by those features.
- Always be ready to entertain and discuss alternative interpretations. Can they be excluded by any observations at hand? What new observations would be needed to test them?

REFERENCE

Milliman, J. D. 1972. *Petrology of the sand fraction of sediments, northern New Jersey to southern Florida.* U.S. Geological Survey Professional Paper 529-J. Washington, DC: U.S. Government Printing Office.

RECORDS OF MORE ANCIENT LANDS

The summit of the Kaibab Plateau is more than six thousand feet above the river and I have mentioned that the summit of the plateau is also the summit rocks of the Carboniferous Age. These beds are about three thousand five hundred feet in thickness and beneath them we have a thousand feet of conformable rocks of undetermined age. This gives us 4,500 feet from the summit the plateau down to the non-conformable beds. Still beneath we have 1,500 feet so that we have more than one thousand five hundred feet of other exposed in the depths of the Grand Cañon. Standing on some rock which has fallen from the wall into the river--a rock so large that top lies above the water--and looking overhead we see a thousand feet of crystalline schists, with dikes of greenstone, and dikes and beds of granite. Heretofore we have given the general name granite to this group of rocks; still, above them we can see beds of hard, vitreous sandstone of many colors, but chiefly dark red. This group of rocks adds but little more than five hundred feet to the height of the walls, and yet the beds are 10,000 feet in thickness. How can this be? The beds themselves are non-conformable with the overlying Carboniferous rocks; that is, the Carboniferous rocks are spread over their upturned edges.

In Illustration 79 [our Figure 11.3] we have a section of the rocks of the Grand Cañon. A, A represents the granite; a, a, dikes and eruptive beds; B, B, these non-conformable rocks. It will be seen that the beds incline to the right. The horizontal beds above, C, C are rocks of Carboniferous Age, with underlying conformable beds The distance along the wall marked by the line x, y, is the only part of its height represented by these rocks, but the beds are inclined and their thickness must be measured by determining the thickness of each bed. This is done by measuring the several beds along lines normal to the planes of stratification; and, in this manner, we find them to be 10,000 feet in thickness.

Doubtless, at some time before the Carboniferous rocks C, C were formed, the beds B, B extended off to the left, but between the periods of deposition of the two series, B, B and C, C there was a period of erosion. The beds, themselves, are records of the invasion of the sea; the line of separation the record of a long time when the region was dry land. The events in the history of this intervening time, the period of dry land, one might suppose were all lost. What plants lived here, we cannot learn; what animals roamed over the hills, we know not; and yet there is a history which is not lost, for we find that after these beds were formed as sediments beneath the sea, and still after they had been folded, and the sea had left them, and the rains had fallen on the country long enough to carry out ten thousand feet of rocks, the extension of these beds to the south, which were cut away, and yet before the overlying Carboniferous rocks were formed as sediments of sand and triturated coral reefs, and ground shells and pulverized bones, some interesting events occurred, the records of which are well preserved. This region of country was fissured, and the rocks displaced so as to form faults, and through the fissures floods of lava were poured, which, on cooling, formed beds of trap, or greenstone. This greenstone was doubtless poured out on the dry land, for it bears evidence of being eroded by rains and streams prior to the deposition of the overlying rocks.

Let us go down again, and examine the junction between these red rocks, with their intrusive dikes and overlying beds of greenstone, and the crystalline schists below.

We find these lower rocks to be composed chiefly of metamorphosed sandstones and shales which have been folded so many times, squeezed, and heated, that their original structure, as sandstones and shales, is greatly obscured, or entirely destroyed, so that they are called metamorphic crystalline schists.

Source: Excerpt from pages 211–214 of *Exploration of the Colorado River of the West and Its Tributaries,* by John Wesley Powell (Washington, DC: U.S. Government Printing Office, 1875).

Dame Nature kneaded this batch of dough very thoroughly. After these beds were deposited, after they were folded, and still after they were deeply eroded, they were fractured, and through the fissures came floods of molten granite, which now stands in dikes, or lies in beds, and the metamorphosed sandstones and shales, and the beds of granite, present evidences of erosion subsequent to the periods just mentioned, yet antecedent to the deposition of the non-conformable sandstones.

Here, then, we have evidences of another and more ancient period of erosion, or dry land. Three times has this great region been left high and dry by the ever shifting sea; three times have the rocks been fractured and faulted; three times have floods of lava been poured up through the crevices, and three times have the clouds gathered over the rocks, and carved out valleys with their storms. The first time was after the deposition of the schists; the second was after the deposition of the red sandstones; the third time is the present time. The plateaus and mountains of the first and second periods have been destroyed or buried; their eventful history is lost; the rivers that ran into the sea are dead, and their waters are now rolling as tides, or coursing in other channels. Were there cañons then? I think not. The conditions necessary to the formation of cañons are exceptional in the world's history.

We have looked back unnumbered centuries into the past, and seen the time when the schists in the depths of the Grand Cañon were first formed as sedimentary beds beneath the sea; we have seen this long period followed by another of dry land--so long that even hundreds, or perhaps thousands, of feet of beds were washed away by the rains; and, in turn, followed by another period of ocean triumph, so long, that at least ten thousand feet of sandstones were accumulated as sediments, when the sea yielded dominion to the powers of the air, and the region was again dry land. But aerial forces carried away the ten thousand feet of rocks, by a process slow yet unrelenting, until the sea again rolled over the land, and more than ten thousand feet of rocky beds were built over the bottom of the sea, and then again the restless sea retired, and the golden, purple, and black hosts of heaven made missiles of their own misty bodies--balls of hail, flakes of snow, and drops of rain--and when the storm of war came, the new rocks fled to the sea. Now we have cañon gorges and deeply eroded valleys, and still the hills are disappearing, the mountains themselves are wasting away, the plateaus are dissolving, and the geologist, in the light of the past history of the earth, makes prophecy of a time when this desolate land of Titanic rocks shall become a valley of many valleys, and yet again the sea will invade the land, and the coral animals build their reefs in the infinitesimal laboratories of life, and lowly beings shall weave nacre-lined shrouds for themselves, and the shrouds shall remain entombed in the bottom of the sea, when the people shall be changed, by the chemistry of life, into new forms; monsters of the deep shall live and die, and their bones be buried in the coral sands. Then other mountains and other hills shall be washed into the Colorado Sea, and coral reefs, and shales, and bones, and disintegrated mountains, shall be made into beds of rock, for a new land, where new rivers shall flow.

Thus ever the land and sea are changing; old lands are buried, new lands are born, and with advancing periods new complexities of rock are found; new complexities of life evolved.

STORIES OF PLACES WE CAN'T SEE BUT CAN HEAR

People have traveled to the Moon and descended into deep ocean trenches, but no one has visited Earth's mantle or core. Humanity's closest approach to Earth's core is in South Africa's 3.9-kilometer-deep Mponeng gold mine, currently the world's deepest mine (Mining-technology.com 2013). At these depths, the core is only 2,886 km away. Miners at Mponeng labor in temperatures that can approach 60°C (140°F)—hot enough that cool air must be forced through the mine for workers to survive (Wadhams 2011). But these temperatures are frigid compared with the core's blistering 6000°C, or about 10832°F (Foley 2013). Suggestions by Hollywood and Jules Verne to the contrary, no one is likely to go there any time soon. Pressures more than a million times that of atmospheric pressure and near-solar temperatures are not suitable for humans, and we have no technology—nor can we even propose a plausible future technology—that can protect us from these conditions.

Even so, every high school student knows about the Earth's core. How can we know that the Earth has a core if we can neither see it nor go there?

Why, we can look it up online, of course!

This seems funny when we realize that, for information to be online, someone had to figure it out. Someone figured out a way to "see" the core without actually visiting it. Sometimes we lose track of the distinction between the practices of science—where our understanding is based on observation—and learning science facts—where our understanding is based on consulting authoritative people or reports. My students and I (Mary) experienced this distinction firsthand in fall 2015 when I began my eighth-grade earth science year by having students look up answers online.

WHEN YOU CAN'T LOOK UP THE ANSWER

My first lesson dealt with map patterns of topography and rock ages. We started with the journey of Lewis and Clark up the Missouri River and over the Columbia Plateau to the Pacific Ocean. "Elevation had to change as they went up the Missouri River," I said. "What's the elevation where they started in St. Louis? What about Great Falls, Montana? What's the elevation at Fort Clatsop near where they reached the sea?"

Students pulled out their devices and got their thumbs moving. They quickly jotted down the elevations for each town—not reading more than the headline blurb. They compiled their numbers on the whiteboard to compare with other students. They found only a small variation in values for St. Louis and Great Falls. But, whoa! The numbers for Fort Clatsop were all over the place.

I provided them with a shaded relief map of the region around each city, showing the low topographic relief at St. Louis and Great Falls, and the high relief at Fort Clatsop. "So, why do you think we got so many different elevations for Fort Clatsop?" I asked.

Despite the maps, this question proved very hard. "How are we supposed to know, if we don't know where to look it up?"

"This is hard," I agreed, "and don't worry if you don't understand right away. But get ready for more of this! Science is about figuring things out and thinking about your observations, not just looking up answers."

Looking up answers in authoritative documents is not different in kind from checking the answer key. We considered the tyranny of the answer key, and its ability to stifle investigation, in Chapter 9. The declaration "It's true because I found it on Google" has its own tyrannical side. Students tend to accept the online result as true without question. If prompted to question the results, they look to the authority to answer the question rather than trying to reason through how the high relief at Fort Clatsop might result in more variable elevation measurements. One student googled the question, "Why are there different elevations at Fort Clatsop?"

The practice of science is not simply a matter of looking up answers. Neither should science in the classroom simply involve looking up answers. Science in the classroom should look more like the process of discovery that gave us the answers to begin with. Jennifer Mesa, Rose Pringle, and Lynda Hayes (2013) said, "When students learn science in ways that mirror how scientists do science, they are better able to understand how scientific knowledge develops and gain deeper understandings of core scientific concepts." [C36]

Ready Set Science!, a precursor document to *A Framework for K–12 Science Education: Practices, Crosscutting Concepts, and Core Ideas* (*Framework;* NRC 2012), proposed that "[t]he way that scientists operate in the real world is remarkably similar to how students operate in effective science classrooms" and that "science learning can be modeled in important ways on how real scientists do science" (Michaels, Shouse, and Schweingruber 2008, pp. 6 and 15, respectively). Likewise, one of the key shifts in science education envisioned by the *Next Generation Science Standards* (*NGSS;* NGSS Lead States 2013) is described in Appendix A, as follows: "K–12 science education should reflect the interconnected nature of science as it is practiced and experienced in the real world" (p. 1).

How is science practiced and experienced in the "real world"? Let's look at the work of one practicing scientist and consider three elements of authentic science that can be part of students *doing* science in your classroom.

SCIENCE IN THE REAL WORLD: WITNESS TO THE UNSEEN

In summer 2015, I (Mary) interviewed Dr. Keith Koper, a University of Utah geophysicist who does forensic seismology. Forensic seismologists use seismic waves to recognize and verify human-made events that release seismic energy—such as explosions and nuclear weapons tests.

In 2000, Dr. Koper was working to understand the differences between the seismic signatures of earthquakes and explosions as a postdoc at the University of Arizona (personal communication 2015). On August 12, 2000, the Russian cruise-missile submarine *Kursk* sank during naval exercises in the Barents Sea, killing all 116 crew members and 2 civilians (Koper et al. 2001; Tyler 2000). Speculation about what happened ran high all around the world. Had the Russian submarine collided with an American submarine spying on the Russian war games? Had it hit a World War II mine?

No one was there to see what happened tens of meters beneath the sea. But thousands of miles away, geologists were *listening*. Seismographs had recorded the shock waves from the event. Dr. Koper and his colleagues asked, "Can that seismic data tell us what happened to the *Kursk?*"

Their question brings us to our first authentic science element: The question arose from Dr. Koper's own curiosity and initiative and was a question he could address with observational evidence.

I imagined that maybe he got some kind of alert when the incident occurred: "Dr. Koper, the U.S. government needs your help to solve this mystery!" I told him, "Somehow in my mind I envisioned, you know, a seismologist having alerts on their computer."

No. No alerts. No crack team of government agents visiting his lab to solicit help. Dr. Koper invented the question because he had something he wanted to understand and an innovative way to understand it—no one else had imagined that these seismic data, free and open to everyone, held answers to all kinds of mysteries, including the mystery of the *Kursk*.

Dr. Koper and his colleagues gathered data from several seismic networks in Scandinavia and compared those data with what they were learning about seismic waves generated by human activity. They focused not on what was already known, but on how scientific exploration might reveal something new.

This brings us to our second authentic science element: Dr. Koper pursued a project to explain something new, driven by a new question—not to prove or reinforce something already reported in a textbook.

At the time of the *Kursk* sinking, seismometers recorded two events; the seismograms of these events differed from the pattern typical of earthquakes, but the two patterns resembled each other, indicating they were of similar origin. The high-frequency whine of a large field of underwater gas bubbles that followed the second event was clearly characteristic of a gas-generating explosion, not a collision. Dr. Koper explained:

> *And so the seismic data, it's kind of technical, but if you have an event underwater, an explosion, you get what's called a bubble pulse, which means the gases and so forth expand and then they contract and expand and contract very rapidly. So it creates sort of a distinctive kind of fingerprint, this rapid oscillation. And we were able to observe that. And that's how we knew it wasn't an earthquake or a collision of some sort. It was actually an explosion.*

Here's our third authentic science element: Dr. Koper's research involved understanding arcane details of seismic waves and how one type of wave differs from another in subtle ways. But he applied that focused study of seismic waves to bigger problems of broad importance.

From the squiggles on the seismograms, Dr. Koper and his colleagues proposed a plausible scenario for the disaster. Two explosions occurred on board the *Kursk.* The first explosion was a small, magnitude 2.2, seismic event. The second was 250 times larger, a magnitude 4.0 event. The seismologists believed that the first explosion, perhaps caused by a torpedo misfiring, breached the hull. The second explosion, coming 135 seconds later with time enough for the sub to sink to the bottom, occurred when a cruise missile exploded with the firepower of a small earthquake. This explosion generated seismic waves heard as far away as Africa.

Their research proved that an explosion, not an underwater collision with an American submarine, caused the accident (Koper et al. 2001). They not only gained new understanding about a fundamental aspect of the natural world—how different seismic events produce different kinds of seismic waves—but also answered a politically sensitive question that had made headlines all over the world.

In June 2002, a year and a half after Koper and his colleagues published their results, the Russian findings were announced. At the conclusion of extensive investigations and salvage operations, Russian official Ilya Klebanov confirmed that a faulty torpedo, not a collision with another sub, was the cause of the explosions (Traynor 2002).

Dr. Koper's work began with a compelling driving question, proceeded through a scientific exploration of something unknown, and yielded an understanding of an issue of wide interest.

Can we make our science classrooms work like Dr. Koper's team? Can our classrooms become places where students make sense of the world in an authentic way—where students ask questions not driven by the teacher or the textbook, where students use the practices of science to make discoveries and construct understanding that go beyond reinforcing already-learned concepts? Can students, like scientists, apply their new understanding to problems in the broader world? We think the answer to all of these questions is "yes."

We suggest an approach for doing this in the "Suggestions for Designing an Activity" section (pp. 253–255). First, in the following text, we offer some ways to help students explore how seismology has been used to see into the Earth.

Authentic Science and the *NGSS*

Dr. Keith Koper's work with the *Kursk* provides a model for students' role in the classroom that parallels the vision of the *Framework* (NRC 2012) and the *NGSS* (as quoted in the following text—we've italicized some words for emphasis).

Dr. Koper asked a question of interest to himself and his colleagues: "At all levels, [students] should engage in investigations that range from those structured by the teacher—in order to expose an issue or question that they would be unlikely to explore on their own ... *to those that emerge from students' own questions*" (NRC 2012, p. 61).

Dr. Koper worked to explain something he didn't already know: "Asking students to demonstrate their own understanding of the implications of a scientific idea by *developing their own explanations* of phenomena ... engages them in an essential part of the process by which conceptual change can occur" (NRC 2012, p. 68).

Dr. Koper knew a lot about seismic waves that he then applied to a new problem within a larger context: "The NGSS focus on deeper understanding of content as well as *application of content*" (conceptual shift 4 in Appendix A; NGSS Lead States 2013).

A JOURNEY TO THE CENTER OF THE EARTH

Like the final minutes of the *Kursk,* the inside of the Earth is hidden from our eyes. Our deepest drilling devices only scratch the outer skin of Earth. Before the early 1900s, Earth's unseen interior was a subject for speculation. For example, Jules Verne's *Journey to the Center of the Earth,* written in 1864, offered a vision of passageways and caverns extending to the Earth's core.

Even today, geology classes are fraught with speculation about what lies underfoot. Students imagine there are caverns deep within planet Earth, filled with water, oil, or lava, or simply empty, awaiting the explorations of the bold adventurer. One eighth-grade student, who realized the "floor" of the ocean is made of rock, asked what would happen if we drilled through the ocean floor: "Would we poke through into empty space and drain the ocean?"

Verne's *Journey to the Center of the Earth* came 16 years before the invention of the seismograph, which brought study of the Earth's interior into the realm of observational science. Although we can't see into the Earth, seismic waves generated by the energy released during an earthquake, a large explosion, or even movement of ocean waves allow us to *hear* into the earth.

After rock ruptures, wave energy speeds through the earth at kilometers per second—about 40 times the cruising speed of a passenger jet. These waves—compressional sound waves being one kind—change velocity, reflect, and bend as they encounter different materials inside the Earth. When the waves return to the surface, we can measure their time of travel and from that reconstruct their pathway and "see" the structures of the Earth's interior. Dr. Koper explains the technique this way:

> *In terms of imaging the geology of Earth's interior, one of the best ways to do it is with seismic waves, with tomography. And the analogy that we use most often is [that seismic tomography is] like a CAT scan. If you go to a doctor you can use electromagnetic waves that make an image of the inside of your body. Or another analogy that is maybe a little bit closer is sonograms. And so, for a fetus, they're doing this imaging, they're basically doing seismic tomography of the fetus. And they use sound waves to do that. (personal communication 2015)*

From tomography, we not only discover the big layers of the Earth—crust, mantle, and liquid core—but we can spot slabs of rock descending into the Earth at the deep ocean trenches. We can see convective structures in the Earth's mantle that drive the movement of Earth's raftlike plates. We can even discover strange islands or mountains at the boundary between the core and mantle.

Understanding *how* we can use seismic waves to "see" the large-scale layering within the Earth is part of the understanding students need to meet *NGSS* performance expectation HS-ESS2-3: "Develop a model based on evidence of Earth's interior to describe the cycling of matter by thermal convection."

A LESSON IN MEASURING WHAT WE CAN'T SEE

Seismic study in real rock is difficult to bring into the classroom. Seismic waves travel fast, and high-speed seismic equipment is sophisticated and expensive. The iSeismometer app, despite its great tech appeal, has a time resolution on the order of tenths of a second, which is not sensitive enough for classroom measurement of seismic wave travel times.

I (Mary) sometimes use Slinky toys to demonstrate compressional waves and old-timey phone cords to demonstrate shear waves. However, visualizing types of waves and wave propagation—a common classroom practice when studying waves in general—is not the same as using seismic wave travel times to reveal information about Earth's interior. The focus of this chapter, and the goal of the activity presented here, is to explore how waves can reveal aspects of the Earth's hidden interior, not to study the nature of waves.

Students sometimes make pie-slice representations of Earth's interior, which are based on seismic studies. But those pictures only represent a model for what we know, not how we figured it out.

Despite the obstacles in doing seismology with secondary students, some simple experiments can be done in the classroom. Students *can* make measurements of unknown and unseen distances that are analogous to measurements made in the Earth's interior.

One seismic technique that doesn't require the processing power of a computer involves measuring the time it takes for a seismic wave to go a distance into the Earth, reflect off a boundary layer, and return (see Figure 12.1). If the velocity of the seismic wave in rock is measured independently of the travel time, then the distance to the layer can be calculated from the travel time and velocity by the following equation:

$$distance = \tfrac{1}{2} \times velocity \times time$$

The division by two accounts for the round-trip distance being twice as long as the distance to the layer.

Students can construct an analogous experiment in the classroom using a string of nonlatex elastic bands and toothpicks like that shown in Figure 12.2. When this contraption is wound up (i.e., twisted) and released, a twisting motion will spiral along the elastic-band chain like a wave. Students can measure the velocity of the wave by measuring the time for a twisting wave to go from one end to the other of an elastic-band string of known length; this is analogous to how researchers measure velocity in rock by measuring travel time through a known distance of rock.

Figure 12.1

Illustration of a seismic wave traveling into the Earth and reflecting off a boundary between two layers located at an unknown depth

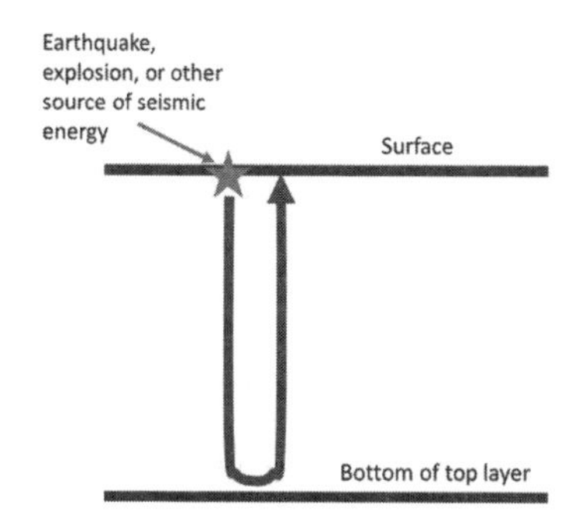

From the time of travel and the velocity of the wave, the distance to the lower layer can be calculated by the equation *distance* = ½ × *velocity* × *time*. Velocity must be measured independently in a separate measurement or experiment.

Figure 12.2

Setup of elastic bands and toothpicks to experiment with wave travel time, velocity, and travel distance in the classroom, simulating measurements with seismic waves in rock

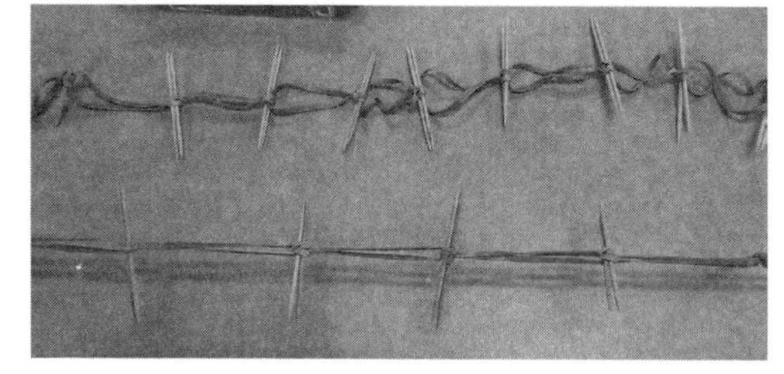

Students can then apply that velocity to determining the length of an unseen string of elastic bands, analogous to measuring the distance from the Earth's surface to subsurface layers.

Safety Note

Have students wear indirectly vented chemical-splash goggles or safety glasses during the whole activity, including setup, hands-on piece, and takedown.

An example lab setup is shown in Figure 12.3, where the long box conceals the true length of a string of elastic bands. To determine the length, students must measure both the velocity of the wave in a separate string of elastic bands whose length they know and the travel time along the string of unknown lengths.

Figure 12.3

Simple lab setup to allow students to use seismic methods to measure an unknown and unseen distance—the length of the nonlatex elastic band string hidden within the long box

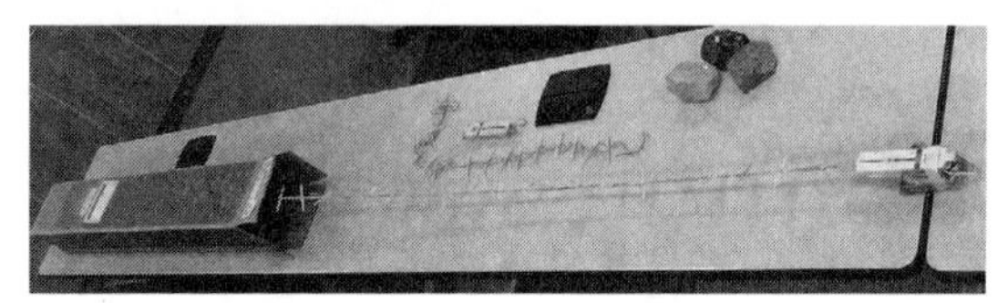

Although the "twisty waves" are not identical to seismic waves, they can be used to determine a distance in a similar fashion. Velocity of these waves depends on the type of elastic bands used, the degree of stretching (thus the spring scale on the right), and the mass and length of the toothpicks inserted into the elastic-band string. Similarly, the velocity of seismic waves depends on the composition, density, and temperature of the rock through which the waves travel.

Students often ask, "What's this separate string of elastic bands for? What am I supposed to do with it?" We need two pieces of information to figure out the depth: the velocity of seismic waves and the travel time of seismic waves. The separate string of elastic bands provides a way to measure the velocity of waves in the string of elastic bands, analogous to lab or field measurement of velocities of seismic waves in rock.

Students also ask, "Is it more accurate to measure the time it takes to go from one end to the other, or should I time how long it takes for the wave to go down and back?" This question provides an opportunity to engage students in discussing the uncertainty in their measurements. There is always some uncertainty in when students start and stop their stopwatch. The longer the travel time, the smaller the proportion this uncertainty is of the total time. We talk about this idea and then I ask, "So which do you think will give better results, the longer or shorter travel time?"

Students can then measure the travel time in the elastic-band string of unknown length, analogous to the measurement of travel times for seismic waves in planet Earth. They have to figure out how to make those measurements. Other measurement challenges might include "It seems like the string is always jiggling—how do I know when a wave comes back?" or "Does the spin direction reverse when the wave comes back?"

Once students perfect their setup and gain a confidence that they can spot the return wave, they can measure the travel time in the "mystery box" elastic-band string.

Elastic-Band Seismicity: Incorporating Engineering Design

You could turn the seismic investigation in Figure 12.3 into an engineering design puzzle by encouraging students to optimize some aspect of the development of the activity:

- Offer a design objective, for example: *Design an elastic-band chain (using nonlatex elastic bands) to optimize how easy it is to see a wave propagate and reflect.*
- Limit the scope of the project by restricting the number of design variables for students to consider, for example: *Choose from among strength of elastic bands, how tightly they are stretched, total number of elastic bands that are connected, number of toothpicks at each node between elastic bands, length of the toothpicks* (or give them a choice between toothpicks or shish kebab skewers).

Students often take nearly half the time available for the seismic elastic-band activity simply learning to see the wave and time its arrival. By engaging them in designing the activity, they have the opportunity both to learn to see the wave and to help create an activity in which the wave is easy to see. I (Mary) fussed with this experiment for quite some time trying to get it to "work." After many design iterations, I tried skewers instead of toothpicks, which slowed the wave sufficiently for me to see it traveling. I felt like an engineer!

From the two measurements—wave velocity and travel time—they can calculate the length of the string of elastic bands. For example, suppose they measure the velocity as 7.2 elastic bands per second. The travel time in the string of unknown length is 8.3 seconds. They calculate that the string is 30 elastic bands long. They can then check inside the box to see how close they got (something that researchers looking into planet Earth can't do).

This experimental classroom investigation includes some elements of authentic science, even though it doesn't directly examine rock in planet Earth. Students, like seismologists, use lab (or field) measurements of wave velocity in a known material coupled with measurements of travel times to determine the length of something they can't see. Like Dr. Koper's study of the *Kursk,* students discover something that they didn't already know.

CONSIDERING DATA FROM A HISTORICAL STUDY: RICHARD OLDHAM AND THE OUTER CORE

Model Development and Revision

Richard Oldham, a pioneering seismologist of the early 20th century, used data from the new array of seismographs to argue for the presence of a core (Oldham 1906). His evidence included the travel times for seismic waves from an earthquake to various points around the globe. By comparing the travel times, he reasoned that Earth's interior had a central area characterized by sharply lower seismic velocities. He inferred the presence of a core.

We present Oldham's averaged data in Figure 12.4 in the form of an activity appropriate for high school students. In this activity, students analyze his data and develop an argument for the existence of a core in an otherwise homogenous Earth. This exercise includes elements of authentic science in that it has students consider real data from a historical study[C9] and provides an opportunity to argue from evidence without constraining your students to a tidy discussion with only a single, reasonable conclusion.

An Untidy Argument From Evidence

Scientific explanations and conclusions are typically presented to students as tidy packages from which the messy work required to gain the hard-won knowledge has been excised. For example, in Figure 12.5 (p. 248), the wrap-up argument for a liquid outer core takes three sentences. This argument appears in one form or another in most geology textbooks that address the Earth's interior.

The problem is, if students encounter scientific research only after it has been sanitized, then their own classroom investigations, which are often messy and confusing, can feel like failure to them. Worse, sometimes we encourage students—either consciously or subconsciously—to believe that the primary goal of their investigations is to confirm and reinforce the textbook answer that we already know is right. Both the crispness of textbook science and our emphasis on confirming answers, can discourage students from diving into confusing results with a will to figure out what their own data mean.

Just as classroom experimental activities often have puzzling or even ambiguous results, initial discoveries in research are often less clear-cut than we make them out to be in wrap-up arguments. This is certainly true with our discovery of the Earth's liquid outer core.

Richard Oldham proposed the existence of a core, but he did not conclude that the core must be liquid. The problem was that he observed the arrival of some S waves in the S-wave shadow zone—although greatly delayed—consistent with presence of a solid core of different composition.

These extra S waves confused initial study of the Earth's interior. Like the shadow of a tree that is not perfectly dark because light can arrive indirectly, the S-wave shadow is

Figure 12.4

High school–level activity based on Richard Oldham's study of Earth's interior

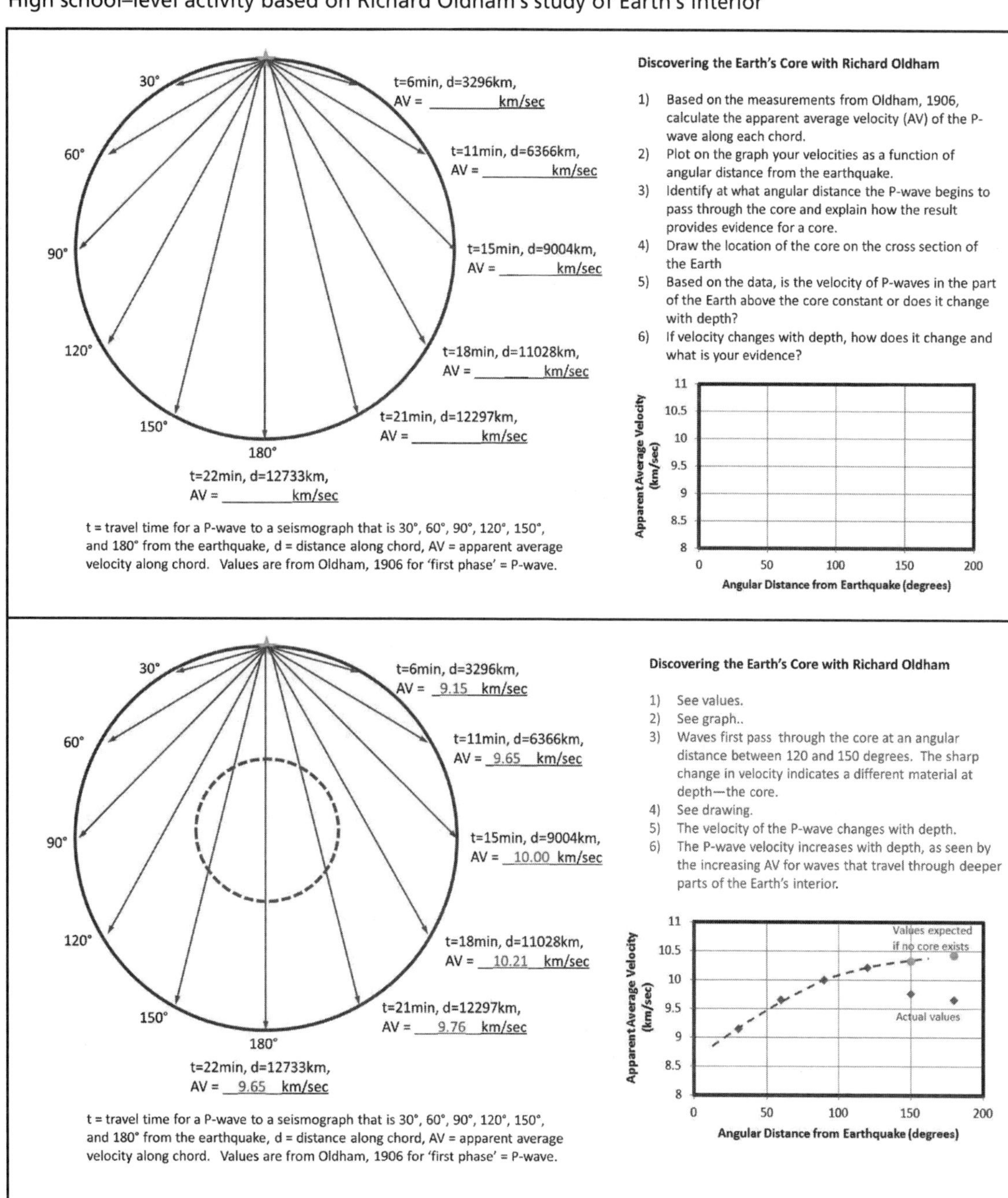

This activity, which uses data from Oldham's study to engage students in a real research investigation, illustrates how new data that do not fit within a preexisting model can be used to modify an old model or create a new one. In this case, the model of a homogeneous Earth gave way to a model of an Earth with a core. The angular values in the figure come from Oldham's original paper. Exact angular values—and size of the core—were refined by later studies.

Figure 12.5

Illustration of the P- and S-wave shadow zones where P and S waves do not show up at seismic stations on the opposite side of the Earth from an earthquake

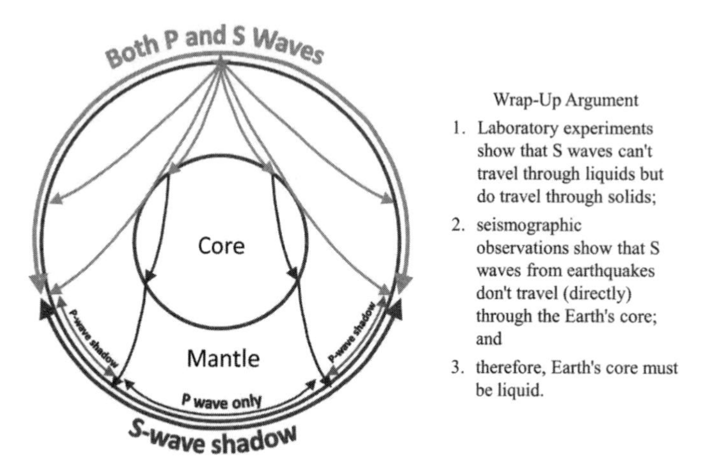

These shadow zones result from a combination of refraction (bending) of the seismic waves as they enter the core and the inability of S waves to pass through the liquid outer core. The wrap-up argument for how we know the Earth's outer core is liquid (from Chapter 9) is included. Like Dr. Koper's investigation of the *Kursk*, and like the simplified tomography measurements in Figures 12.1–12.3, our inference that the Earth's outer core is liquid is based on making two different kinds of measurements (lab experiments and seismic observations) and on applying smaller-scale studies to larger-scale problems.

not perfectly devoid of S waves. S waves can reflect off surfaces, like the bottom of Earth's crust, arriving in the shadow zones in a series of skiplike reflections. What's more, P waves, on entering any solid material such as the inner core or the mantle, partially convert into S waves, generating S waves where we might not otherwise expect them.

Oldham understood that the "second phase" waves he observed in the seismographic record from the shadow zone might not be secondary shear waves (S waves). But he argued that if that second phase was not an S wave, then the seismographic record showed no S wave at all. Either way, the S wave was being affected by passing through a region of significant change in both P and S waves—a core.

He favored the view that the waves did not disappear entirely. Only now in retrospect do we know that the S waves he saw in the seismograms were, in fact, S waves that arrived indirectly through processes such as reflection.

At the time, other workers had proposed a liquid in the Earth's interior, and the presence of S waves that Oldham reported seemed contradictory to that. However, Oldham did not deny the possibility of a liquid core. He argued that the presence of an S wave did not preclude a liquid core because liquids under the immense pressure of the Earth's core might indeed transmit S waves even when materials at the Earth's surface do not. So, the arguments presented in Figure 12.5 were not universally convincing back in 1906 because researchers couldn't be sure that S waves behaved the same in the Earth's core as at the surface.

Discovery of the Earth's core was a very significant moment in the history of the study of Earth's interior, but that understanding did not emerge fully formed in its modern state. The arguments at that time were much closer to arguments from the fog than they were to wrap-up arguments. It took more researchers and more years of research, to confirm that S waves did not arrive directly in the shadow zone, and that they in fact do not travel through liquids even at the high pressure of the Earth's core.

READING SCIENTIFIC LITERATURE: INGE LEHMANN'S PESKY P WAVES AND THE INNER CORE

Understanding science as scientists report it is key to one of the *NGSS* science and engineering practices, Obtaining, Evaluating, and Communicating Information. As noted in the Appendix F matrix for this practice, students in grades 6–8 should "[c]ritically read scientific texts adapted for classroom use to determine the central ideas and/or obtain scientific and/or technical information"; similar wording is used for students in grades 9–12.

Although the original 1936 paper by seismologist Inge Lehmann is beyond most high school students, the following account is within reach:

> *In 1929 a large earthquake occurred near New Zealand. Danish seismologist Inge Lehmann—"the only Danish seismologist," as she once referred to herself—studied the shock waves and was puzzled by what she saw. A few P-waves, which should have been deflected by the core, were in fact recorded at seismic stations. Lehmann theorized that these waves had traveled some distance into the core and then bounced off some kind of boundary. Her interpretation of this data was the foundation of a 1936 paper in which she theorized that Earth's center consisted of two parts: a solid inner core surrounded by a liquid outer core, separated by what has come to be called the Lehmann Discontinuity. Lehmann's hypothesis was confirmed in 1970 when more sensitive seismographs detected waves deflecting off this solid core.* (www.amnh.org/explore/resource-collections/earth-inside-and-out/inge-lehmann-discoverer-of-the-earth-s-inner-core; *see also Mathez 2001)*

An illustration modified from Lehmann's original paper is shown in Figure 12.6 (p. 250). Comparing the text above to this model can provide opportunities for students to interpret the evidence, understand the connection between evidence and conclusion, and argue from evidence in their own words. Students might want to refer to the model shown in Figure 12.5 for reference. Note that students need to figure out which of the observations—made at the points where the numbered paths encounter the surface of the Earth on the model in Figure 12.6 (p. 250)—moved Lehmann to propose a model of

Figure 12.6

Illustration modified from Figure 1 in Inge Lehmann's 1936 report addressing the arrival of P waves in the P-wave shadow zone

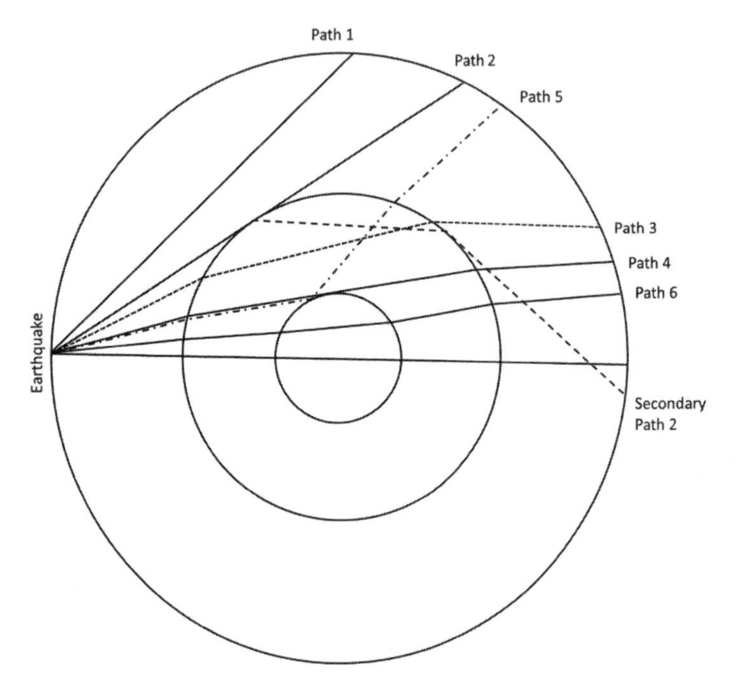

The illustration shows her interpretation of the paths of seismic waves through the Earth that result in P-wave arrivals at different locations on the opposite side of the Earth from the New Zealand earthquake of 1929.

an inner and outer core (the value measured at the end of path 5).

Example prompt questions for students to help guide them through the process, might include the following:

- Where on the diagram from Lehmann's technical report is the area where the unexpected P waves were observed?
- Why weren't the P waves expected there?
- Can you put into your own words what evidence Lehmann found for the presence of an inner core?

In this activity, students don't simply paraphrase the evidence. The text provides an incomplete explanation. Students need to pull in the idea of waves reflecting off of a boundary that marks a density change. They have to compare a textual explanation to a graphical explanation. They have to apply their understanding of wave refraction such as that illustrated in Figure 12.5.

As with the previous illustrations and activities in this chapter, this study provides an example of taking a small bit of information—in this case a deviation from the expected pattern of P waves—and applying it to understanding a bigger picture. Almost all the data fit the model of a liquid core—except for some pesky P waves where they shouldn't be! Lehmann's contribution was to realize that those minor P waves revealed something big—yet another layer in the Earth's inner structure.

FINAL THOUGHTS

Seismic waves join stratigraphy and the stories that rocks tell as key languages by which we read the earth and sky. Using small-scale, seemingly arcane studies of waves, we can answer big questions that previously seemed unanswerable: What's at the Earth's core? Where does lava come from? What makes tectonic plates move?

On Discovering the Earth's Core: Leaving the Realm of Speculation

Richard Oldham wrote the following in his 1906 publication:

> *The object of this paper is not to introduce another speculation [about Earth's interior] but to point out that the subject is at least partly removed from the realm of speculation into that of knowledge by the instrument of research which the modern seismograph has placed in our hands. Just as the spectroscope opened up a new astronomy by enabling the astronomer to determine some of the constituents of which distant stars are composed, so the seismograph, recording the unfelt motion of distant earthquakes, enables us to see into the earth and determine its nature with as great a certainty up to a certain point as if we could drive a tunnel through it and take samples of the matter passed through. The subject is yet in its infancy and much may ultimately be expected of it; already some interesting and unexpected results have come out which I propose to deal with in this paper.*

With seismic data, we can detect nuclear weapons tests, estimate the yield of explosives used by terrorists, look for untapped oil reserves, or find abandoned gasoline storage tanks. With seismic data we can detect thunder, ocean waves, and the shifting of glaciers with the change of climate. Seismic waves have even been used to probe the depths of our Sun.

What's more, seismic data records all kinds of human activities and behaviors. For example, seismic data record variations in background vibrations, which might be used to monitor urban activity or truck traffic, with variations reflecting time of day, weekday/weekend differences, or the changing of seasons. Unexpected variations might indicate new economic activity or preparations for war.

And, of course, seismic data can solve mysterious submarine disasters. An interesting aspect of seismic data that we take advantage of in the "Suggestions for Designing an Activity" section is that a lot of data are freely available to the public, including to classroom students. Dr. Koper conveyed a bit of the excitement that he felt in using publicly available seismic data to solve the puzzle of the *Kursk:*

> *[W]e leave the seismometers on 24/7, because we don't know when earthquakes are going to happen. They're very sensitive and so because they're very sensitive and they are always on, they record all sorts of things, you know, other than earthquakes, that*

we can study and learn about. I think there's just a lot of people that didn't know that and so it's kind of exciting to get that across.

REFERENCES

Foley, J. A. 2013. Earth's core temperature a hellish 6,000 degrees Celsius, a new study confirms. Nature World News. *www.natureworldnews.com/articles/1606/20130426/earths-core-temperature-hellish-6-000-degrees-celsius-new-study.htm.*

Koper, K. D., T. C. Wallace, S. R. Taylor, and H. E. Hartse. 2001. Forensic seismology and the sinking of the *Kursk. Eos* 82 (4): 37–46.

Lehmann, I. 1936. P. *Union Geodesique et Geophysique Internationale, Serie A. Travaux Scientifiques* 14: 87.

Mathez, E. A., ed. 2001. *Earth: Inside and out.* American Museum of Natural History Book. New York: New Press.

Mesa, J. C., R. M. Pringle, and L. Hayes. 2013. Show me the evidence! Scientific argumentation in the middle school classroom. *Science Scope* 36 (9): 60–64.

Michaels, S., A. W. Shouse, and H. A. Schweingruber. 2008. *Ready set science! Putting research to work in K–8 classrooms.* Washington, DC: National Academies Press.

Mining-technology.com. 2013. The top ten deepest mines in the world. *www.mining-technology.com/features/feature-top-ten-deepest-mines-world-south-africa.*

National Research Council (NRC). 2012. *A framework for K–12 science education: Practices, crosscutting concepts, and core ideas.* Washington, DC: National Academies Press.

NGSS Lead States. 2013. *Next Generation Science Standards: For states, by states.* Washington, DC: National Academies Press. *www.nextgenscience.org/next-generation-science-standards.*

Oldham, R. D. 1906. The constitution of the interior of the Earth, as revealed by earthquakes. *Quarterly Journal of the Geological Society* 62: 456–475.

Traynor, I. *Guardian.* 2002. Moscow Finally Accepts Fuel Leak Sparked Kursk Disaster. July 1. *www.theguardian.com/world/2002/jul/02/kursk.russia.*

Tyler, P. E. *New York Times.* 2000. Russians Point to a Collision in Sub Sinking. August 18. *www.nytimes.com/2000/08/18/world/russians-point-to-a-collision-in-sub-sinking.html.*

Verne, J. 1864/1991. *Voyage au centre de la terre [Journey to the center of the Earth].* Translated by Lowell Blair. New York: Bantam Books.

Wadhams, N. 2011. Digging for riches in the world's deepest gold mine. Wired. *www.wired.com/2011/02/st_ultradeepmines.*

SUGGESTIONS FOR DESIGNING AN ACTIVITY

AUTHENTIC QUESTIONS AND REAL SEISMIC DATA

In this chapter, we identified several important aspects of authentic science, including asking questions, figuring out problems that you don't already know the answer to, and putting your results into the context of questions of wider interest. Most scientists recognize the value of doing authentic science in the classroom. For example, Dr. Koper's colleagues ran a master's program at the University of Utah, where science teachers did authentic research projects at the university.

> *We had them write theses and give a defense and go through the whole thing. They didn't necessarily have a science background so we had to pick the projects carefully. But nevertheless they got the sense, I think, of what it's like to gather data and interpret data and try to figure something out that you don't know ahead of time. (Keith Koper, personal communication 2015)*

"[T]ry to figure something out that you don't know ahead of time." Wow. That's quite a different experience from doing experiments to reinforce what you've learned from a textbook. But can students and teachers really ask questions about things that they don't know the answer to—or maybe no one knows the answer to—and then pursue the answer in a meaningful way? How can students get the kind of data needed to address real science questions about the Earth?

When talking about his study of the *Kursk*, Dr. Koper took note of how seismic data are available to everyone:

> *One of the interesting things at the time was the openness of the data, the transparency. This was part of a military exercise that was going on at the time this accident happened. There was sophisticated military monitoring of everything that was going on, but that data was classified, not openly available. And so one of the neat things is we were able to get a tremendous amount of information about the event from the openly available data that anyone, really, could download and use. (personal communication 2015)*

In the following activities, we start with an authentic question from a student (the authentic question from your students might be different) and consider how to approach the question in the classroom using earthquake data available online.

EXAMPLE AUTHENTIC QUESTION 1

In 2011, my students and I (Mary) watched video footage of a Pacific Ocean tsunami rolling over the east coast of Japan. As the water pushed houses, ships, and cars inland like bits of driftwood, questions tumbled out like the flotsam of the tsunami. Tevon asked with immense interest, "Are we having more and bigger earthquakes than we used to?" What an interesting question. Dr. Koper thought so, too. "I totally agree with your student," he said. "That's an interesting question."

By using earthquake data that are available online, you and your students can address the question. Join them in a journey of real discovery, prompted by an authentic question, looking for an answer that neither you nor anyone else has the answer to. Put that small question into the context of a bigger picture, such as high-level disciplinary core ideas such as Earth's Systems and Earth and Human Activity and *NGSS* performance expectation MS-ESS3-2: "Analyze and interpret data on natural hazards to forecast future catastrophic events and inform the development of technologies to mitigate their effects."

To answer this question, students can gather historical earthquake data and make graphs that test a variety of possibilities. For example, they might plot number of earthquakes of various sizes by year for several decades. Is there a trend in the number of earthquakes of magnitude 8 or larger, 7 or larger, 6 or larger? What if you plot the ratio of number of earthquakes over magnitude 7 versus earthquakes of magnitude 5–6 for that same range of years? Is there a trend? What if you average the data over periods of 10 years, providing for less scatter in the data? Do you see a trend now? Is the trend real or a statistical fluke? What other innovative approaches or graphs can you and your students come up with?

EXAMPLE AUTHENTIC QUESTION 2

After spending time observing the patterns in earthquake epicenters and focal depths, one of my students, Jonah, asked a different question, "Are deeper earthquakes generally stronger?" Students can address this question, too. For example, your students could put earthquake data into statistical bins according to depth: earthquakes more than 300 km deep, those 200–300 km deep, those 100–200 km deep, and those less than 100 km deep. They might plot the ratio of the number of earthquakes greater than magnitude 7 to those less than magnitude 7 as a function of depth. Is there a trend? Again, what other graphs or methods can you and your students come up with?

SUMMARY CHECKLIST FOR TEACHER AS PRACTITIONER OF SCIENCE

Find a source of data. For example, the U.S. Geological Survey maintains a searchable database from which the data of interest can be extracted and tabulated into the form you need to address your question: *http://earthquake.usgs.gov/earthquakes/search.*

The purpose of an investigation of this sort is to pursue questions you or your students are interested in and then to step into the fog of what you don't yet know. Enjoy the journey. Gather data, ask more questions, propose possible explanations, and support your ideas with evidence and reasoning. Students love to do this. Your job is part mentoring scientist and part project manager. With no more suggestions or steps for how to proceed, you can figure out an interesting and valid approach. *You and your students can do it together.*

13 STORIES OF PLACES WE CAN'T GO BUT CAN SEE

Distance learning has a long history. For ages, humans have traveled the universe with nothing but light for our spaceship and a mind to guide us. Eighteen hundred years before Columbus, the Greeks not only knew the shape of the Earth but had measured its size (Ferguson 1999). They understood—based on light and shadows—the cause of the phases of the Moon and why the planets seem to wander against the backdrop of stars. They knew roughly how far the Moon and Sun are from Earth. By 250 BC, people knew that Earth was a very small sphere in a big universe.

And, they learned it with light.

Aristarchus of Samos weighed into this great exploration by comprehending the immense size of the Sun and our distance from it (Ferguson 1999). How did he do it? First, he considered the lunar eclipse. He watched as the edge of blackness drifted slowly across the Moon's face. Then, he watched how long the Moon remained sheltered in the coppery darkness of Earth's shadow. The Moon, he saw, was completely darkened for about the same length of time as it took for the leading edge of the shadow to first pass across the Moon's face. This signified to Aristarchus that Earth's shadow was twice the size of the Moon. Using his expertise in geometry, and realizing that the apex angle of the cone of Earth's shadow had to equal the angular size of the Sun in the sky (an assumption that is true if the Sun is much larger than Earth and its distance from Earth is much greater than the Earth's diameter), he calculated that the Moon is a little less than 2.8 times smaller than the Earth. Then, considering the way that objects appear smaller with distance, and knowing the Moon's true size, he calculated the distance to the Moon as about 9.5 times greater than the size of Earth itself (about a third of the modern measurement).

This was an astonishing realization in a world where the Earth was the center and foundation of all things and thus, surely, must fill up most of the space that exists.

Using similar reasoning, he found the Sun to be at least 18 times farther from the Earth than the Moon and more than 7 times larger than the Earth. So the Earth was no longer the largest of all things. It was a small part of a large universe.

Some years back, a friend asked, "Why doesn't the space shuttle run into the stars?" Our friend knew—had he paused to think—that the Sun is our nearest star and is about 93 million miles away, much too far away for the shuttle to run into it. Knowing distances, however, isn't the same as apprehending the relative sizes and spacing of objects in our solar system, or galaxy, or universe.

The question "Why doesn't the space shuttle run into the stars?" is a fundamental and important one, the beginning of science investigation. *How do we know* that stars are far away? How do we know anything about places we can't go? How do we know that the Moon orbits the Earth or that the Earth orbits the Sun? How do we know what stars are made of, or how galaxies wheel through the universe?

The clues are found in light, and what we can see and figure out at a distance. For most of human history, we couldn't travel beyond the Earth. Even today, most of the universe lies far beyond our fastest spaceships. Since we can't travel out to the universe, the universe has to travel to us, in the light we can see with our naked eye, in the light we can see with telescopes, and in the light we can see with spectroscopes.

ENLIGHTENED: READING THE SKY WITH OUR NAKED EYE

Objects move in our solar system, and that movement affects what we see in the sky—one of the significant discoveries made by our "spaceships of light." Movements in space cause the phases of the Moon, which have influenced the development of our calendar and the life cycles of plants and animals. Movements in space cause the seasons, which influence agriculture, commerce, and our parsing of time. These movements are called out in performance expectation MS-ESS1-1 in the *Next Generation Science Standards* (*NGSS;* NGSS Lead States 2013): "Develop and use a model of the Earth-sun-moon system to describe the cyclic patterns of lunar phases, eclipses of the sun and moon, and seasons."

Students can often describe the appearance of the Moon in the sky and explain the pictorial model for the Earth-Moon-Sun system, but they have a harder time connecting the descriptive appearance to the model. We suggest that classroom experiences need to intentionally relate observations of the sky to our conceptual models for motions in space. Aristarchus considered bodies in space from both a ground perspective (he observed a lunar eclipse) and from the distant viewpoint of a space alien watching from above Earth's North Pole (he imagined the cone of Earth's shadow). Likewise, students need to visualize Earth's place in the cosmos both from our earthbound perspective and from the perspective of a distant observer in space.

There is no shortage of great ways to visualize what causes the phases of the Moon, ranging from illustrations in a textbook to online animations. One of our favorites is the classroom classic of using an overhead projector for the Sun (if you can still locate such antiquated equipment!), Styrofoam balls on sticks for the Moon, and the students' own heads for the Earth. If it's not cloudy, students can go outside and use the Sun itself as the Sun. Students can see that the Sun always lights up half the Styrofoam ball, but yet the lit portion as seen from the Earth varies depending on the Moon's location in orbit around their head.

Because indirect lighting keeps the shaded side of the Styrofoam ball from being perfectly dark, students often need some guidance to notice the shadowed and lit sides in this particular model. To cue them in on what to look for, I (Mary) often do a pre-activity in which students walk around stationary half-black/half-white Styrofoam balls. Students draw what they see from different perspectives as they go around the "Moon." Half the ball is always black, but, depending on their perspective, students see a variety of different proportions—phases—of light and dark.

Once they know what to look for and how to interpret what they see, students can see the phases in the Styrofoam balls lit by the overhead projector, or the Sun, and see how the phase changes as the Moon orbits the Earth. Many will quickly encounter lunar eclipses when their head inadvertently comes between the ball and the Sun. A few prompt-questions can help them connect their space-alien perspective (the pictorial model) to their earthbound perspective (observational data). For example, at what position in its orbit, relative to the Earth and Sun, will the Moon appear half lit? At what position does the Moon appear fully lit? At what position do you see only a tiny crescent of the lit side?

Despite the abundance of approaches to visualizing the Moon's phases, students often fail to connect their understanding of orbital motion from the space-alien perspective to their earthbound observations of the sky. For example, in our experience, incoming college students often don't understand the difference between phases and eclipses or how the Moon's phase, its location in the sky, and the time of day must all be related. In addition, they often have misconceptions born out of the models themselves. Pictorial models always have some shortcomings that tend to confuse students. This is not unexpected given that models do not "correspond exactly to the more complicated entity being modeled" and tend to "bring certain features into focus while minimizing or obscuring others" (NRC 2012, p. 56).

Teaching Models in the Classroom

Joe Krajcik and Joi Merritt, in their 2012 article "Engaging Students in Scientific Practices," had this to say about teaching models:

> *Perhaps the biggest change the modeling practice [of NGSS] brings to classroom teaching is the expectation for students to construct and revise models based on new evidence to predict and explain phenomena and to test solutions to various design problems in the context of learning and using core ideas. ...*
>
> *Often in science class, students are given the final, canonical scientific model that scientists have developed over numerous years, and little time is spent showing them the evidence for the model or allowing them to*

> *construct models that will explain phenomena. As a result, often learners do not see a difference between the scientific model and the phenomena the model is predicting and explaining, or the value of the model for explaining and finding solutions. (p. 7)*

As we've talked about in previous chapters, using models involves more than memorizing a picture and then regurgitating it. Memorizing a picture instead of developing, using, and revising a model often leads to hidden misconceptions that become difficult to root out. I (Russ) often ask my education majors to combine a physical model of the Earth-Moon-Sun system with a picture model of Moon phases (e.g., Figure 13.1) and use those two forms of model to interpret or predict earthbound observations of changing phases. I ask them questions about the location of the Moon in the sky at different phases and times of the day.

Figure 13.1

Common illustration of the phases of the Moon, as viewed from above the Earth's North Pole (sizes and distances of bodies are not to scale)

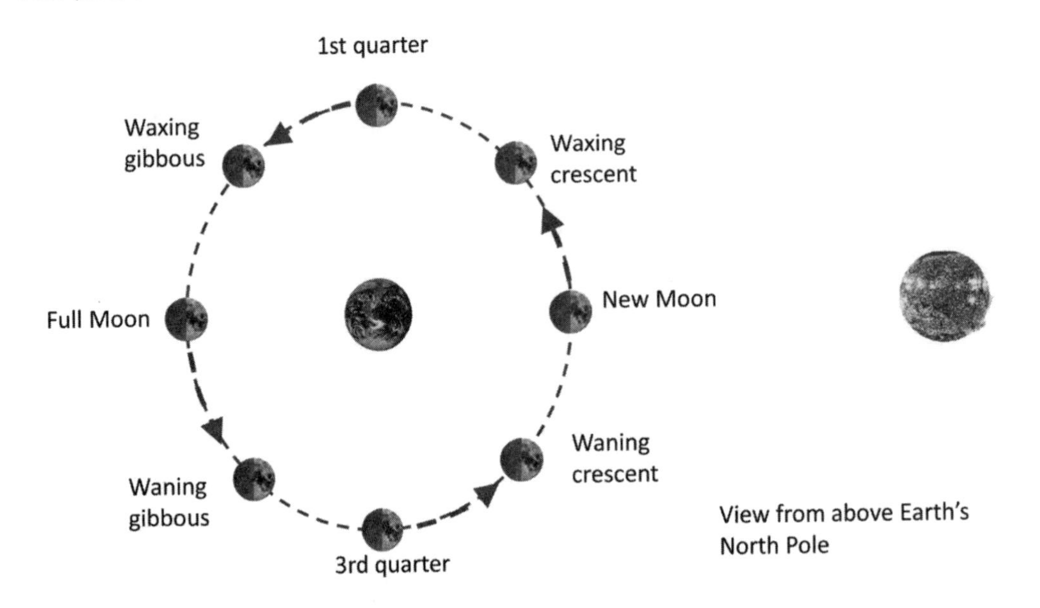

Images of the Earth, Sun, and Moon are taken from near the plane of the ecliptic, not from above Earth's North Pole, and so the details of the surfaces of these bodies do not appear as they actually would if viewed from space.

Source: Images of the Earth, Sun, and Moon courtesy of NASA.

After a number of years, I realized that one wrong answer was very common. The question was "Where in the sky would you look to see the first quarter Moon at midnight?"

At midnight, the first quarter Moon will just be setting in the west (roughly, on average). However, a common answer was "overhead to the north." Since my students and I live in the Northern Hemisphere, "overhead to the north" is not one of the places that the Moon goes. I didn't understand my students' misconception.

Then, one of the confused students pointed at our model on the blackboard—similar to Figure 13.1—and said, "But how can you say I got the question wrong when I can see the first quarter Moon here in the north part of your picture!"

I got it. If you think the picture of the model is a map, then up will be north, and it would seem to be "overhead" as well. Thus, the first quarter Moon appears overhead to the north. So the picture itself was causing confusion. There is no north, south, east, or west in this space-alien view of the Moon phases in which one looks down on the Earth's North Pole. But if students presume that the drawing is a map with a north, south, east, and west, then they develop hidden misconceptions.

It's important to remember that a scientific model is, in part, a conception in our head, and a physical or pictorial portrayal will only be an approximation of that model told in symbols. If students don't understand the symbols and what they represent, then memorizing the picture will not help them derive meaning from the model.

Dangers of Pictorial and Physical Models

Most teachers have experienced cases in which students misunderstood the pictorial or physical models used to convey the idea of a scientific model. One time, I (Mary) had a student who thought that the Earth somersaults pole to pole because of an illustration of the Earth's orbital motion that included an arrow showing the polar orbit of the Landsat satellite.

Kim Kastens (2011) has pointed out that some illustrations of the phases of the Moon show the various *positions* of the Moon as viewed from above the North Pole (like Figure 13.1) but show the *phases* of the Moon as viewed from the surface of the Earth, introducing a source of confusion. This kind of model doesn't make physical sense and is confusing if students don't recognize the symbolism of the two perspectives.

One of the biggest pitfalls of pictorial models is the problem of scale. Students get the wrong idea about the relative sizes of the Sun, Moon, and Earth and about their relative distances from each other. Textbook models are never to scale because, if they were, we wouldn't be able to see the key features that we need to see to understand the model. Suppose that we make the Moon as small as we can and still distinguish the shadowed side and the lit side, say about ¼ inch. Put that in the center of an 8.5 × 11 inch sheet of paper. The Earth would be about the size of a quarter and off the page, over 2 feet away. The Sun would be a giant beach ball nearly 9 feet wide and nearly three football fields away, probably off the school property.

> All pictorial and physical models are merely conceptual illustrations of our mental models, which are, in turn, abstract constructions of reality. Some aspects of pictorial or physical models will be wrong. We teachers need to be aware of what models do and do not show, and help students avoid creating new misconceptions even as they dispel their old ones.

For this reason, we suggest engaging students in a variety of exercises to help connect their conceptual model with reality. These exercises might begin with a question about observational data. For example, why do we see the crescent-shaped phase in the morning or evening but not at midnight? Or, why is the crescent phase seen at sunset or sunrise only near the horizon? In addressing these questions, students connect their mental model of orbital motion, and their diagrammatic representation of the model, to real-world observations.

Another approach to connecting model to observational experience is for students to make testable predictions based on their model. For example, given their model, when will the next first quarter Moon occur? This testing of a model through prediction is in itself an important practice of science. If the model predicts something contrary to observation, then the model—at least our version of it—must be wrong.

In Figure 13.2, we offer a few example applications of the Moon-phase model, put into the format of a murder mystery. Suppose that you're a detective trying to solve a series of murders. Each suspect tells a story of their whereabouts at the time of each murder. One murder took place during a full Moon, another during waxing crescent, and a third during the third quarter Moon. Based on your model for the phases of the Moon (Figure 13.1 and Figure 13.3 [p. 264]), which of the suspects' stories must be false?

Making predictions based on models and then testing those predictions against observations encourages students to realize that their model of reality is a work in progress. Iterative modification of models is an important practice of science. As stated in Appendix F of the *NGSS*, "Students can be expected to evaluate and refine models through an iterative cycle. ... When new evidence is uncovered that the models can't explain, models are modified."

Phases of the Moon are only one part of understanding movements in space. Seasons, the trek of planets across the background of stars, and how the stars seem to swirl around the North Star are other observations that you might ask students to explain with their models of the Earth-Moon-Sun system.

In the activities discussed earlier, we started with the model and then applied and tested that model in a variety of ways, consistent with *NGSS* performance expectation MS-ESS1-1 (quoted earlier in this chapter) and Appendix F (cited two paragraphs previously). In contrast, in original research, people start with the observations and infer the

Figure 13.2

Murder mystery puzzle applying the Moon phase model

MURDERS	ALIBIS		
	SUSPECT 1	SUSPECT 2	SUSPECT 3
MURDER 1: COMMITTED DURING A FULL MOON	I was coming out of the bar at midnight. I remember because the Moon was just rising and I stopped for a moment to watch it, all big and orange.	I was still out on the lake fishing at five o'clock that morning. I was watching the Moon, how the rabbit turns upside down as it crosses the sky. By the time we left it was approaching the western horizon.	I was on a flight headed to New York. We got some sunlight coming in the windows from behind us about sunset, and ahead of us I could see the Moon sitting on the horizon.
MURDER 2: COMMITTED DURING A WAXING CRESCENT MOON	I was with my girlfriend out at the lake, sitting on the dock watching the sunset. After it started to get dark, we noticed that there was a thin sliver of the Moon still hanging over the horizon, and we talked about how sometimes it holds water and sometimes it doesn't.	I was with my boyfriend at the lake, and we'd been out fishing all night. The Sun hadn't risen yet, but the crescent Moon was just poking above the horizon when we headed back to shore.	I came out of the bar around midnight and headed home and it was really dark. I remember wondering why I couldn't see the Moon, because the sky was clear and I'd heard that the new Moon was over a few days before.
MURDER 3: COMMITTED DURING A THIRD QUARTER MOON	We camped all night that night back in the park and then hiked out the next day. It was like noon before we got back to the parking lot. I happened to glance over to the hills in the west, and there was the Moon sitting on the top of the hills. It was weird having the Moon up during the day, like it didn't know that it was daytime and had forgot to go to bed.	We quit fishing early that night. About three o'clock I'd guess. The Moon was just rising, but we were in a hurry because I had to be at work early that day and so we couldn't stop to watch.	I was already home by sunset that night. I remember because my porch has a view toward the east and I stopped for a while to watch the Moon rise.

This puzzle has three murders with three suspects for each murder, and each suspect gives an alibi for each murder. Three of the alibis in the puzzle contain factual errors in reference to the phase and location of the Moon. Can you figure out which stories are wrong, based on your model for the phases of the Moon, and reveal which suspect is guilty of each murder? Visit the Extras page (*www.nsta.org/learningtoread*) to see if your choices are correct.

Figure 13.3
Illustration of a model for the phases of the Moon, showing the time of day at different locations on the Earth (e.g., midnight on the side of Earth away from the Sun and noon on the side of the Earth toward the Sun; sizes and distances of bodies are not to scale)

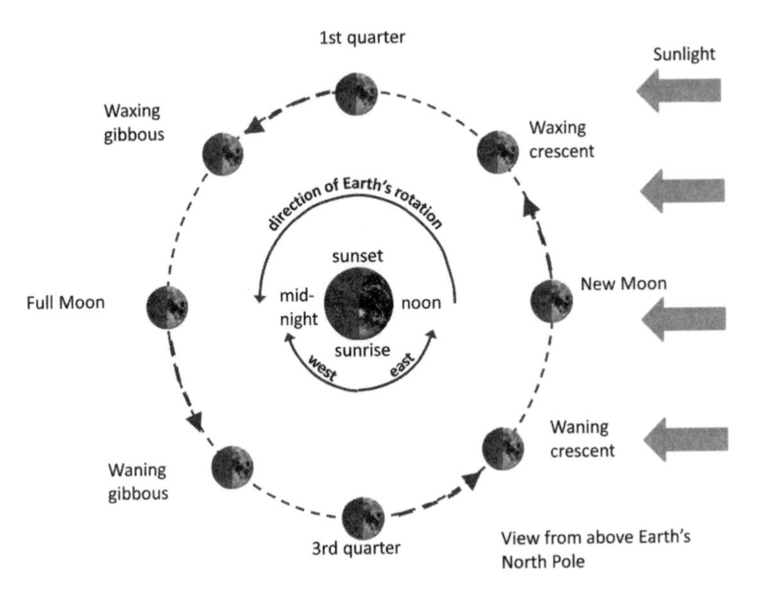

Although there is no north, south, east, or west in this picture, there is an east and a west as viewed from the Earth's surface, and these directions are indicated by the arrows. As viewed from above the North Pole, Earth rotates counterclockwise, or toward the east. Analogously, "toward the west" on the Earth's surface corresponds to a clockwise direction. This model can aid in visualizing where in the sky the Moon will be during different phases and times of the day. For example, at midnight on the side of the Earth away from the Sun, the first quarter Moon will just be setting because it will appear on the western horizon as the Earth rotates toward the east. If the Moon were full at this same time of day, the full moon would appear overhead, halfway through its journey across the sky. A new Moon would be on the other side of the Earth.

Source: Images of Earth and Moon courtesy of NASA.

model. Starting with the observations and constructing a model engages students with a different set of practices: identifying patterns and conceiving how to explain all the observations.[C37] We offer a way to start with observations in building a model for phases of the Moon in the "Suggestions for Designing an Activity" section (pp. 272–274).

ENLIGHTENED: READING THE SKY WITH A TELESCOPE

Most of us probably know that Galileo proved that the Sun is at the center of our solar system and that the Earth and other planets revolve around it. We know that Galileo spent the last few years of his life under house arrest because of this belief, a punishment decreed by the Catholic Church. Some of us may even know that it took until 1992 for the Catholic Church to fully recant its actions (Cowell 1992).

We think we're so much wiser now, knowing and accepting that Galileo proved the Earth orbits the Sun. Yet most of us don't know how he did it. If we accept that the Earth revolves around the Sun, yet we don't know how we know that, then our belief is based as much on faith as was the former stance of the Catholic Church.

Developing and Using Models is one of the *NGSS* science and engineering practices, and many of the performance expectations involve explaining or describing some aspect of the world in terms of models. However, to truly understand a model or theory, we need to understand the practices of science that led to our acceptance of it. Those practices of science begin with observation, not with memorizing and using models.

One proof that the Earth orbits the Sun comes from the parallax of the stars, which we talked about in Chapter 9. However, parallax is not visible to the naked eye, and it was not even measurable by telescope until gradual improvements in telescopes led to the first measurements of parallax in 1838 (Bessel 1838). So Galileo didn't have that kind of evidence. But, Galileo's telescope was powerful enough to see the phases of Venus. His observations of the phases of Venus became one of his key lines of evidence that Venus orbits the Sun, not the Earth. Let's go through his reasoning.

Nicolaus Copernicus had proposed a Sun-centered model of the solar system in the early 1500s, but at the time of Galileo the Ptolemaic model for the solar system was still widely believed (Ferguson 1999). The Ptolemaic model put the Earth at the center of the solar system, with the planets and the Sun orbiting it.

Figure 13.4 shows a common modern misconception of the Ptolemaic (Earth-centered) model for the motion of the Sun and Venus around Earth. Even the ancient Greeks knew this model was wrong. Think for a moment to see if you can figure out how the Greeks could know this, even without a telescope.

Figure 13.4

Illustration of a common misconception of the Ptolemaic (Earth-centered) model of the solar system, with only the Earth, Sun, Venus, and the Moon shown (sizes and distances of bodies are not to scale)

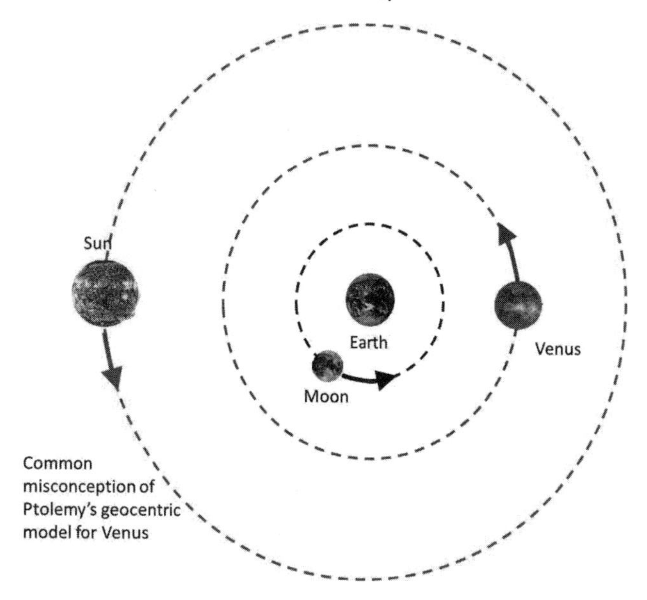

Even without a telescope, the ancient Greeks could exclude this model on the basis of observation. Can you think of a simple observation that disproves this model?

Source: Images of Earth, Sun, Venus, and Moon courtesy of NASA.

Figure 13.5

Illustration of the Ptolemaic model of the solar system, showing only the Earth, Sun, Venus, and Moon (sizes and distances of bodies are not to scale)

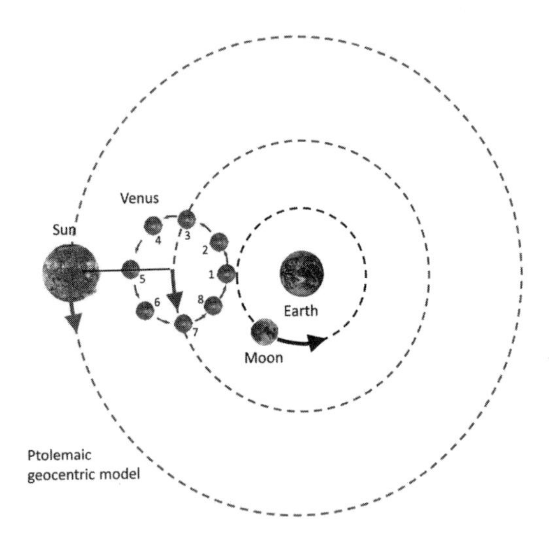

The orbits of the Sun and Venus are linked so as to always be on the same side of the Earth. The Ptolemaic model was complex, with odd, looping patterns to the orbits of Venus and other planets. The complexity makes this model less appealing to modern minds. However, without the use of a telescope, the general features of the model were difficult or impossible to disprove on the basis of observation.

Source: Images of Earth, Sun, Venus, and Moon courtesy of NASA.

Figure 13.6

Simplified illustration of the Copernican (Sun-centered) model of the solar system, showing the Earth, Sun, Venus, and Moon (sizes and distances of bodies are not to scale)

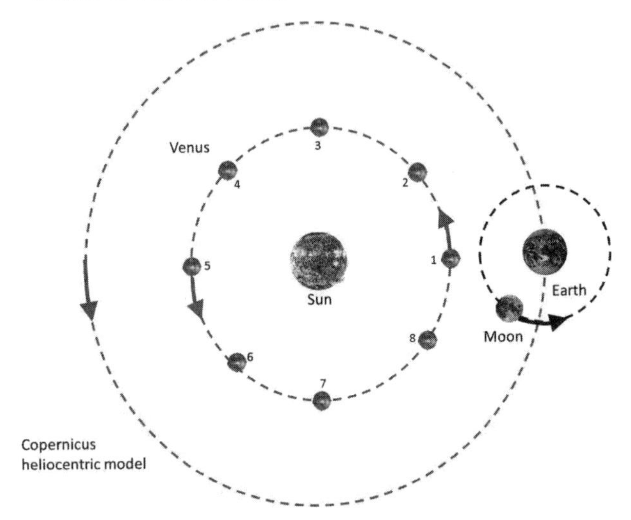

Source: Images of Earth, Sun, Venus, and Moon courtesy of NASA.

In the model shown in Figure 13.4, Venus is sometimes on the far side of the Earth from the Sun. Thus, it would appear high overhead during the middle of the night. Yet, Venus was never observed high overhead in the middle of the night. Venus is only seen near the Sun, and thus is only visible (to the naked eye) near sunset or sunrise when the bright Sun has set or not yet risen (thus, we call Venus the evening star or the morning star).

The real Ptolemaic model is better represented by Figure 13.5. We can compare this model, as Galileo did, to the heliocentric model of Copernicus, shown in Figure 13.6.

For each model, it's possible to predict how the phases of Venus should vary with time. For example, you might encourage your students to draw predicted phases at each of the positions for Venus shown by the numbers 1–8 in the two models in Figures 13.5 and 13.6. Students might draw something like what's shown in Figure 13.7. The actual phases, observed by Galileo, are shown in Figure 13.8.

Figure 13.7

Example illustration of the phases of Venus predicted from the geocentric and heliocentric models of the solar system shown in Figures 13.5 and 13.6

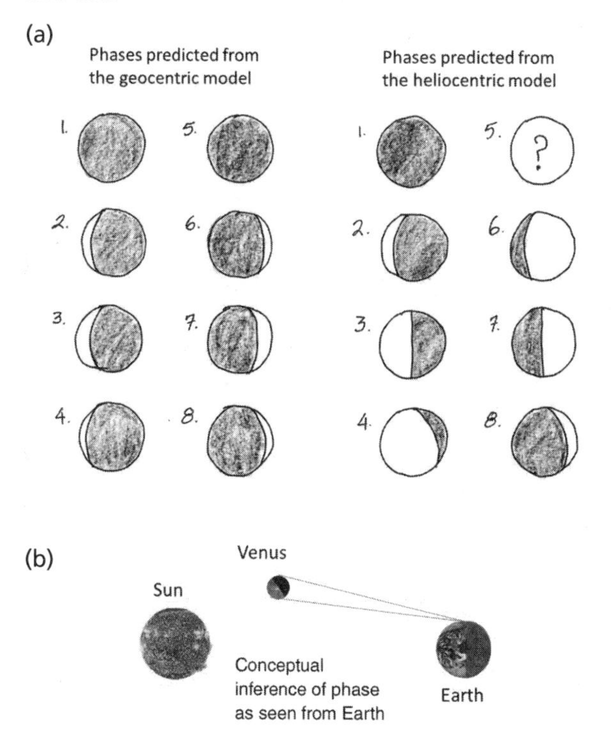

Source: Images of Earth, Sun, and Venus courtesy of NASA.

Figure 13.8

The phases of Venus as illustrated by Galileo (top) and as photographed by Efrain Morales Rivera (bottom)

Galileo's illustration is based on his telescopic observations (pre-camera), published in *Il Saggiatore* (Rome: Giacomo Mascardi, 1623). The photos on the bottom were taken by Rivera, an astronomy photographer who viewed the phases of Venus through a telescope.

Source: Galileo illustration, *http://library.lehigh.edu/omeka/exhibits/show/heavenlyspheres/galileo/saggiatore;* Rivera photographs, *www.jaicoa-observatory.com.*

Based on the phases of Venus predicted by each model, and the phases observed by Galileo and later researchers, students can argue from evidence to address the question "Why are we convinced that Venus orbits the Sun instead of the Earth?"

The size of Venus as observed by Galileo also changes with the phases. You might encourage students to explain how the size change makes sense in terms of the heliocentric model.

ENLIGHTENED: READING THE SKY WITH A SPECTROSCOPE

In the mid-1800s Auguste Comte, a French philosopher who believed that science must be based on observation, made the claim that it was certain that we could never know what stars are made of because we can't go there (Ronan 1971). Comte's claim seemed

Figure 13.9

Model of the concept behind spectroscopy

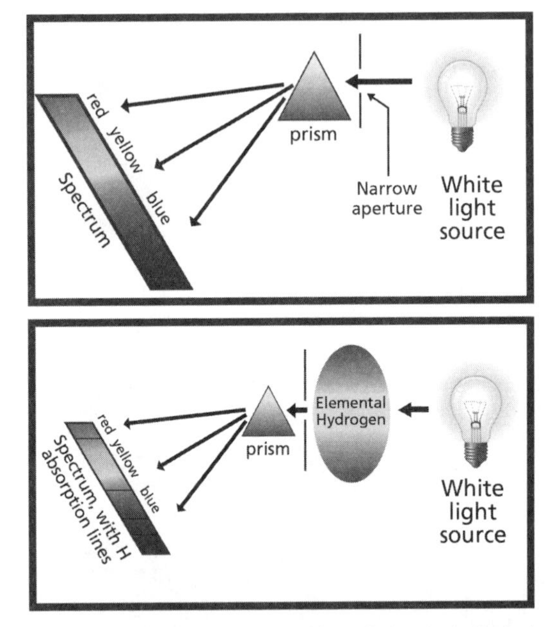

The electromagnetic spectrum can be split into its individual components of different wavelengths. Different materials absorb or emit specific wavelengths of light, allowing us to identify which elements are present in a particular material or star.

reasonable at the time—since we can't go to the stars, we can't measure the composition of the stars. But the development of spectroscopy was already laying the foundation for bringing the stars to us. Because the spectrum of a star is composed of the unique spectral character of each element in the star, we gained the ability to analyze the composition of stars and other objects that are too far away to ever sample.

The concept of spectroscopy is fairly straightforward: Light is composed of many different wavelengths of an electromagnetic spectrum, and we can split off those wavelengths to study them individually. Each element absorbs and emits very specific wavelengths, and so by studying light that has passed through the atmosphere of a star, or through some other material, we can determine what elements are in the star or the material. This concept is shown in Figure 13.9.

The invention of the spectroscope not only brought the heavens to us but it provided the means to study and understand many other aspects of the earth and sky. Spectroscopy is used to analyze the composition of rock, water, soil, and air—for example, the Curiosity rover on Mars carries a variety of spectroscopes for examining the composition of rock and gas. Without an understanding of the spectrum and how different wavelengths of light are either absorbed or not absorbed by the atmosphere, we would have no means to understand or predict the "greenhouse effect" and its possible impact on global climate.

HOW WE KNOW IS MORE IMPORTANT THAN *WHAT* WE KNOW

Sometimes the complexity of scientific theories and models obscures the fact that science is ultimately about making sense of what we see in the world around us. We don't need to know quantum mechanics to understand that light is absorbed differently by different materials. For example, we know that visible light does not pass through a slab of rock. We know this because we can't see through even a few centimeters of rock. However, we know that light does pass through air because we can see through miles of air.

In an effort to connect these simple observations with the broader scientific model of the electromagnetic spectrum, I (Russ) sometimes pose the following question to students: "What is one part of the spectrum that is not absorbed by our atmosphere?" The question follows discussion of the electromagnetic spectrum and experimentation with flashlights, lasers, and colored filters to understand the ideas of light transmission and light absorption. The goal of the question is not to prompt a memorized answer but rather to encourage students to think about the implication of the question and what simple observations might bear on the answer. All of my students know from their own personal experience that visible light passes through our atmosphere; otherwise none of us would be able to see!

Even so, my students often have no answer to this question. They think that I'm asking for some "memorized fact" type of answer, and they don't have one. That expectation reveals how we often test science learning—asking for memorized facts or explanations based on models.

Rather than asking students to recall scientific information—such as what are stars made of or what's an element found in the atmosphere of a distant planet—we can engage students in exploring how we figure out those answers with spectroscopic data. The two short activities in the following subsections might provide inspiration for how you can do that in the classroom.

Example Spectroscopy Puzzle 1: Measuring the Composition of a Star

An example stellar spectroscopy puzzle is shown in Figure 13.10; students are asked to figure out which elements are in the star Sirius. The blue-white star Sirius has among the simplest spectral signatures, particularly if we limit ourselves to only the strongest spectral lines and only those lines in the visible light range as in Figure 13.10. You can develop much more complex puzzles if you choose. Most stars will contain more elements than Sirius because they formed from the debris of multiple generations of precursor stars, which produced many elements by nucleosynthesis.

Figure 13.10
Example spectroscopy puzzle 1

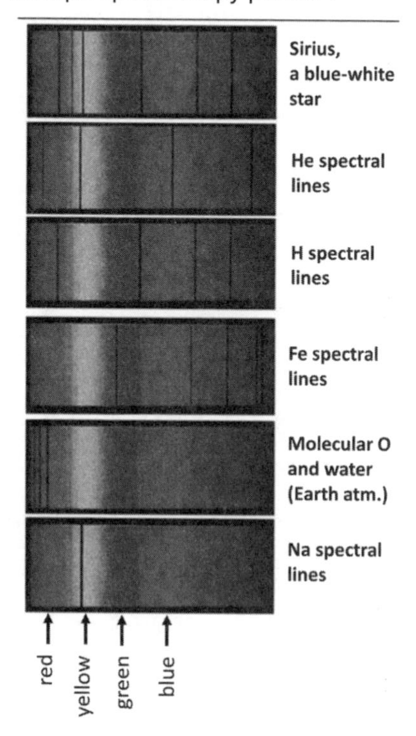

This illustration shows the spectrum of Sirius, a blue-white star, compared with the spectra for a few elements. Can you figure out which elements are in the star Sirius?

Sources: The spectrum for Sirius is modified from a lithograph published in 1870, as found in the *Cambridge Illustrated History of Astronomy*, edited by Michael Hoskin (Cambridge, U.K.: Cambridge University Press, 1997). Spectral lines are simulated from data in the *CRC Handbook of Chemistry and Physics: A Ready-Reference Book of Chemical and Physical Data*, 63rd ed., edited by Robert C. Weast (Boca Raton, FL: CRC Press, 1982), and from *Stars and Their Spectra: An Introduction to the Spectral Sequence* by James B. Kaler (Cambridge, U.K.: Cambridge University Press, 1989).

Figure 13.11

Example spectroscopy puzzle 2

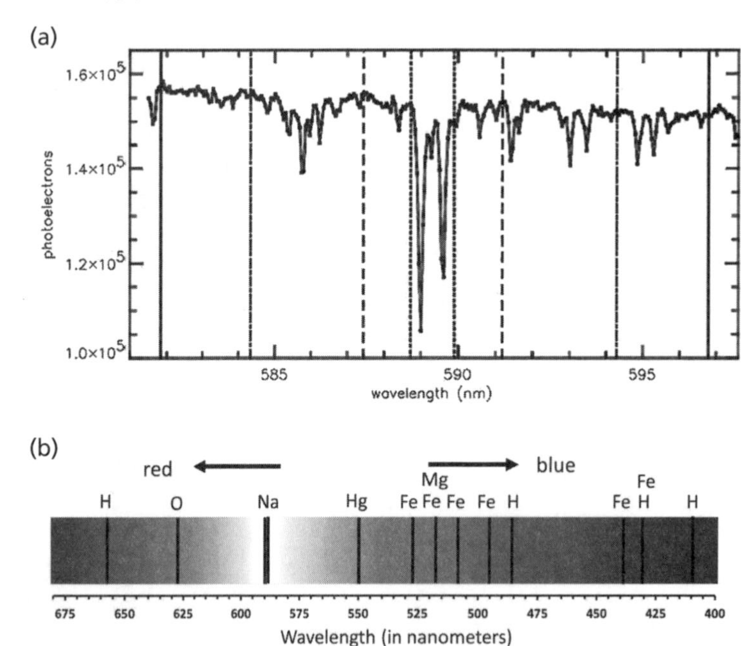

The graph in the top part of the figure comes from the paper "Detection of an Extrasolar Planet Atmosphere" by David Charbonneau, Timothy M. Brown, Robert W. Noyes, and Ronald L. Gilliland (2002). The *x*-axis plots wavelength of light in nanometers (billionths of a meter), and the *y*-axis plots a measure of the amount of light arriving on Earth at each wavelength. The graph is centered on the characteristic wavelength of the element that they studied. The bottom part of the figure shows a spectrum with selected elemental lines.

Can you figure out which element Charbonneau and his colleagues discovered in the atmosphere of a distant world?

To solve this simple science reasoning puzzle, students need to pay attention to the scale on each of the figures and identify which element's absorption band is at the same wavelength as the absorption bands at the center of the graph. The strongest absorption in the center of the graph occurs at just under 590 nm wavelength. By comparing it to the spectrum, that absorption is from sodium (Na).

Example Spectroscopy Puzzle 2: First Measurement of the Atmosphere of an Exoplanet

Exoplanet is the name we give to a planet outside our own solar system (that is, beyond the eight planets plus several dwarf planets that orbit our Sun). Despite the worlds of *Star Wars* and *Star Trek* and many other science fiction stories, humans had never observed a world outside our own solar system until the mid-1990s.

It took humans a bit longer to make measurements of the atmosphere of an exoplanet. The first spectroscopic measurements of an element in the atmosphere of a planet outside our solar system were reported by David Charbonneau and his colleagues in 2002. The puzzle in Figure 13.11 has been simplified from that research; students are asked to figure out which element Charbonneau and his colleagues discovered.

REFERENCES

Bessel, F. W. 1838. Bestimmung der Entfernung des 61sten Sterns des Schwans. *Astronomische Nachrichten* 16: 65–96.

Charbonneau, D., T. M. Brown, R. W. Noyes, and R. L. Gilliland. 2002. Detection of an extrasolar planet atmosphere. *The Astrophysical Journal* 568 (1): 377–384.

Cowell, A. *New York Times.* 1992. After 350 years, Vatican Says Galileo Was Right: It Moves. October 31. *www.nytimes.com/1992/10/31/world/after-350-years-vatican-says-galileo-was-right-it-moves.html.*

Ferguson, K. 1999. *Measuring the universe: Our historic quest to chart the horizons of space and time.* New York: Walker Books.

Kastens, K. 2011. Bad diagrams. *Earth and Mind: The Blog. http://serc.carleton.edu/earthandmind/posts/bad_diagrams.html.*

Krajcik, J., and J. Merritt. 2012. Engaging students in scientific practices: What does constructing and revising models look like in the science classroom? *Science Scope* 35 (7): 6–10.

National Research Council (NRC). 2012. *A framework for K–12 science education: Practices, crosscutting concepts, and core ideas.* Washington, DC: National Academies Press.

NGSS Lead States. 2013. *Next Generation Science Standards: For states, by states.* Washington, DC: National Academies Press. *www.nextgenscience.org/next-generation-science-standards.*

Ronan, C. A. 1971. *Discovering the universe: A history of astronomy.* New York: Basic Books.

SUGGESTIONS FOR DESIGNING AN ACTIVITY

CONSTRUCTING A MOON PHASE MODEL FROM OBSERVATION

The study of the phases of the Moon provides an opportunity to engage students in using a model to explain how things work and predict future observations. We offered some examples of this approach earlier in this chapter. An equally valid approach to learning phases of the Moon is to allow students to make initial observations and use those observations to construct a model.[C37] The following activity can give students an understanding of how people thousands of years ago figured out the Moon's phases and what these phases tell us about our universe.

Students can construct a model by making a few simple observations of the phases of the Moon over a period of a month. Students might observe and document the following:

- *Which phase is the Moon in?* This observation can be recorded by drawing a picture that shows how much of the visible Moon is lit on each particular day.
- *Where is the Moon in the sky?* This observation can be reported by drawing a half circle to represent the horizon and sketching the Moon at its observed location. The example data record on page 274 shows only the east-south-west position of the Moon and not its angle, or altitude, above the horizon.
- *Where is the Sun?* This observation can be recorded analogous to the position of the Moon in the sky. In the example data record, observations of the Moon are only made near sunrise or sunset to simplify the location of the Sun.
- *Which side of the Moon is lit—the side toward the Sun or away from the Sun?* This of course won't change—only the side toward the Sun is lit. However, by making this observation, students can come to understand how ancient people figured out that the Moon reflects the Sun and does not provide its own light.
- *When you make your observation, is the Sun to the right of the Moon or to the left of the Moon as you face toward the south?* If you are facing toward the south, then the position of the Moon relative to the Sun will be the same as in the

Moon phase model diagram as viewed from above the North Pole, as in Figure 13.1 (p. 260). If you live in the Southern Hemisphere, you might face toward the north to see the Moon, and these relationships will be reversed.

- *What is the angle between the Moon and the Sun?* This can be determined by pointing an arm at each celestial body and measuring the angle between your arms. Younger students (eighth grade) might simply estimate the angle using a protractor. Older students (high school) might determine the angle using trigonometry, as shown in Figure 13.12.

An example data record beginning September 9, 2015, is shown in Figure 13.13 (p. 274). A full lunar phase model can be made by putting all of the Moon positions shown in the data record into one drawing. That model will resemble Figure 13.1 for observers in the Northern Hemisphere.

Figure 13.12

Using measurements and trigonometry to determine the angle between the Moon and the Sun

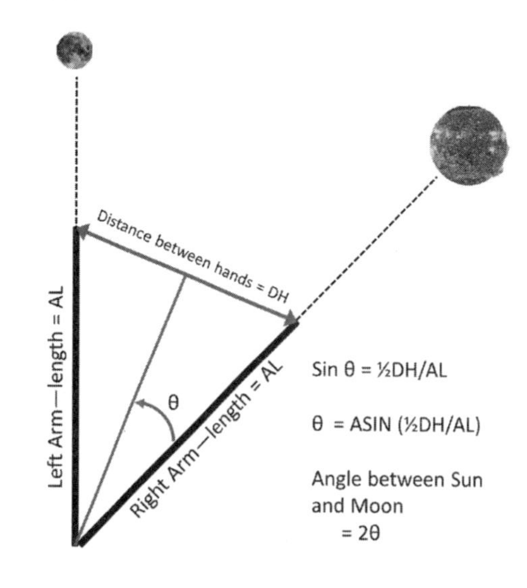

Source: Sun and Moon images courtesy of NASA

Figure 13.13

Example observational data record from which a lunar phase model can be constructed

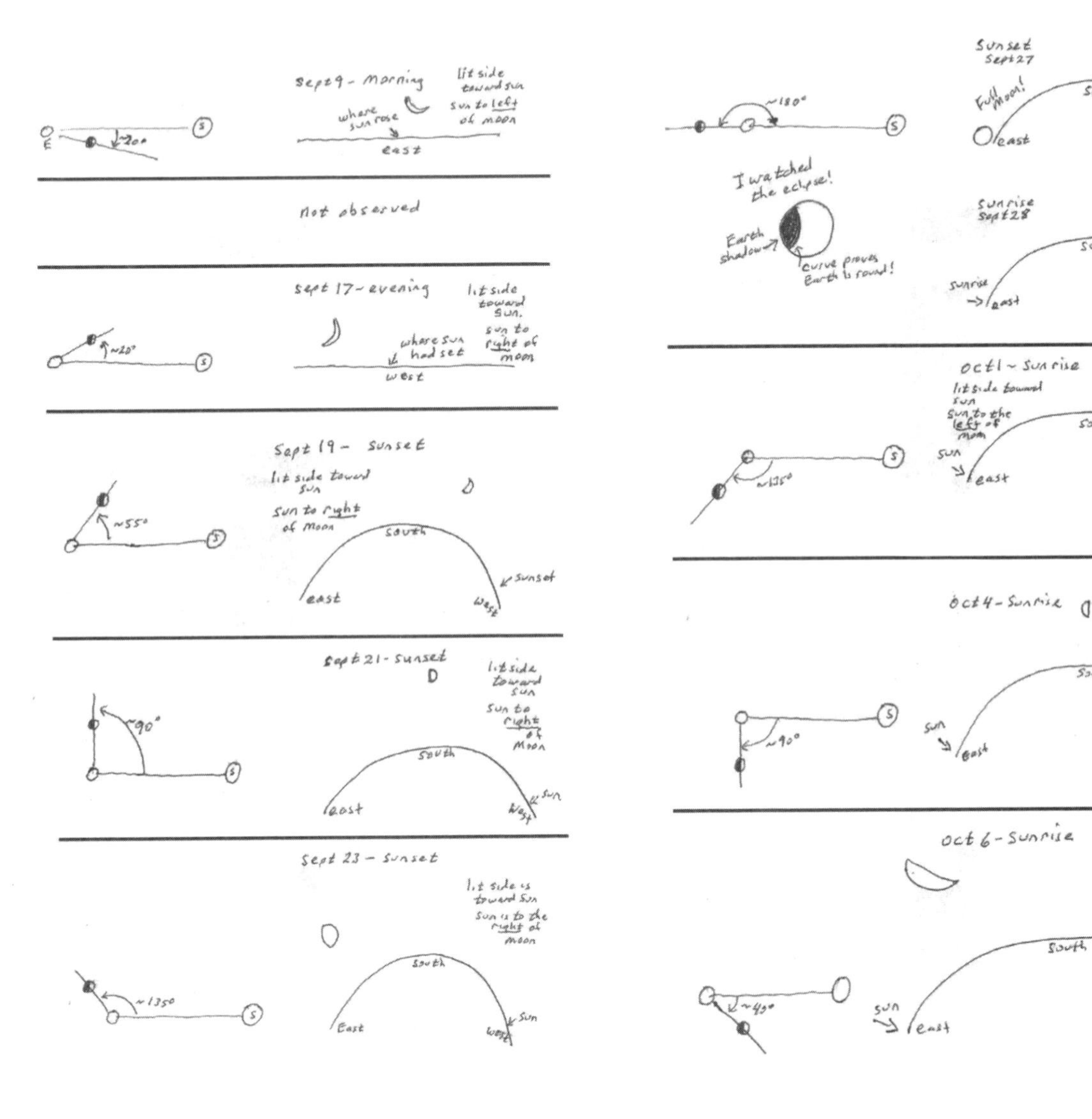

STORIES TOLD BY ATOMS

One February day, I (Mary) was out on a frozen pond with a group of Minnesota high school students drilling a hole through nearly 3 feet of ice to sample the water underneath. Our goal was to explore questions that we didn't know the answer to ahead of time, as part of a community learning center project to connect classroom learning to the outside world. One of those questions was "How does the quality of water in the regional ponds change during the winter freeze-up?" We sampled water from the fall through winter as the ice grew progressively thicker. By late February, the ice had reached the limits of our 3-foot ice auger.

One of the water quality measurements that jumped out for students was the progressive increase in the concentration of calcium in the water. Where was the extra calcium coming from? How could calcium be flowing into the pond when everything was frozen over?

Of course, calcium wasn't flowing into the pond. The amount of calcium was staying the same, but it was being increasingly concentrated in the liquid because the ice rejected the calcium as it froze. "Look how the concentration changes," one student observed, pointing at our graphs back in the lab. "It doesn't change with air temperature, or which month it is, but the thicker the ice is, the higher the calcium concentration!"

Geochemical cycling, the idea that atoms in Earth systems move and get redistributed (often unevenly) is an important part of the foundation for the *Next Generation Science Standards* (*NGSS;* NGSS Lead States 2013), as shown in this grade band endpoint for ESS2.A in *A Framework for K–12 Science Education: Practices, Crosscutting Concepts, and Core Ideas* (*Framework;* NRC 2012):

> *All Earth processes are the result of energy flowing and matter cycling within and among the planet's systems. This energy is derived from the sun and Earth's hot interior. The energy that flows and matter that cycles produce chemical and physical changes in Earth's materials and living organisms. (p. 181)*

The *Framework* also says that the "ability to examine, characterize, and model the transfers and cycles of matter and energy is a tool that students can use across virtually all areas of science and engineering" (p. 95).

Geochemical cycling is readily adapted to classroom-friendly experimental and field activities. Even so, we know of few teachers who deal with it in middle or high school earth science. When taught, it's often limited to a pictorial portrayal of the carbon cycle, usually as part of a life science lesson.[C38]

CHAPTER 14

FINGERPRINTS OF THE EARTH

Geochemical cycling and its companion process, geochemical differentiation, have an impact on far more than just the biosphere. They bear on understanding how ore deposits form, what drives ocean circulation, how pollutants migrate, and how we can learn about planetary interiors and processes based on the composition of lava that reaches the surface. Using the unique geochemical signatures of materials—the "fingerprints" of the materials—archaeologists can map the migration of ancient people based on tooth chemistry, environmental geologists can track down who dumped a pollutant into the air thinking the wind would hide their guilt, and a detective can figure out where a suspect was prior to a murder based on the composition of the mud on his boots.

Geochemical differentiation has even had an impact on the politics of human societies, both modern and ancient. When people in the Fertile Crescent (where the food was grown) had to import copper from the southern Arabian Desert and tin from the Taurus Mountains of Turkey in order to manufacture bronze, it forced cooperation between groups of people who might have otherwise ignored each other or been at war. One might argue that the entire idea of cooperation between distant groups of people found its origins in the uneven distribution of resources. The resulting trade encouraged other types of interactions as well, providing an avenue for exchange of a wide range of culture and ideas.

This dependence of humans on the geochemical cycling and differentiation of earth materials is called out in the *NGSS* disciplinary core idea progressions (NGSS Lead States 2013); for example:

- Humans depend on Earth's land, ocean, atmosphere, and biosphere for different resources, many of which are limited or not renewable. Resources are distributed unevenly around the planet as a result of past geologic processes. (Appendix E, ESS3.A for grades 6–8)
- Resource availability has guided the development of human society and use of natural resources has associated costs, risks, and benefits. (Appendix E, ESS3.A for grades 9–12)

These disciplinary core ideas encompass the entirety of global geochemical cycling, a rather high level to begin a classroom investigation. Although teachers could engage students in arguing from evidence or interpreting models as they focus on these large-scale disciplinary core ideas, we argue in this book, and in the following text, that it's better to begin at a more basic level, by asking questions, designing and doing experiments, graphing, interpreting results, and so on.

Writers of the *NGSS* recognized and discussed the idea that for students to develop an understanding of planetary-size cycles, teachers would need to engage students in activities at the other end of the spectrum—the classroom-size experiments and explorations

Geochemical Differentiation and Cycling

Geochemical differentiation and cycling emphasize different aspects of the complex processes involved in movement of matter on a planetary scale. The word *differentiation* comes from the idea of "make different" and emphasizes the ever-changing character of a differentiating world. In contrast, the word *cycle* emphasizes the steady-state nature of geochemical processes. Both processes can be understood through partitioning and mass balance, as explored in this chapter. A schematic illustration of each of these ideas is shown below.

One of these ideas is not "wrong" and the other "right," but rather each focuses our attention on a different aspect of geochemical processing. On shorter time frames, many chemical components maintain a rough steady state, like water cycling from ocean to mountains and back again to the ocean, or carbon cycling from atmosphere, to life, and back to the atmosphere through metabolic processes. On longer time frames, compositions change, like the changing composition of the Great Salt Lake during cycling of water discussed in this chapter, or the change in carbon concentration in the atmosphere and other reservoirs that contributes to climate change as discussed in Chapter 4.

The different emphases call to mind the *NGSS* crosscutting concept Stability and Change: Cycling emphasizes stability, differentiation emphasizes change.

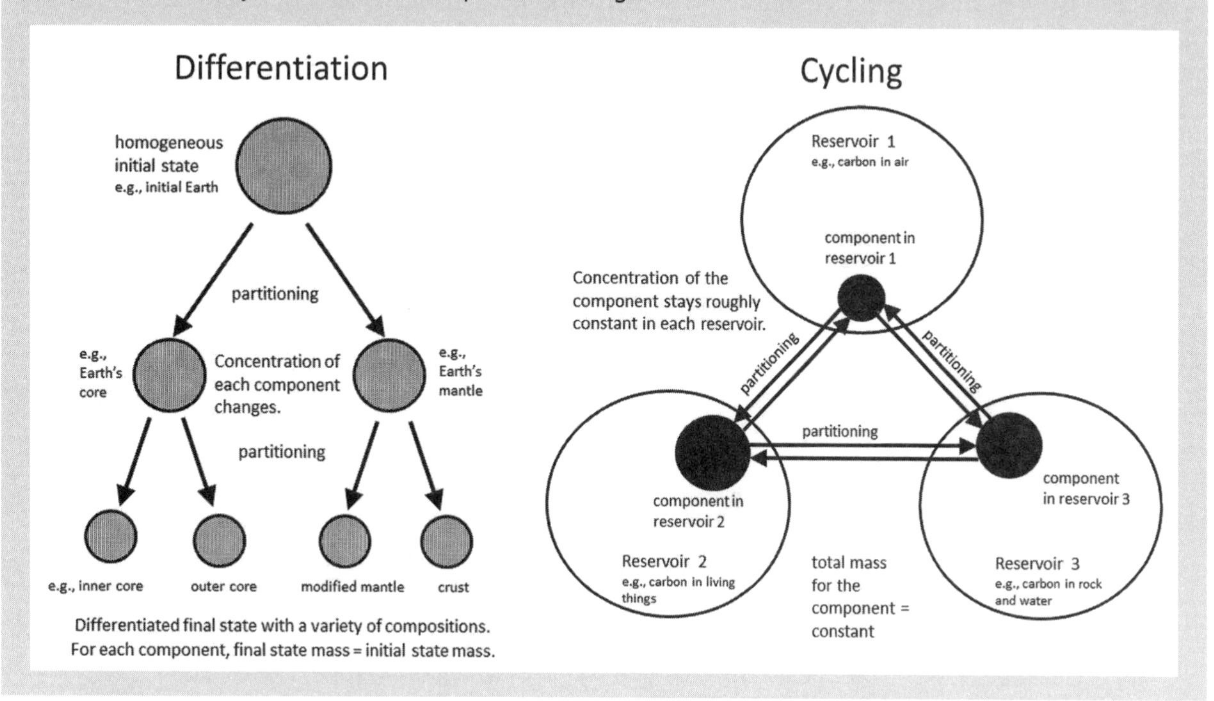

that help students understand how the large-scale models work and how we figured them out. As we have pointed out in previous chapters, this is also how scientific research proceeds. Scientists investigate some aspects of the cycling of elements through the systems of planet Earth, and those investigations add to a growing understanding of how the larger system works. Trying to start students out with the big cycles of the Earth will, in the end, offer them only a grand model to memorize and little insight into how we came to understand that model or how it works in nature.

In the following text, we consider some activities that can engage students at the "classroom-friendly" end of the learning spectrum. Understanding geochemical cycling and the differentiation of matter in planetary systems rests on two fairly simple classroom-size ideas—mass balance and chemical partitioning—which we address in this chapter.

CLASSROOM-SIZE IDEAS: MASS BALANCE

Non-nuclear chemical processes do not create or destroy elements but rather redistribute the elements already present; we call this idea *conservation of matter.* Conservation of matter means that most of the atoms on Earth today were present in the ancient past. For example, consider your next breath. You might inhale some atoms that were inhaled by a tyrannosaur some 65 million years ago. In fact, you probably will!

Given that mass is conserved, we can keep track of particular elements—account for them—as they get redistributed through Earth systems. We call this accounting process *mass balance.* The variations in calcium concentration in Minnesota ponds that we talked about previously did not involve creating or destroying atoms of calcium, or even the movement of new calcium into the ponds. Instead, most of the seasonal variation in calcium is explained by the freezing of the water, leaving calcium behind in the shrinking mass of liquid water underneath the ice.

Physical and Chemical Changes

Chemical reactions are often divided into two types in the chemistry classroom: *chemical change* and *physical change.* Chemical change occurs when a reaction forms new compounds, such as the reaction of baking soda and vinegar to form carbon dioxide gas. Physical change occurs when a reaction doesn't form new compounds, such as phase changes (freezing water) or dissolution reactions (dissolving salt).

Some people might find it surprising that a physical change (freezing) results in a change in chemistry. With pure substances, such as in the chemistry laboratory, this would not occur. However, with the nonpure substances of the earth science world, changes in phase do result in changes in composition. Thus, the distinction between physical and chemical change from the chemistry classroom is often not helpful in thinking about geochemical differentiation

Further confusion results from the fact that physical and chemical changes in the chemistry classroom are not parallel to physical and chemical weathering in earth science. *Chemical weathering* is weathering by chemical reaction, which includes both chemical change (such as the formation of clay minerals from feldspars) and physical change (such as the dissolution of evaporate deposits). *Physical weathering* refers to a process of mechanically breaking rock into smaller pieces, which is a process unrelated to physical change as used in chemistry.

That mass is truly conserved during chemical processes is not always obvious or easy to measure in the classroom. As part of my unit on weathering, I (Mary) engage my eighth graders in experiences with physical and chemical weathering. One of those experiences involves the reaction between baking soda and vinegar. Using balances, students measure the loss of mass during the reaction. Did the mass change? Did some mass "disappear"? If not, where did it go?

Common Geochemical Misconception

The ideas of mass balance and conservation of mass are not widely understood. For example, a survey of adults reported by Smith, Piburn, and Reynolds in an August 1999 *Geotimes* article indicated that 85% of people incorrectly answered the following true/false question: "Elements are created and consumed by chemical processes in the earth." Eighty-five percent answering incorrectly is much worse than random guessing. Some of this misunderstanding might be related to not remembering the difference between elements and chemical compounds, and some might be related to mixing up the concepts of amount and concentration, ideas that we explore in this chapter.

By putting the reactants into a sealed bag, students can see the bag expand during the reaction, getting a sense of the generation of a gas. This helps them realize that the loss of mass does not correspond to atoms disappearing but rather to the atoms recombining to form a gas.

Safety Note

Have students wear indirectly vented chemical-splash goggles or safety glasses during the whole activity, including setup, hands-on piece, and takedown.

One year, a group of my students invented their own experimental procedures to test how the proportions of reactants influence the extent of the reaction. They hoped to figure out how they could generate the most gas possible, so they did a series of experiments with different proportions of baking soda and vinegar and measured the mass loss for each experiment. However, none of these experiments really showed that the total mass stays the same during a chemical reaction. Trying to measure the mass inside a sealed bag runs into the problem of buoyancy—as the bag expands with the generation of gas, buoyancy will increase and so the measured weight of the bag decreases during the reaction, even though all the material is still inside. My students and I measured the apparent mass loss by comparing the mass when the reaction was done in a sealed bag with the mass when it was done in open air. We thought we detected a slightly greater mass when the experiment was done in a sealed bag (expected because carbon dioxide [CO_2] is somewhat denser than air and because the bag may pressurize above 1 atm), but we didn't do enough experiments to see if the difference was greater than our experimental uncertainty.

For older students, or students who've studied buoyancy (see Chapter 6, "Modeling, Part 1: Conceptual Modeling"), the conservation of mass can be more positively demonstrated by taking buoyancy into account. Students can remove the air from their sealable plastic bag, combine the baking soda and vinegar to start the reaction, and then reseal the bag. The mass of 1 gallon of CO_2 at standard temperature and pressure (STP) is about 8 g. However, the buoyancy at STP will also be close to the equivalent of 8 g, the weight of 1 gallon of displaced air. This means that as the 8 g are lost from the reactants and become part of the gas, the measured weight of the closed bag and its contents will decrease. By including the weight of the displaced air (the buoyancy force) in the calculation, the products and reactants can be shown to have the same mass:

Mass of reactants = mass of products

Apparent weight of the reactants = apparent weight of the products + buoyancy force

Mass Balance and Math

When students apply mass balance to Earth systems to understand some aspect of geochemical cycling, they engage in mathematical thinking as both a practice of science and an interdisciplinary connection between math class and science class. They also engage in a robust way with the crosscutting concept of Scale, Proportion, and Quantity: "The ideas of ratio and proportionality as used in science can extend and challenge students' mathematical understanding of these concepts" (NRC 2012, p. 90).

Let's apply the idea of mass balance to calcium in Minnesota ponds to predict how we would expect the concentration of calcium to change. If half of the pond water freezes to ice and all of the calcium remains in the liquid, then the concentration in the liquid has to double. *Concentration = amount of calcium/amount of water,* where the amount of water is cut to half but calcium remains the same. This calculation engages students not only with some simple math but with the ideas of *amount* (a quantity) and *concentration* (a proportion).

The amount of a material and the concentration of a material are not the same thing. A friend who taught middle school math gave us a puzzle in mass balance and concentration one year: Suppose that a watermelon is composed of 99% water. It sits out for a while and at a later time is only 98% water. Is it still edible?

We were stumped. How would we know if it was still edible? Clearly, the question included concentrations of water, but how did it bear on mass balance?

Being a good math teacher, she gave us a hint that still required us to work through the problem. Suppose that you have 99 units of water and 1 unit of everything else. That's 99% water, right? How many units of water do you have to lose to get to 98% water?

Aha. With mass balance, the number of non-water units stays constant at 1, while the water units evaporate. To get 2% non-water, we need to have 49 units of water and 1 unit of everything else, 1/50 = 2%. The water plus non-water is reduced to 50 units. Which means that 50 of the water units—more than half of the original 99—had to evaporate.

That watermelon would be a shriveled puck half its original volume. Not something most of us would want to eat!

Figure 14.1
Illustration of a modeling puzzle in geochemical evolution

(a)

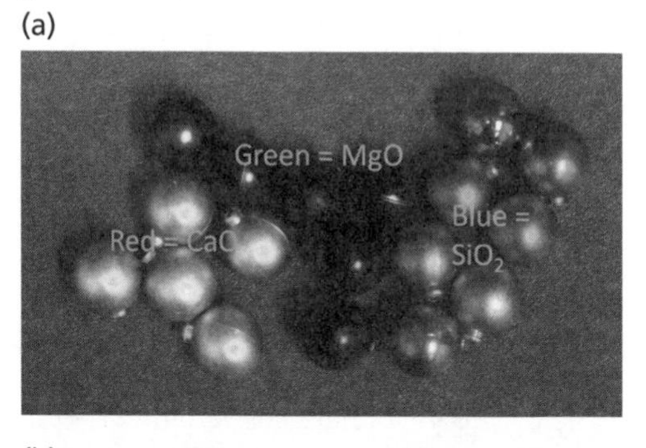

(b)

Suppose that you start with a simplified magma composed of three chemical components as symbolized by the Christmas tree ornaments in the top photo: calcia (CaO), magnesia (MgO), and silica (SiO_2). Then consider that some of that magma crystallizes to form olivine made of MgO and SiO_2 in the proportions shown in the bottom photo. How will the composition of the remaining magma change?

Mass Balance Modeling and Application to Real Data: Example Stories Told by Atoms

On a family trip to Hawaii some years back, we were a bit startled to discover deposits of volcanic ash on the Big Island. Our too-simple mental model for "hot spot" volcanism suggested that Hawaiian lava should be poor in the chemical components silica (SiO_2) and water, leading to quiet eruptions, not the more explosive eruptions that produce volcanic ash. Did some process change the quiet, basaltic lava typical of a hot spot volcano into more explosive SiO_2 rich lava?

To address this question, let's consider a classroom-friendly model for how the composition of the liquid part of a magma might change as minerals crystallize from it at depth. Suppose that we consider a body of magma residing underground beneath Puʻu ʻŌʻō (pronounced poo-oo oh-oh), an active vent near the Kīlauea caldera. Let's say the magma contains the chemical components magnesia (MgO), calcia (CaO), and SiO_2, represented by Christmas tree ornaments in Figure 14.1.

When considering natural geochemical processes such as changes in the lava at Puʻu ʻŌʻō, it's important to understand that chemical components are not the same thing as chemical compounds. A chemical compound is a single substance, in other words, a phase. In contrast, chemical components are the constituents that make up a phase. In the preceding two paragraphs, we used the words silica, magnesia, and calcia to distinguish the chemical components SiO_2, MgO, and CaO from their analogous chemical compounds silicon dioxide, magnesium oxide, and calcium oxide. The liquid part of magma might consist of the chemical components SiO_2, MgO, and CaO but would not be described as containing those chemical compounds. To engage in geochemical modeling, one needs to consider this key idea: that phases—single substances—are made up of many chemical components. Identifying those chemical components allows us to engage in the mass balance accounting that lies at the heart of geochemical processes.

Consider that magma is slowly cooling in the magma chamber beneath Puʻu ʻŌʻō and that the mineral olivine crystallizes from this magma as it cools. A simplified olivine contains only MgO and SiO_2 (as shown in the bottom half of Figure 14.1), thus the remaining melt (liquid part of the magma not including the crystallized olivine) will have less MgO and SiO_2, but the same amount of CaO.

How will the concentrations of MgO, SiO_2, and CaO be different in the final melt, after olivine crystallized, compared with the starting melt? Will the concentrations of each be lower? Higher? The same? A quick calculation gives us the information shown in Table 14.1.

Table 14.1

Calculation of the concentration change for magnesia (MgO), calcia (CaO), and silica (SiO_2) in a sample of magma as olivine crystallizes from it

Chemical Component	Starting Amount	Starting Concentration	Final Amount	Final Concentration	Concentration Change
MgO	6	6/18 = 33.3%	2	2/12 = 16.7%	decrease
CaO	5	5/18 = 27.8%	5	5/12 = 41.7%	increase
SiO_2	7	7/18 = 38.9%	5	5/12 = 41.7%	increase slightly

The starting concentrations in the initial magma have an effect on the final result. For example, consider the starting composition shown in Figure 14.2. If the same amount of olivine is removed, how will the final concentrations change? Calculation results are shown in Table 14.2.

Table 14.2

Calculation of the concentration change for magnesia (MgO), calcia (CaO), and silica (SiO_2) in a sample of magma as olivine crystallizes from it

Chemical Component	Starting Amount	Starting Concentration	Final Amount	Final Concentration	Concentration Change
MgO	6	6/18 = 33.3%	2	2/12 = 16.7%	decrease
CaO	6	6/18 = 33.3%	6	6/12 = 50.0%%	increase
SiO_2	6	6/18 = 33.3%	4	4/12 = 33.3%	no change

Figure 14.2

Illustration of a modeling puzzle in geochemical evolution similar to that in Figure 14.1, but starting with a different magma composition

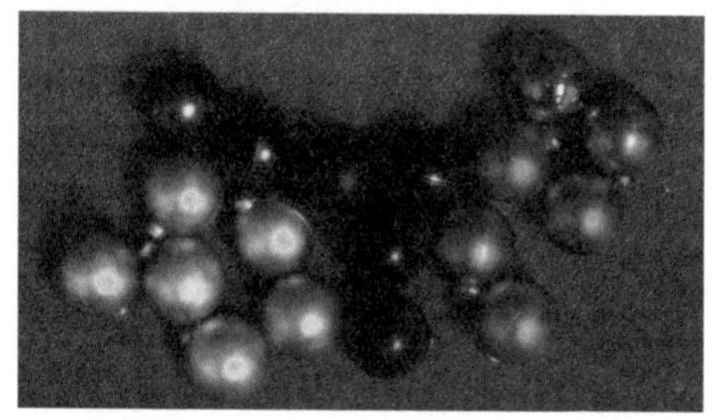

How will the composition change as olivine is removed from it?

As minerals crystallize, the direction of change in the concentrations of components in the melt is not always intuitive to students. They expect that if you remove something from the melt, the concentration of the component left in the melt will decrease. However, that's not true for SiO_2 in either of the cases. Some students expect that if you leave the amount the same, like CaO, the concentration will stay the same. In both example magmas, CaO wasn't removed, but its concentration in melt increased as olivine crystallized. Modeling exercises such as this can tune students' intuitive feel for how the great diversity of rock compositions have developed in the earth as matter cycles and differentiates through natural processes.

Figure 14.3

Variations in the composition and temperature of lava erupted from Pu'u 'Ō'ō on Hawaii from 1998 through 2001

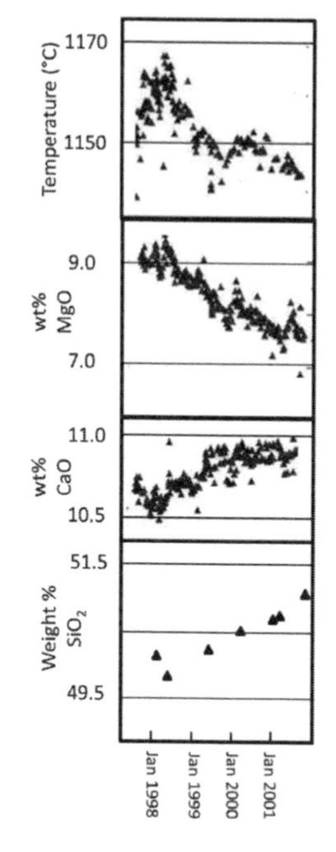

These variations can be understood as arising from the crystallization of olivine in the magma chamber that feeds the volcano, as explained in the text. Scatter and jumps in the data reflect the influence of other processes.

Sources: Data for temperature, MgO, and CaO are from Thomber (2003). Data for SiO_2 are estimated from Thomber (2003) and Marske et al. (2008).

Now let's apply our modeling results to actual compositions of basalt erupted from volcanoes in Hawaii and consider whether our model matches what we see and whether that variation might result in more explosive styles of eruption.

Composition data for lava erupted from Pu'u 'Ō'ō from 1998 through 2001 are shown in Figure 14.3. The successive layers of basalt, cooled from each new lava flow, tell the story of composition changes in the magma chamber feeding the vent.

Are the data shown in Figure 14.3 consistent with olivine crystallizing at depth? You may need to prompt students to think about two things. First, the flows of lava at the surface form from molten material erupted from depth. Second, the crystallization of olivine from the magma at depth will cause the composition of the remaining melt to change in the way that we modeled with the Christmas ornaments. Is the variation over time in the three components—SiO_2, CaO, and MgO—consistent with this interpretation?

Since the concentrations of SiO_2 and CaO increase with time, and the concentration of MgO decreases, the direction of the trends is consistent with crystallization of olivine at depth. The values of the concentrations don't match the model exactly because real lava contains many components other than SiO_2, CaO, and MgO, and thus their detailed concentrations will differ.

Does the temperature change in the manner expected for increasing amounts of olivine crystallized? Lower temperatures should result in more crystallization of olivine and thus lower concentrations of MgO and higher concentrations of CaO, as is observed.

Conceptually, is the starting composition at depth in the magma chamber that feeds Pu'u 'Ō'ō more like that in Figure 14.1 or that in Figure 14.2? Since the concentration of SiO_2 increases rather than staying constant, the starting composition must be, conceptually, more like that in Figure 14.1.

Even though we can't go down into the magma chamber, or take samples directly from it, we can put constraints on its composition and determine what minerals are crystallizing from it by considering mass balance constraints. Our observational data include how the composition of the lava erupted to the surface changes with time. The variations in magma composition with olivine crystallization give us a way to understand why some eruptions might be more enriched in SiO_2 and water than others and thus more explosive, producing ash deposits.

Mass Balance in the News

BIOSPHERE 2

While driving back from a South Dakota Science Teachers Association meeting in February 2014, I (Mary) listened to a TED talk on Biosphere 2—an experiment on how to live on other planets carried out in the early 1990s near Oracle, Arizona (for more information, go to *http://biosphere2.org*). Like many of the TED talks, I found it riveting, and more important, found in it a classroom-size puzzle in mass balance. Biosphere 2 was in the news again in the September 2015 issue of *Earth*, which provided more details (Rosen 2015).

In 1991, eight intrepid souls entered Biosphere 2, a 3-acre sealed, self-sustaining environment with ocean, desert, marsh, rainforest, and savanna. The goal was to spend two years in complete isolation from any inputs from outside except for sunlight. As with any good experiment, unexpected issues arose: some species of insect became dominant and wiped out others, the water cycle inside the biosphere delivered too much condensation "rain" to the desert regions, and reflection of sunlight off the glass roof limited agricultural production more than expected.

Failed Experiments Are the Best Kind

Public support for the Biosphere 2 experiment waned as the two-year project proceeded. People viewed the need to pump in oxygen to keep the environment safe for the eight participants—an action that was initially kept secret from the public—as a failure. In addition, one participant left for medical reasons and she returned with a bag of supplies whose unknown contents raised public suspicion. Combined with other "failures," such as the inability to grow sufficient food, the collapse of the insect ecosystem, and so on, these problems made people ask—why did we spend so much money on something that didn't work?

Some 20 years have given us a different perspective—experiments that teach us things that we didn't know are the best kind. In fact, if everything had gone exactly as predicted in Biosphere 2, and so we learned nothing new, then that would have been the true waste of money.

Sometimes in the classroom, when an experiment doesn't go as predicted, it can feel like a failure to us and we ask, "What did we do wrong?" A better question might be "What can we learn from this?"

However, an interesting mass balance puzzle arose from a mysterious drop in the molecular oxygen (O_2) level in the domed habitat. Biosphere 2 was one of the most tightly sealed environments ever built. Once the door closed, nothing went in or out except the light passing through the glass walls and ceiling—a great experiment in mass balance. Even so, the O_2 concentration in the air fell from a normal 21% to 14% over a period of 16 months—a 33% decrease. Eventually, it was necessary to pump in extra O_2 to keep the participants healthy, even though this undermined the goal of a truly sealed environment. Given the size of the facility and the amount of air in it, this loss of O_2 corresponded to a loss of roughly 7 tons of oxygen.

Whoa. Where did all that oxygen go? Mass balance constraints meant that it couldn't just disappear or get used up. Was there a leak somewhere?

Coincident with the decrease in O_2 was an increase in CO_2, which increased from the normal 0.035% to around 0.2%–0.4% in the same time period—a 900% increase. One possible explanation for the oxygen loss was that the O_2 in the Biosphere 2 atmosphere was reacting with the organic humus in the soil to produce CO_2—thus decreasing O_2 in the air and increasing the CO_2.

One classroom-size puzzle, appropriate for high school students, is this: Does the 900% increase in CO_2 account for the 33% decrease in O_2?

Let's think through the problem. Suppose that we model the air in Biosphere 2 with 100,000 air molecules, with 21% of them, or 21,000, being O_2 and 0.035% of them, or 35, being CO_2. After 16 months, Biosphere 2 had lost a significant amount of O_2 so that there was only 14% O_2 remaining in the Biosphere 2 atmosphere. That means that there was a decrease in the number of O_2 molecules in the air. We can solve for the amount of lost O_2 with some simple algebra:

$$14\% = 0.14 = \text{amount of } O_2/\text{total molecules}$$

$$= (21{,}000 - \text{lost } O_2)/(100{,}000 - \text{lost } O_2)$$

Solving for lost O_2 gives us

$$14{,}000 = 21{,}000 - 0.86 \times \text{lost } O_2$$

and

$$\text{lost } O_2 = 8{,}136 \text{ molecules}$$

So, considering that one molecule of O_2 is needed for each molecule of CO_2, does the increase in CO_2 match the decrease in O_2? Do we have a gain of 8,136 molecules of CO_2? Using the model of 100,000 molecules, the 900% increase in CO_2 corresponds to an increase from 35 molecules (0.035%) to 350 molecules (0.35%), an increase of only 315, nowhere near enough to account for the 8,136-molecule drop in O_2.

The problem in Biosphere 2 was a bit more complex than this because CO_2 was being scrubbed from the air to keep the atmosphere healthy. But, even after accounting for the scrubbed CO_2, there wasn't enough CO_2 to account for the O_2 loss. Most of the O_2 had to be going somewhere else.

Maybe the O_2 was going into the concrete walls. Calcium and magnesium in concrete can react with CO_2 in the air (which contains O_2) to form calcium and magnesium carbonates. After the experiment ended, researchers tested the walls, finding a carbonate-rich layer on the inside of the concrete walls, supporting this conclusion.

Oahu: Dissolving From Within

Another mass balance problem in the news was found in the *Live Science* article "Hawaiian Island Dissolving From Within" (Pappas 2012), based on original research by Nelson, Tingey, and Selck published in 2013.

We all know that landscapes are washing away to the sea. We learn this in earth science lessons ranging from weathering to the cycling of tectonic plates. John Wesley Powell saw evidence for multiple cycles of erosion in the rocks of the Grand Canyon, which we explored in Chapter 11, "Stories in Rock Layers." The role of water in shaping the surface of Earth through weathering and erosion is a core idea in both the *Framework* and the *NGSS*.

But how does bedrock wash away? Does the rock mostly dissolve and then become part of the water that runs to the sea, or do tiny particles bouncing down the rivers to the sea carry most of the mass? The black basalt sand beaches of the Hawaiian Islands might suggest that particle erosion is the key process, but is that true?

Nelson, Tingey, and Selck (2013) addressed this question with a study readily translated into the high school classroom. They measured the concentration of silicon in groundwater in south-central Oahu. Silicon is the most abundant element (after oxygen) in the basaltic rock of the island, and the authors used it to understand how the rock as a whole was behaving. They also measured how much groundwater ran to the sea each year, and they measured how much silicon is in the top meter of rock on the island. From these measurements, they calculated how much depth of rock is carried to the sea each year by groundwater and compared this result with a similar calculation of how much rock was being carried to the sea as particles in surface water.

An example data set for students is shown in Figure 14.4 (p. 288), along with the calculated solution. Students can figure out that the amount of sediment carried to the sea in solution (dissolved in river and groundwater) is 10 times greater than that carried as sediment particles.

You can adjust the difficulty of this problem by how much you process the raw data for the students, how many of the unit conversions you do for them, and how thoroughly you set up the mass balance calculation for them. If you set up the sequence of calculations entirely, the problem becomes straightforward arithmetic, not a mass balance problem in which students have to figure out that the mass of silicon washing to the sea must equal the mass of the silicon in the eroded landscape. We suggest allowing students to grapple with using the data to address the investigative question: How much erosion is due to dissolving of the rock? Most of the students' time and class discussion will focus on how to set up the calculation, not on doing the calculation itself.

Also, you may not want to do all the unit conversions for your students because science and engineering practice 5 of the *NGSS*, Using Mathematics and Computational Thinking, specifically calls out unit conversions for grades 9–12: "Apply ratios, rates, percentages,

Figure 14.4
Mass balance modeling activity from research in the news

How Oahu Washes Away to the Sea
A Mass balance problem!

Driving questions:
- How much does the south central part of Oahu erode away each year?
- Is most of the erosion due to movement of solid particles or dissolved elements?

Observational Data:
- Concentration of Si in ground water = $57.51 \frac{\text{milligrams}}{\text{liter}}$
- Amount of groundwater flowing to the sea = $4.964x10^{11} \frac{\text{liters}}{\text{year}}$
- Si in each meter depth of rock in south central Oahu =

$1.319x10^{3} \frac{\text{kilograms } Si}{\text{cubic meter basalt}} * 6.745x10^{8}$ square meters land area $= 8.899x10^{11} \frac{\text{kilograms}}{\text{meter depth}}$

Comparison Data:
- Depth of landscape erosion due to particulate erosion = $3x10^{-6} \frac{\text{meters}}{\text{year}}$

Calculated dissolution erosion rate:
- The calculation is set up by recognizing that for the mass to balance, the amount of Si washed to the sea each year must equal the amount of Si in the rock eroded each year.
- Depth of landscape erosion due to solute erosion =

$$\frac{\text{kilograms Si}}{\text{liter groundwater}} \times \frac{\text{liters groundwater}}{\text{year}} \times \frac{\text{meters depth}}{\text{kilogram Si}} = \frac{\text{meters eroded}}{\text{year}} = 3.2x10^{-5} \frac{\text{meters}}{\text{year}}$$

Does most of Oahu wash away as particles or as material dissolved in water? One mathematical solution to the problem is shown by the "Calculated dissolution erosion rate" at the bottom of the figure. The activity is based on research by Nelson, Tingey, and Selck (2013).

and unit conversions in the context of complicated measurement problems involving quantities with derived or compound units (such as mg/mL, kg/m^3, acre-feet, etc.)."

CLASSROOM-SIZE IDEAS: CHEMICAL PARTITIONING

Chemical partitioning refers to how an element or other chemical component is distributed between two different materials in contact with each other in the natural environment. For example, in the puzzle of calcium concentration in Minnesota ponds described at the beginning of this chapter, the calcium partitioned preferentially into the liquid phase of water, not into the ice, resulting in its concentration in the liquid as more water froze. Alcohol partitions preferentially into the vapor phase when we cook food with wine. CO_2, formed from the reaction between O_2 in the atmosphere and the soil organics of Biosphere 2, partitioned into both the air and the concrete walls.

Although partitioning involves chemical reactions—which depend on temperature, pressure, oxygen availability, acidity, reaction kinetics, compositions of the materials involved, and many other parameters of the systems—earth scientists often simplify partitioning by considering a fixed ratio of concentrations between two materials in contact with each other. This ratio is called a partition coefficient and is often symbolized by the letter *D*. This simplification is often a good enough approximation for geologists to explain differentiation in even fairly complex systems and allows for conceptual and mathematical modeling appropriate for middle and high school students.

We can think of the partitioning of calcium (Ca) between liquid water and ice from the puzzle at the beginning of this chapter: D_{Ca} = (*concentration of Ca in liquid water*)/(*concentration of Ca in ice*). Because the calcium goes mostly into the liquid, the partition coefficient will be a large number. A small number would signify that the element goes mostly into the material (usually a phase) that you put on the bottom of the ratio. A number close to 1 signifies that the element partitions equally between the two materials (phases).

This concept is highly useful in addressing geochemical cycling and differentiation with secondary students of all ages. For example, at the start of the last school year I (Mary) engaged my eighth-grade students with my Tapestry of Time and Terrain map activity (see Chapter 5, "Analyzing and Interpreting Data, Part 2: Maps and Cross-Sections")—a broad survey of landforms and geological processes that I use as a foundation and springboard for many later activities. We were looking at the Great Basin and a student asked, "What's this black splotch?" "That's the Great Salt Lake," I said.

"Is it salty?"

"Yep, saltier than the ocean."

"So, how did it get so salty?"

A great driving question for our next unit on weathering. During weathering reactions, sodium partitions from the rock into water. The water runs in streams into the Great Salt Lake. During evaporation from the lake, sodium partitions into the liquid water instead of the vapor, leaving the sodium increasingly concentrated in the lake. As more stream water enters the lake, bringing more sodium, the lake gets increasingly salty until it reaches the salty state we see today.

Each of the steps for this process is something that students can test in the classroom—they can measure the change in total salts in water when rocks weather (at least if the rock chosen contains highly soluble minerals), they can calculate and model how sodium concentration will change as water evaporates, and they can do experiments to test the model calculations. We offer some suggestions for experiments of this sort in the "Suggestions for Designing an Activity" section (pp. 297–300).

Example Experimental Partitioning Activity

Johnny Honest has been spraying for green bugs in the north forty for most of the day, using the pesticide DEATH-X. The day has been long and hot, and the reserve of water in his thermos has long since been used up. Each time he loops the tractor and trailing sprayer past the south end of the field, he looks longingly toward Grandma's well and the tank of water where the cattle come to drink. On each pass, his gaze becomes a bit more longing until at last he turns the tractor toward the tank, parks, and dismounts. He skims back the layer of algae growing on the surface of the water tank, leans over, and takes a long, satisfying drink. Ahh!

He returns to the tractor and moves to reengage the power take-off to start up the sprayer again but accidentally bumps one of the hydraulic levers, which upends the DEATH-X tank, emptying most of its contents onto the ground beside the cattle tank and well (see Figure 14.5).

Figure 14.5

Setup for the problem with the pesticide spill at Grandma's well

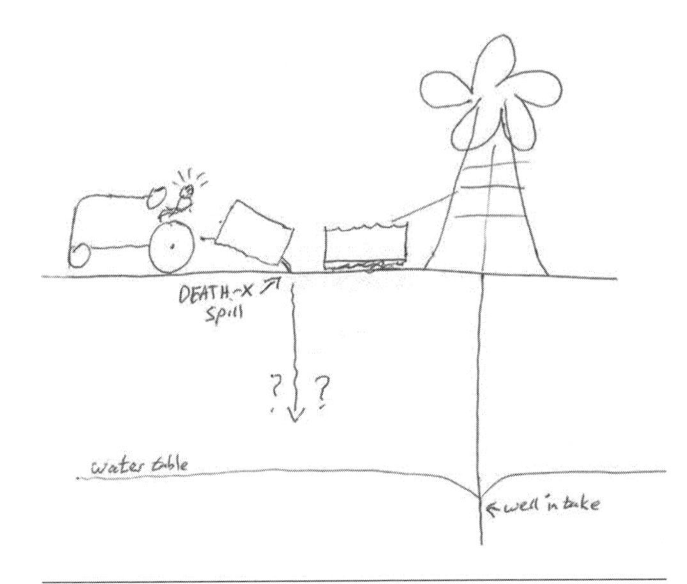

Will the pesticide seep down through the sediment to the water table and get into Grandma's well? Or will the pesticide partition into the sediment, effectively filtering the water and leaving Grandma's well safe?

Oh no! He really doesn't want to have to admit his mistake to Grandma. However, he doesn't want Grandma to get poisoned when the pesticide leaks down through the ground to her well-water intake.

Aargh! What to do?

He reasons that he doesn't have to tell her anything if he can be sure that the DEATH-X won't get down to the water table and into her well. That might be the case if the DEATH-X partitions into the sediment instead of into the water. Then, as rainwater washes the DEATH-X downward, the pesticide will partition into the sediment, and the sediment will clean the water like a filter.

However, if the DEATH-X partitions into the water, then rain will wash it straight down to the water table and into Grandma's well.

Only one thing to do: He has to carry out experiments to test how the DEATH-X partitions. Taking a sample of DEATH-X that didn't spill, he heads home where he gathers some sediment samples and containers for the experiments. He does two different experiments, using two different kinds of sediment—clay and sand—hoping that one of them will match the sediment above Grandma's well (Figure 14.6).

Figure 14.6

Two experiments with two different kinds of sediment that address the question of whether the spilled pesticide will get into Grandma's well

Grandma's Well Experiments

Using red food coloring as our proxy for the deadly insecticide DEATH-X, and using cornstarch as a proxy for clay, we can do two experiments to equilibrate water + sediment + DEATH-X and measure the partition coefficients that result.

5 drops red food coloring stirred with 20 ml loose clay and 50 ml water

After sediment settles, use a syringe to extract 20 ml of water from each experiment for chemical analysis.

5 drops red food coloring stirred with 20 ml loose sand and 50 ml water

Safety Note: Have students wear indirectly vented chemical-splash goggles and aprons during the whole activity, including setup, hands-on piece, and takedown.

In each experiment, he mixes 50 ml of water, 20 ml of sediment, and 5 drops of DEATH-X. He stirs each to ensure that the experiment approaches equilibrium, then lets the sediment settle. Using standards that he made with the leftover DEATH-X (see Figure 14.7), he measures the concentration of DEATH-X in the water and calculates the concentration in the sediment (see Figure 14.8).

Figure 14.7

Manufacture of a set of standards by which the composition of water samples from the Grandma's well experiments can be measured

Figure 14.8

Illustration of the use of standards to analyze the concentration of red food coloring in the experimental water samples for the Grandma's well experiments

Grandma's Well Experiments—Analysis of Compositions

Samples of water from the two experiments are compared to the standards in order to determine the concentration of DEATH-X (red food coloring) in the water that equilibrated with sediment.

Concentration in clay (cornstarch) calculated by difference:

There is 1 drop of red food coloring in 50 ml of water, leaving 4 of the original drops in 20 ml of clay, for a concentration of 4 drops/20 ml or 0.2 drops/ml.

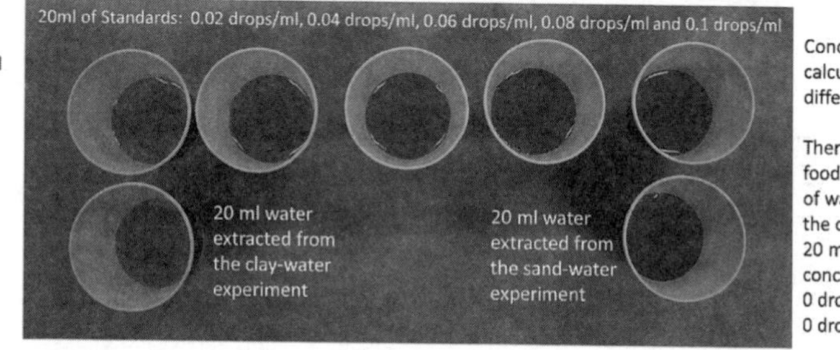

Concentration in sand calculated by difference:

There are 5 drops of red food coloring in 50 ml of water, leaving 0 of the original drops in 20 ml of sand, for a concentration of 0 drops/20 ml or 0 drops/ml.

The concentration in water from the clay (cornstarch) experiment is determined to be 0.02 drops/ml, and the concentration in water from the sand experiment is 0.1 drops/ml. The concentration of red food coloring in the sediment (clay or sand) cannot be measured directly by this method but can be calculated by difference as shown in this figure. These values for concentration in clay and sand are 0.2 drops/ml and 0 drops/ml, respectively.

Based on the results of his experiments shown in Figure 14.8, and assuming that the sediment above Grandma's well is clay-rich, will he have to tell Grandma? What if the sediment above the well is mainly sand?

The experiments shown in Figures 14.6–8, yield the following partition coefficients:

> *$D_{Death\text{-}X}$ sand/water = concentration of DEATH-X in the sand divided by the concentration in the water = (0 drops/ml) / (0.1 drops/ml) = 0: This number is much less than 1, meaning that the DEATH-X partitions almost entirely into the water rather than sand. A partition coefficient of 0 is not theoretically possible, but because the standards go in increments of 0.02 drops/ml, students are unable to distinguish the result from 0 in this case.*
>
> *$D_{Death\text{-}X}$ clay/water = (0.2 drops/ml) / (0.02 drops/ml) = 10: This number is much greater than 1, meaning that the DEATH-X partitions into the sediment rather than the water.*

Thus, the clay is likely to filter the DEATH-X, whereas the sand will not.

Johnny can check the sediment at the well, and, depending on whether it is clay-rich or sand, decide how likely the spill is to reach the water table. Of course, Johnny tells Grandma regardless of the outcome of his experiments. His name is Johnny Honest, after all.

This activity can be simplified for middle school students by skipping the making of standards and the quantitative calculation of partition coefficients. Students can still evaluate the partition coefficients qualitatively as shown in Figure 14.9. The qualitative partition coefficients can be used to determine whether Johnny has to tell Grandma.

Figure 14.9 ✪

Simplified version of Grandma's well experiment from Figure 14.6 for younger students or shorter duration

Grandma's Well Experiments—Qualitative Estimate of Partitioning

Results of experiments are estimated qualitatively without making standards or analyzing compositions
Might be more appropriate for middle school students

Light red color indicates that most food coloring stayed with the clay (cornstarch).

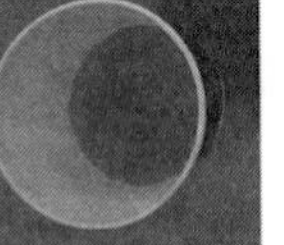

20 ml water extracted from the clay-water experiment

20 ml water extracted from the sand-water experiment

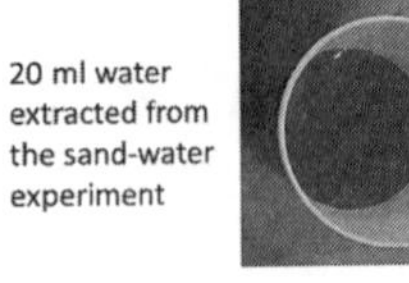

Dark red color indicates that most food coloring stayed with the water.

D sand/water < 1

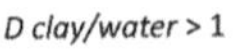

D clay/water > 1

This shorter version starts with the two experiments shown in Figure 14.6, but skips the step for making standards. Students evaluate the partition coefficients qualitatively rather than calculating a numerical value. Students can still evaluate whether the DEATH-X partitions into sediment or water for each of the sediment types and determine if Johnny has to tell Grandma about his mistake.

Clay adsorbs many elements, so a clay-rich sediment will tend to filter pollutants, "scrubbing" the water and retaining the pollutants. Sandier sediment will filter less, leaving the pollutants in the water. The activity in Figures 14.6–9 uses cornstarch instead of clay for convenience. However, if you have access to it, disaggregated clay will also work. Also, the activity uses red food coloring instead of DEATH-X; the organic red food coloring acts like many organic, water-soluble pollutants and is a good analogue.

One advantage of using the non–instrument-based analytical approach shown in Figure 14.7 is that students are engaged with the chemical analysis at a fundamental level. Students sometimes see an analytical tool as a magic black box and miss the basic idea that chemical analysis depends on comparing something that we don't know to something that we do know by means of an observed or measurable characteristic.

This activity engages students in thinking about how different sediment types will affect pollutant migration. It engages them in an experimental investigation of partitioning and in a calculation of partition coefficients that gives students practice linking proportional thinking to amount and concentration. It exposes them to the idea of chemical analysis in which unknown samples are compared to a standard. Students apply the results of their experiments to a bigger problem, the mobility of pollutants in the environment. They also gain an insight into the partitioning and mass balance processes by which all of Earth's geochemical cycles work.

APPLICATION OF MASS BALANCE AND PARTITIONING CONCEPTS TO LARGE-SCALE GEOCHEMICAL MODELING

Geochemical processes appear in the *NGSS* crosscutting concepts as well as the disciplinary core ideas, reflecting their importance. Crosscutting concept 5, Energy and Matter: Flows, Cycles, and Conservations, is described in *NGSS* Appendix G as follows: "Tracking fluxes of energy and matter into, out of, and within systems helps one understand the systems' possibilities and limitations" (NGSS Lead States 2013).

How to apply the classroom-size concepts of mass balance and partitioning to the global-scale flows of matter implicit in the *NGSS* disciplinary core ideas and crosscutting concepts may not be obvious. In the "Digging Deeper" section (pp. 301–303), we consider how the classroom-size ideas discussed in this chapter might be applied to global-scale questions such as "Why is there so much gold in South Africa?"

REFERENCES

Marske, J. P., M. O. Garcia, A. J. Pietruszka, J. M. Rhodes, and M. D. Norman. 2008. Geochemical variations during Kīlauea's Pu'u 'Ō'ō eruption reveal a fine-scale mixture of mantle heterogeneities within the Hawaiian plume. *Journal of Petrology* 49 (7): 1297–1318.

National Research Council (NRC). 2012. *A framework for K–12 science education: Practices, crosscutting concepts, and core ideas.* Washington, DC: National Academies Press.

Nelson, S. T., D. G. Tingey, and B. Selck. 2013. The denudation of ocean islands by ground and surface waters: The effects of climate, soil thickness, and water contact times on Oahu, Hawaii. *Geochimica et Cosmochimica Acta* 103: 276–294.

NGSS Lead States. 2013. *Next Generation Science Standards: For states, by states.* Washington, DC: National Academies Press. *www.nextgenscience.org/next-generation-science-standards.*

Pappas, S. 2012. Hawaiian island dissolving from within. *Live Science.* December 27. *www.livescience.com/25838-hawaiian-island-dissolving.html.*

Rosen, J. 2015. Benchmarks: September 26, 1991: Crew sealed inside Biosphere 2. *Earth* 60 (9): 105–107.

Smith, M., M. Piburn, and S. Reynolds. 1999. Research for earth science learning. *Geotimes* 44 (8): 27, 35.

Thomber, C. R. 2003. Magma-reservoir processes revealed by geochemistry of the Pu'u 'Ō'ō-Kūpaianaha eruption. In *The Pu'u 'Ō'ō -Kūpaianaha eruption of Kīlauea volcano, Hawai'i: The first 20 years,* ed. C. C. Heliker, D. A. Swanson, and T. J. Takahashi, 121–136. Professional Paper 1676. Reston, VA: U.S. Geological Survey.

SUGGESTIONS FOR DESIGNING AN ACTIVITY

EXPERIMENTAL TEST OF THE MATHEMATICAL MODEL FOR GEOCHEMICAL DIFFERENTIATION

Understanding geochemical cycles and systems rests on the mathematical modeling of those systems. Although the math gets complex when all the components of interacting systems are put together at once, the mathematics of mass balance for one partitioning component in a system is simpler.

In science, every mathematical model needs to be tested against observational evidence. This can be done for geochemical differentiation by means of an experimental investigation into how the concentration of salt changes in water as the water evaporates. Salt partitions into the liquid rather than the vapor phase. This results in the progressive concentration of salt in a basin of evaporating water, such as an aquarium or the Great Salt Lake. This is the process by which the oceans became salty. This process also contributes to ocean circulation as we talked about in Chapter 6. It's analogous to the process for concentrating calcium in Minnesota ponds and the magmatic process for concentrating gold addressed in the "Digging Deeper" section (pp. 301–303).

CASE 1: CONCENTRATION OF SALT IN WATER AS THE WATER EVAPORATES

The mathematical model for the change in salt concentration can be constructed from the concept of mass balance analogous to modeling activities earlier in this chapter. For example, take a 1,000-unit model for water. If 10 of those units are salt (NaCl), then the concentration is 10 ppth (parts per thousand).

If 100 units of water evaporate, and virtually all of the salt stays with the liquid, then the new concentration of salt in the water will be 10/900 = 11.1 ppth. If 200 units of water evaporate, then the concentration of salt will be 10/800 = 12.5 ppth. If half of the water evaporates, 500 units, then the concentration will be 10/500 = 20 ppth, or double the original concentration.

This mathematical model can be tested by putting water of known salinity, say 10 ppth, in a basin and letting the water evaporate. We suggest using a small (2–3 gallon) rectangular or cylindrical aquarium—the simple shape makes the changes in water volume easy to calculate. Salinity can be controlled at the start of the experiment by mixing a measured amount of salt into a measured amount of distilled water. You might include an aquarium heater to keep the water warm so it will evaporate more quickly. By marking and measuring water levels on the side of the tank, you can keep track of how much water has evaporated. Salinity can be measured using a simple hydrometer from the local pet store, which works on the principle of buoyancy and typically can measure salinities from 4 ppth o 40 ppth.

Safety Notes

1. Have students wear indirectly vented chemical-splash goggles and aprons during the whole activity, including setup, hands-on piece, and takedown.
2. Remind students to immediately wipe up any spilled water to avoid a slip and fall hazard.

Salinities measured in the experiment can be graphed along with the modeled values (e.g., plotting both against the fraction of water evaporated) to see if the model matches observation. If the values don't match, you can explore why they don't (e.g., the observed salinities may trail the model values slightly because of precipitation of salt around the heater or on the sides of the tank).

CASE 2: CONCENTRATION OF SALT IN WATER WHERE NEW WATER IS ADDED TO REPLACE EVAPORATED WATER

It might seem intuitive that as long as you keep adding freshwater to a basin, it won't get salty. That intuition would be wrong. The Great Salt Lake got salty even as freshwater rivers continued to flow into it. Those of us with fish tanks know that we can't keep adding more water from the faucet as water from our tanks evaporates. If we do, the water will get progressively saltier until first our plants die and eventually our fish die.

We can construct a mathematical model to predict how the salt concentration will change for this situation using a similar approach to that taken above. Suppose that we again consider a 1,000-unit model for water that includes 10 units of salt (10 ppth).

If 100 units of water evaporate, and then we add back in another 100 units of water that contains the original concentration of salt (10 ppth), the concentration of salt will be 11/1000, or 11 ppth. Ten of the units of salt come from the original water. The 11th comes from the 100 units of water added to bring the water level back up to where it started. If another 100 units of water evaporate, and we add in another 100 units of water, the concentration rises to 12 ppth. Another 100 units and the concentration rises to 13 ppth, and so on. Unless the added water contains no salt at all, we will always increase the concentration of salt by this process.

As in the case 1 experiment, you can test the model by experimenting with your aquarium, being sure to add back water whose composition is the same as the initial water. Because adding more water back into the tank dilutes the concentration of salt, the rise in concentration will be slower than for the case where the water simply evaporates from the aquarium. Again, you might encourage your students to graph the measured values versus the modeled ones to see how closely the model predicts the measured values.

Safety Notes

1. Have students wear indirectly vented chemical-splash goggles and aprons during the whole activity, including setup, hands-on piece, and takedown.
2. Remind students to immediately wipe up any spilled water to avoid a slip and fall hazard.

CASE 3: PARTITIONING OF SALT BACK INTO THE SOLID PHASE

The origin of the salt in Great Salt Lake, and in the oceans, is sodium and chlorine from weathering of rocks. The salt partitions into the water and runs into the sea. This makes the sea saltier as the water evaporates and the salt remains behind in the liquid water. However, this process can't go on forever. For example, the saltiness could never exceed 100% salt. In fact, the limitation is lower than this. Water will stop getting saltier at the point where salt crystals begin to crystallize, thus partitioning sodium and chlorine back into the solid phase. At this point, the saltiness will tend to stabilize as a balance is reached between the partitioning of salt into the liquid during evaporation and the partitioning of salt into the solid.

It's possible to experimentally confirm this stabilization of concentration by growing salt crystals in a jar. Put a string with something on it, perhaps a paper clip, into an open jar of water that is close to salt saturation (about 36 g of salt per 100 g of water at room temperature). Leave the jar open so that water evaporates. Periodically measure the salinity of water and the amount of water evaporated. Does the salinity increase? Can you make a qualitative connection between the amount of water that evaporates and the amount of salt that crystallizes?

Safety Notes

1. Have students wear indirectly vented chemical-splash goggles and aprons during the whole activity, including setup, hands-on piece, and takedown.
2. Remind students to immediately wipe up any spilled water to avoid a slip and fall hazard.

The pet store hydrometer probably won't work to measure the salinities in this experiment because they are too high. One work-around is to dilute the water by a known factor, such as a factor of 10, and then measure the salinity with the hydrometer, remembering to multiply the result by 10. Or you can splurge for a salinity refractometer, which can measure the salinity more quickly and accurately. A hand refractometer runs between $50 and $100, as of this writing.

This balance between crystallization of salt and evaporation of water is why the oceans don't continue to get saltier and is how the oceans have produced most of the world's deposits of salt.

We think that restricting the math to problems that involve relatively simple proportional thinking is important for students in grades 6–12. Reasoning through partitioning puzzles can be done with straightforward proportional thinking in cases where the partition coefficient is very different from 1, as in this chapter. For example, salt partitions almost entirely into the liquid phase in preference to the gas phase, meaning that the partition coefficient for salt, D_{salt} *liquid/gas,* is much greater than 1. For values of D closer to 1, the mathematical expressions become exponential and are perhaps more appropriate for college students.

DIGGING DEEPER

Gold Ore in South Africa

The importance of geochemical processes is identified in *NGSS* performance expectations such as MS-ESS3-1: "Construct a scientific explanation based on evidence for how the uneven distributions of Earth's mineral, energy, and groundwater resources are the result of past and current geoscience processes" (NGSS Lead States 2013). A middle school science teacher from Idaho posted the following question about this performance expectation on an NSTA listserv:

> *This [performance expectation] got me to thinking that I am not sure that I could give a very satisfactory answer if a student were to ask me why, from a geology standpoint, does a country like South Africa have such overwhelming deposits of gold and diamonds. Can anyone help me answer this? All I can find online are historical accounts of South Africa's gold and diamond industries from the human perspective. I want to know what geological process or features of that area caused that particular region to be so rich in precious metals and stones while other areas of the world have none.*

What a good question!

The direct answer to the question is fairly straightforward, albeit unhelpful: South Africa has a long history of magmatic activity, mountain building, and plate tectonic rifting that provided the energy to drive many differentiation events, which, through mass balance and partitioning, resulted in the concentration of a variety of materials in that region. The formation of ores and minerals in South Africa was the consequence of many processes happening over billions of years of time.

This answer is not much more classroom ready than the original performance expectation on which the question was based. The problem is, you can't bring billions of years of geochemical processes into the classroom. The answer was developed from thousands of small-scale studies carried out by many researchers that have been collected into a global model of "a long history of magmatic activity, mountain building, and rifting."

In scientific research, and in the classroom, it's important to keep conclusions—even grand-scale conclusions—tied to laboratory- and field-size observational evidence. When we (Russ and Mary) were graduate students at the University of Tennessee, faculty and grad students alike referred to scientific reports that became too distant from the underpinning evidence as "arm waving." By that we meant that much of the report focused on telling a cool story with insufficient reference to the evidence that supported the story.

Classroom activities that start with the high-level question of why South Africa has more ores than other places will probably degenerate into arm waving with little scientific underpinning unless you first build an understanding of how mass balance and partitioning support model development of geochemical differentiation and cycling. Fortunately, geochemical processing, even billions of years of it, can be understood by breaking it down into simpler steps of partitioning and mass balance.

Let's consider gold as an example. One step in the process of concentrating gold into an ore might be magmatic, similar to the mass balance activity that we did with the Christmas ornaments (see Figure 14.1, p. 281). During crystallization of olivine from a highly mafic magma, gold partitions into the liquid phase, not into solid (see Figure 14.10A). Gold becomes increasingly concentrated in the liquid in the same way that CaO was concentrated in the liquid in our activity with the Christmas ornaments.

Figure 14.10

Model of a magmatic process and a hydrothermal process to concentrate gold

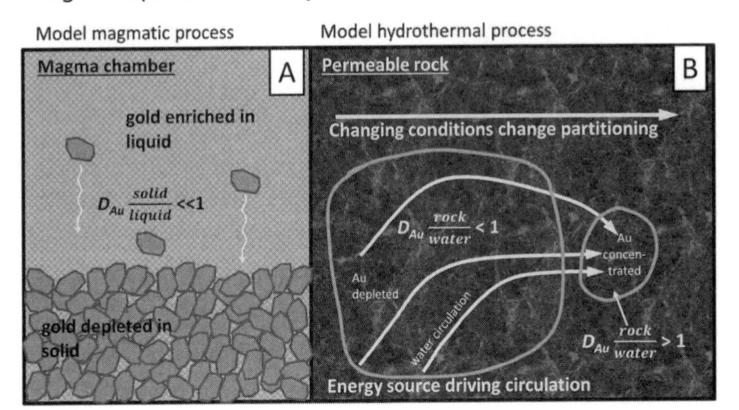

Understanding why South Africa has lots of mineral resources requires an understanding of specific examples of how mass balance and partitioning might have produced particular ores. This figure illustrates models for two processes for concentrating gold that fit within the classroom-size experiments and calculations that we've done in this chapter. (A) The concentration of gold (Au) in the magma chamber can be modeled by students in the same way that we modeled the changing concentration of CaO in the Christmas ornament activity in the text and Figures 14.1 and 14.2. D_{Au} *solid/liquid* is the partition coefficient for gold between minerals and the liquid part of magma, and it equals the concentration of gold in minerals divided by the concentration of gold in liquid with which the minerals are in chemical equilibrium. (B) The hydrothermal process can be modeled by students by comparison to the Grandma's well experiments in Figures 14.6–14.9, as explained in the text. D_{Au} *rock/water*, equals the concentration of gold in rock divided by the concentration of gold in water with which the rock is in chemical equilibrium.

Suppose that the concentration of gold in a magma is 1 part per billion (ppb)—one part gold for every billion parts of total melt. If we crystallize 90% of the magma, what will be the concentration of gold in the remaining liquid? Analogous to the Christmas ornament problem, the *amount* of gold in the final melt will still be 1 part (since the gold partitions into the liquid). We crystallized 900,000,000 parts (90% of a billion), leaving 100,000,000 parts of melt. Thus, the gold concentration is 1/100,000,000 = 10 ppb, a 10-fold increase in concentration.

Typically, magmatic processes are not sufficient to concentrate gold enough to make an ore (*ore* being the name we give a material that can be mined at a profit). Usually, other processes are needed to concentrate the gold even further.

Another process for concentrating gold might be hydrothermal. Depending on conditions of temperature, composition, acidity, and so on, widely dispersed gold in one region of rock might partition into water. Energy from mountain building or magmatic activity can circulate that water through rock to a place where the gold partitions back into the rock (Figure 14.10B). In this way, gold from a large region that is relatively depleted in gold can be concentrated into a small region.

Let's do a thought experiment based on the partitioning activity with Grandma's well earlier in this chapter. In the region on the left side of Figure 14.10B, the gold is loosely bound to the rock (partitions into water) like the red food coloring was loosely bound to sand. In the region on the right side of Figure 14.10B, the gold partitions into the sediment, like the red food coloring partitioned into the cornstarch. Imagine red food coloring dried on the surface of sand in the left-hand side of a tube. Cornstarch without food coloring is in the right-hand side of the tube. As water flows from left to right through the tube, the red food coloring will dissolve from the sand into the water, be carried to the right and precipitate on the cornstarch, analogous to the movement of gold in the hydrothermal system. If there is much more sand with red food coloring than there is cornstarch, then the effect is to concentrate the food coloring.

The small-scale processes of partitioning and mass balance can be used to understand the processes for concentrating gold shown in Figure 14.10, which in turn can be used to understand how billions of years of various geochemical processes have made South Africa particularly gold-rich.

When you examine geochemical cycling in the classroom, remember that the bigger the scale of the cycling, and the more divorced from calculation and experiment it is, the more the discussion becomes arm waving rather than arguing from evidence. We suggest that it's better to do a small-scale, specific study appropriate for the classroom first and then fit that into an understanding of large-scale cycles and systems.

PART III

YOU CAN DO IT!

As soon as I saw these concretions that are 70 million years old, and you open them up, and the sunlight, you know, for the first time it sparkles off of these, I mean, it's just fire off the ... shells! ... I was hooked!"

—Joyce Grier

PART III

Not having learned it is not as good as having learned it; having learned it is not as good as having seen it carried out; having seen it is not as good as understanding it; understanding is not as good as doing it. The development of scholarship is to the extreme of doing it, and that is its end and goal. He who carries it out, knows it thoroughly.

—Xunzi (312–230 BC), translated by Homer Dubs in *The Works of Hsüntze*, 1928

The *Next Generation Science Standards* propose that earth science be taught as a practice rather than a body of factual knowledge. That practice is not a fixed set of steps but rather a fluid and dynamic activity that will be somewhat different in each new situation and application. The practice of science in the classroom involves student initiative in asking questions and creating investigations to answer those questions. It connects results of those creative investigations to broad ideas that exceed the limits of the classroom.

This type of learning experience—fluid, flexible, and creative—cannot be captured by a curriculum, regardless of how good the curriculum might be or how well aligned to the standards. Success of this type of learning experience depends on the participation of the teacher. The teacher is needed to connect the spontaneous questions and creative investigations that arise in the classroom to the bigger ideas of science. The teacher is also needed to prompt and guide creative investigations so that students learn from an expert mentor how to focus and pursue their questions. To do this, the teacher needs to depart from set activities whose primary purpose is to reinforce factual learning, and truly explore with students.

This final section of *Learning to Read the Earth and Sky* examines the importance of the teacher in (1) narrating the link between spontaneous and creative explorations of the classroom and the core ideas of the discipline, (2) initiating creative explorations from where you are in place and understanding, and (3) mentoring student explorations as a fellow scholar and practitioner of science.

TEACHER AS CURRICULUM NARRATOR

In this book, we've argued that science is not about what we know, or think we know, so much as it is about how we know it. For that reason, the practices of science held center stage in Chapters 1–14. We've argued that the person who *knows how we know* is the one who participates in science. We've told stories about students doing science at the classroom level rather than students *knowing* explanations, models, and theories. But this vision of students doing science is not quite complete. Scientists, after all, don't practice science to learn the practices of science. Rather, they practice science to learn something of wider interest and importance.

In Chapter 12, we considered Keith Koper's use of seismic data to study the tragic events surrounding the Russian submarine *Kursk*. Dr. Koper's conclusion emerged from the practice of science, but the product of his investigation was the conclusion—a story of wide interest—not the investigation itself.

The practices of science are at the heart of scientific investigation. But the story, the wider picture, is the *product* of the investigation, without which the practice of science becomes pointless.

The Heart of Scientific Investigation

Some people characterize the crosscutting concepts of the *Next Generation Science Standards* (*NGSS*) as themes that cut across all of the science disciplines. Others see them as a lens through which to view problems in all of the science disciplines. We consider the crosscutting concepts to be, in some ways, an alternative characterization of the practices of science shared by all of the disciplines.

For example, every scientific study requires the identification and interpretation of *patterns*—patterns in data, patterns in maps, patterns in structures. *Cause and effect* is important in interpreting experimental results or correlations observed in nature. *Scale and proportion* are important in considering representations of phenomena from atoms to galaxies. The idea of *flows and cycles of matter and energy* provides a framework for understanding and interpreting observations in all of the sciences.

Both the *NGSS* crosscutting concepts and the science and engineering practices can be thought of as classification schemes for the *practices of science*, and thus both are at the heart of scientific investigation.

CHAPTER 15

DISCIPLINARY CORE IDEAS: THE PRODUCT OF DOING SCIENCE

The disciplinary core ideas (DCIs) of the *Next Generation Science Standards* (*NGSS;* NGSS Lead States 2013) address the products of scientific investigation and summarize our present understanding of the universe. For students to truly understand how science builds understandings of larger importance, students need to apply the results of their own investigations to these larger ideas.[C39]

The problem is that the core ideas of a discipline encompass vastly more ideas, conclusions, and theories than could possibly be "covered" in a single year of study, or 10 years, or a lifetime. The danger is that so much effort might go into teaching students the products of scientific investigations that *doing* investigations will be given insufficient time for students to develop an understanding of the practices by which science proceeds.

We science teachers, both secondary and college, dispense facts, theories, and core ideas with great facility. Our familiarity with the ideas makes the information easy to understand, and we have a stockpile of illustrative activities to help students understand.

Teaching the theories is not only more familiar to us, but it feels like we and our students gain momentum as we "move through the material." I (Mary) have a colleague who teaches high school physics and has expressed frustration with investigative practices: "I let my students explore and experiment with balanced forces using carts and rails, but it took so long that I couldn't get through all the material."

It's true. There isn't enough time to cover all the material. Nor is there enough time for students to investigate everything.

Teachers must make choices about how much time to spend on a particular investigation or how thoroughly to address a particular core idea. The classroom experience comprises a selection of investigative activities, partial investigative activities, non-investigative activities, arguing from evidence, opportunities to work through short science reasoning problems, opportunities to see the teacher work through problems, and so on, all linked to each other and to the big conceptual ideas and theories of science.

Linking all those parts together into a coherent whole—the narrative of the curriculum—is one of the teacher's main jobs.[C40] As Paul Strube said in his 1994 article, "Narrative in the Science Classroom," narrative "offers a structure that allows scientific concepts to be more easily integrated into other conceptual understandings … and more easily structured and ordered in the mind." Through activities, discussions, and questions, a teacher prompts students to make connections among curricular elements: "Here's what we've done today, here's how it connects with what we did yesterday and what we're going to do tomorrow. Here's how our small experiments tie in to the big ideas of science. Here's how the big ideas of science explain our universe and predict the effects of human actions on it."

The interconnected nature of the scientific enterprise means that it is possible to start with almost any question, investigate the question through practices of science, and end up at a DCI from the *NGSS*. Figure 15.1 illustrates an example of how a curricular narrative—a story line—might guide students from a question (in this example, "Does convection hold up hot-air balloons?") to a DCI. A story line includes how the teacher ties one driving question, activity, or investigation to the next, and then the next, and finally to the bigger picture of science including the *NGSS* DCIs. The story line is what helps motivate curiosity and creates a comprehensive and connected fabric of understanding;

Figure 15.1

Illustration of the unifying nature of narrative starting with a single driving question

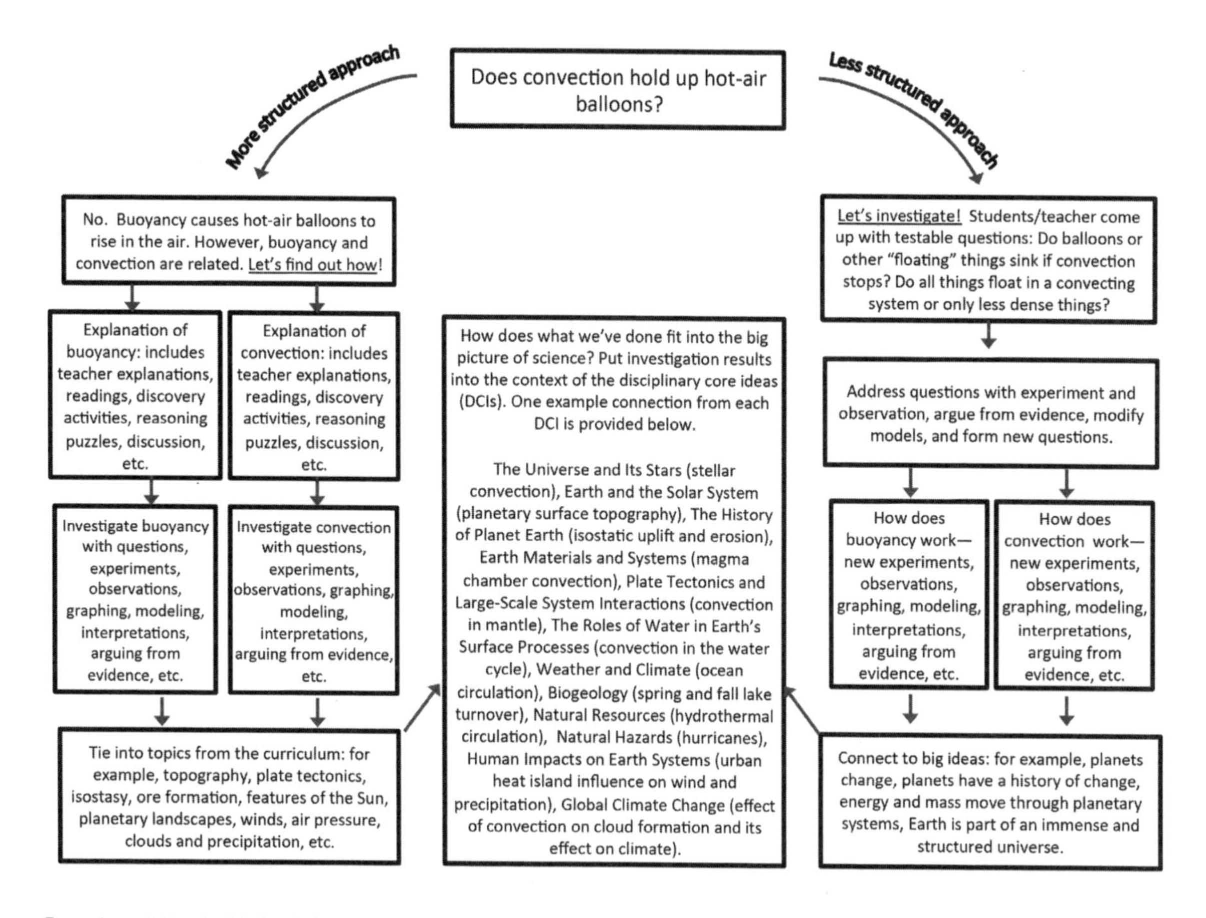

Example activities in this book that address issues relevant to buoyancy and convection include plate tectonics (its relationship to mantle convection is illustrated in Figure 5.6, p. 83), global air circulation (its relationship to atmospheric convection is illustrated in Figure 5.7, p. 85), modeling air pressure (Chapter 6), buoyancy and isostasy (Chapters 6 and 7), and crystal settling and hydrothermal circulation (both illustrated in Figure 14.10, p. 302). In this illustration, the arrows represent the curricular narrative, or story line.

it can take one of several directions and tie together investigative activities that can be more or less structured, depending on student initiative and the choices the teacher and students make. In either case, the work done by students can be woven into a narrative that connects to any number of core ideas for the discipline.

The narrative, represented by arrows in Figure 15.1, links the components of the story lines to show how classroom work supports the core ideas. For example, I (Mary) link my students' experimental investigation of uneven heating of land and water (which we talked about in Chapters 2 and 9) to the core idea of global circulation through several intermediate steps. After interpreting and discussing our experimental results, we model convection by observing convective circulation in a beaker (with pieces of oatmeal to mark the circulation). We apply the convective model to an interpretation of coastal breezes (I give them observational data for shifting winds in the morning at Padre Island in Texas, and they construct convective circulation models from sea to land and land to sea). Finally, we scale the model to global-size circulation, considering convergence near the equator and mean wind directions in the tropics. We often expand the idea of convection to consider convection in the mantle when we start our unit on plate tectonics.

As you considered the many investigative activities in this book, you might have thought, "Oh, my, I don't think I could fit all of this into a single year and still cover all the core ideas of the discipline." You're right. You can't. There are far more suggested investigations in this book than anyone can do in a class year. And this is only a small fraction of the teaching resources available for the classroom.

The good news is that the variety gives you options when a student's question, or your own curiosity, takes a class in an unexpected direction.

It isn't necessary to cover every possible idea of science (an impossible goal in any case). What is necessary is that you engage students in the practices of science and tie those classroom practices to core ideas, such as the *NGSS* DCIs, through a classroom story line.

You are the narrator in the classroom. Your role as curriculum narrator is varied and complex. It includes facilitating classroom discourse, thinking out loud with your students, and guiding students to make connections between different investigations and to the bigger ideas of science. The narration includes not only what you say, but also the way that you sequence classroom observing, questioning, investigating, modeling, and arguing. Much of this narration is improvised in live time, depending on what students ask, say, or do.

STORY HOOK: AWAKENING THE INNER SCIENTIST

The teacher steps to the front of the classroom. Student hearts pound with excitement as they anticipate the next unit of study. "Today, we're going to develop a model to illustrate how Earth's internal and surface processes operate at different spatial and temporal scales to form continental and ocean-floor features." (See *NGSS* performance expectation [PE] HS-ESS2-1.)

Isn't your heart pounding already?

A hand goes up. "Why are we going to study that?"

"Because it's the next thing in the *Next Generation Science Standards*! Plus we're going to be engaged in developing and using models, analyzing and interpreting data, using mathematics and computational thinking, and more!"

Wow. What a thrill ride it's going to be!

Pointedly, our (Russ and Mary's) hearts do pound with excitement at such giddy, seductive language. But the emotional engagement of most students probably rests on a stronger hook, something to draw them in and help them see the broader context and significance of the study.

The "hook"—the invitation that gives a fiction reader a reason to care—has long been a subject of discussion in creative writing classes. It's known for its intangible, undefinable character. "You know it when you see it" is a common, if unhelpful, description for the beginning writer.

So, why should students care in the science classroom? What provides the hook? Probably not "Today, we're going to develop a model to illustrate how Earth's internal and surface processes operate …." However, the hook has to be more than entertaining activities or nonscience questions from pop culture. The hook has to be tied to the actual science you plan to pursue as well as to student interests and previous understanding. Somewhere between "Today, we're going to learn about metamorphic rocks" and "Hey! Let's eat brownies and play games!" lies a sweet spot for engaging student interest in earth science.

Like hooks in literature, effective science hooks will probably be different for different people. Taking a canned hook from a textbook is probably going to be less effective than tailoring a hook for a particular class, situation, and moment. The hook might be something intriguing or unexpected that can be explored through the practices of science; for example:

"Hey, look how the layers in this rock are bent!"

"How can you bend a rock? Wouldn't it break first?"

Or the hook might link an observation or claim to bigger concepts that address the core ideas of the discipline; for example:

"A few billion years ago, a range of towering mountains once stretched across central Minnesota."

"Whoa! How can we possibly know that?"

Sometimes, simply turning a statement into a question can change the tone and create a hook. Brian Reiser has talked about the importance of a driving question as a way to initiate investigative activities that address a particular problem or observed phenomenon (see Reiser 2014).

Even the teacher's reaction to an activity can be influenced by whether it's posed as an assignment or a question. One fall, I (Mary) engaged my students in thinking about two different landscape features, Devils Tower in Wyoming and Ayers Rock in Australia, and why they are different. I first posed this challenge: "Develop a possible explanation for why the shapes of these two landforms are different even though the relief on each is the same." Later, I posed the assignment as a question: "Shouldn't landforms with the same relief have the same shape?" When I reframed the exercise as a question, it took on a new feeling in my mind. It presented a testable hypothesis that engaged my curiosity in a different way. It opened up possibilities for exploration instead of merely giving instructions for a task.

Let's consider how a statement might be turned into a question. A statement might be something such as the following: "Our next topic is phases of the Moon. We're going to learn what causes them." The implied question in this statement is "Why does the appearance of the Moon change from one night to the next?"

We might inject a bit of the unexpected by reformulating the question: "How did people thousands of years ago figure out that the Moon is a sphere orbiting the Earth and lit by the Sun?"

We might take an idea from the Moon phase activity in Chapter 13 to create a different kind of challenge: "What if a murder suspect told a story about seeing the Moon at a time and place where it couldn't be? Could you catch him or her in the lie?"

A driving question or other type of hook doesn't need to start with an *NGSS* DCI or PE. In fact, it's better if it doesn't. The best hooks start with small-scale phenomena that can be explored in the classroom through the practices of science. Thus, "What causes global warming?" is less effective than "What makes my can of soda go flat, and how does that bear on global warming?" Similarly, "What causes planet Earth's cycles of uplift and erosion?" is less effective than "What keeps ice cubes from sinking in my pop when I can feel they have weight, and what does that have to do with mountain ranges?"

The small-scale questions lead naturally into consideration of the bigger questions. For example, a question about ice cubes leads into an investigation of buoyancy like that in Chapter 6. Buoyancy, in turn, can be linked to cycles of uplift and erosion, plate tectonics, landscape, or other big ideas and DCIs through modeling and other curricular narrative.[C41] As explained in *Ready Set Science!* (Michaels, Shouse, and Schweingruber 2008), while teachers identify and promote long-term goals and connections to core concepts

> *they must also define shorter term goals for students that involve more immediate understanding. At each grade level, teachers will need to aim for teaching specific intermediate ideas, with an eye to how these will connect with and inform the more sophisticated concepts that students are building toward understanding. (p. 60)*

Small-scale questions usually create more *memorable* hooks than global-scale questions. I (Mary) remember a question posed by a physics professor during a dinner for new freshmen when I first entered Allegheny College in Pennsylvania. He asked the group at our table, "Which will cool off faster, a coffee cup with a spoon in it or one without a spoon?"

Wow. So much better than "Today, we're going to study heat transfer by radiation, conduction, and convection." I've remembered the driving question of the coffee cup for some 35 years!

Like the coffee cup question, a hook can arise from curiosity about phenomena that are part of students' everyday lives. Pressure in the building at the Horizon Middle School where I (Mary) teach often causes the doors to blow open. The question "Why do the doors in the middle school keep blowing open?" can be a hook for explorations of pressure, which can lead to a unit on weather.

A good hook—or driving question—gives students a compelling reason for doing science, which includes the following: observing, experimenting, graphing, interpreting, modeling, arguing from evidence, and asking more questions.

BETWEEN THE HOOK AND THE ENDING: THE HEART OF THE STORY

Much of the story of science lies in the messy middle[C42] between the initial hook (such as a driving question) and the final product (such as a core idea of the discipline). It is in this middle part of the story where students can dig into the crosscutting concepts and practices of science. It is in this middle part where students develop explanations for what they observe.

We teachers may be tempted to jump from the initial questions to the end of the story—the *NGSS* DCIs—in part because those ideas can seem like the most important content and in part because they may appear to be the focus of many of the *NGSS* PEs. This skips the messy middle of the story where students, like scientists, face the challenge of trying to figure something out that they don't already know the answer to. Jumping to the end allows students to apply models to understanding the universe, but they apply models that they are given without the opportunity to explore the foundational basis for them.

We suggest that it's more consistent with the vision of the *NGSS* to start with narrow questions such as the hooks proposed earlier in this chapter and use the practices of science and crosscutting concepts as appropriate to answer questions through classroom-scale investigation. Then, the results of those investigations can be fit into the overarching umbrella of bigger ideas, such as the *NGSS* DCIs.

Consider the story in Chapter 10 of the leaf inset into the snow (p. 197). That investigation did not start out to study heat transfer, differential absorption, or climate change. It began with an observation that caught the students' interest. They noticed something unusual, and they wanted to understand and explain what was going on in the real world. The investigation arose out of students' curiosity and was expanded on by a teacher who was able to leverage the curiosity so that the question connected to heat transfer and absorption. That study could proceed in the classroom through experimental activities, graphical interpretations, arguing from evidence, and so on. The teacher might connect the study with related activities to establish a wider understanding of heat transfer, absorption, or climate change.

In this way, the classroom narration emerges from the interaction between teacher and students. Figure 15.2 offers an example of how a unit of study might develop from an initial investigation into subsequent investigations, creating a story arc narrated by the teacher. In this imaginary unit on weathering and erosion, the teacher has two planned activities: the initial conservation of mass investigation and a concluding sediment deposition/transport experimental investigation leading into a study of point bars. The diagram shows conceptually how a teacher might develop a narrative to allow students to pursue spontaneous questions while also working toward the planned concluding activity—two different routes to the concluding activity are shown. Other related investigations could arise during a unit like this, such as the investigation of salt derived from the weathering of rock and the formation of Great Salt Lake or the mass balance exercise in Hawaii, both discussed in Chapter 14. The teacher as narrator helps keep the corresponding bigger ideas in the minds of students, ideas such as weathering and erosion, cycles and conservation of matter, or building and destruction of mountains.

As an example, I (Mary) engaged my students in making a variety of observations and interpretations of a map of the age of rocks in North America (the Tapestry of Time and Terrain activity that we talked about in Chapters 5 and 14). One of my students, after reading a news article about a 94-million-year-old fossil find in Dallas, Texas, asked, "How do we know how old rocks are?" That hook became the driving question for a variety of related investigations. Did the age of the fossil found in Texas match the age of the rock on our map? How do fossils form and what kinds of fossils are there? Do we find fossils in any age of rock, or does each type of fossil exist only in rocks of a particular age?

From our fossil investigations, we moved into a set of activities that specifically addressed the student question—how do rock layers and the crosscutting relationships

Figure 15.2

Diagram for an example unit of study on weathering and erosion illustrating how a variety of planned and spontaneous classroom investigations might arise from student questions

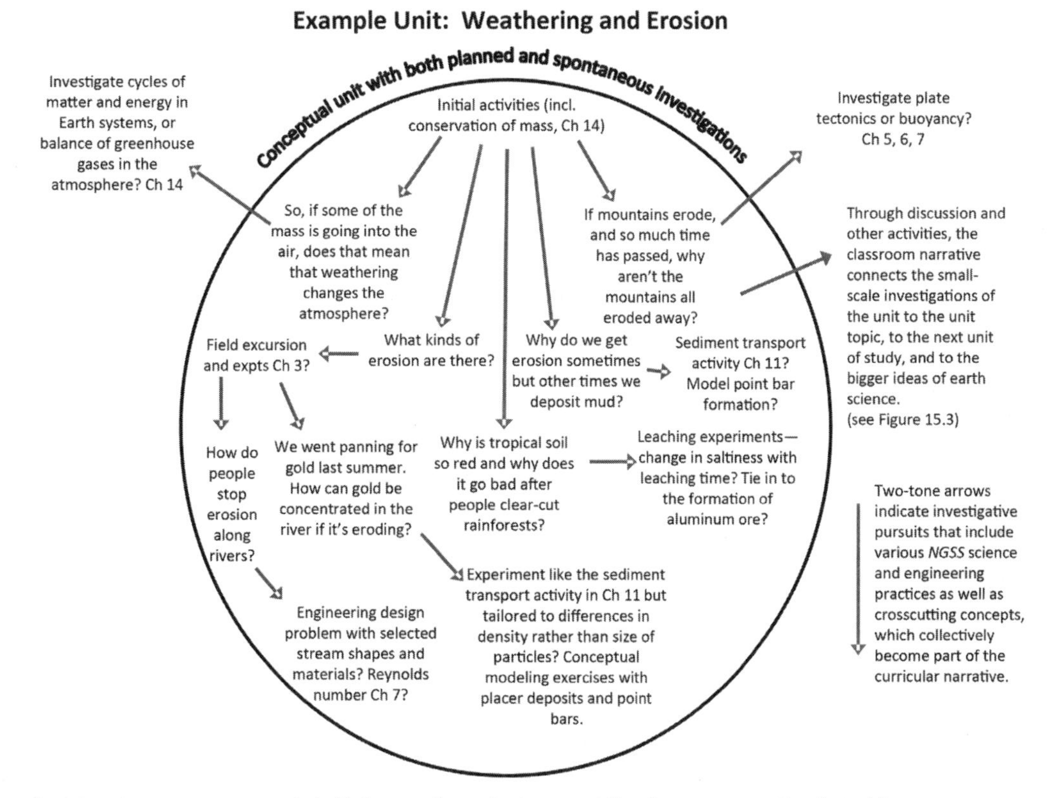

Each two-tone arrow represents both the practices of science and the classroom narrative that addresses one question and spawns the next. Note that each question and its arrow can be compared to the story line for a single question as in Figure 15.1.

reveal the relative ages of rocks—and from there back to my next planned unit of study on stratigraphy and the Grand Canyon (which I shortened from the activity we presented in Chapter 11). All of these investigations collectively addressed the bigger ideas of the "history of planet Earth" and "planets have a history of change," which I kept in students' minds by posting them on the wall and periodically engaging students in discussion of how what we were doing fit into those ideas.

PLANNING AND ORGANIZING THE STORY

Structure and Classification

In planning my curriculum for the school year, I (Mary) consider five or so big ideas that lend a structure to the work my students and I do. Wiggins (2010) offered a vivid description of the usefulness of big ideas:

> *An idea is big if it helps us make sense of lots of otherwise meaningless, isolated, inert, or confusing facts. A big idea is a way of usefully seeing connections, not just another piece of knowledge ... a big idea is a powerful intellectual tool, from which we can derive more specific and helpful understandings and facts. ... A true idea doesn't end thought, it activates it. It has the power to raise questions and generate learning. So, build your unit around one idea with power, an idea that helps learners make sense of otherwise isolated content and which cannot help but bring inquiry to the fore.*

My big ideas are worded in student-friendly language and fit within the *NGSS* DCIs, although they are not identical to them. I think about what the students need to understand (the "little ideas") to get to each big idea and then develop a series of units that engage students in the practices of science, the crosscutting concepts, and the core ideas that collectively address a particular big idea. An example set of six units addressing a big idea ("planets have a history of change") and the *NGSS* performance expectation MS-ESS1-4 ("Construct a scientific explanation based on evidence from rock strata for how the geologic time scale is used to organize Earth's 4.6-billion-year-old history") is shown schematically in Figure 15.3. Each unit includes a set of driving questions and investigations like those shown in expanded form for one unit in Figure 15.2, as well as reminders of factors that influence the narrative, such as student questions, teacher prompts, news items, and suggestions from the curriculum. Each unit connects to other units, and to the big idea, through the curricular narrative developed by the teacher.

Notice that the objective of the PE cited in Figure 15.3 (MS-ESS1-4) is not to "learn" a DCI, or a particular science or engineering practice, or a crosscutting concept. Rather, the PE is addressed collectively by the entire suite of learning activities: science and engineering practices, crosscutting concepts, little ideas, big ideas, and the DCI. Even as students consider a focused question such as "How do we know ancient mountains eroded away at the Grand Canyon?" (one of the ideas explored in Chapter 11), they can be thinking about bigger questions such as "What causes the Earth to change?" (a big idea) or "How does matter and energy cycle through both time and space on Earth?" (see *NGSS* DCI ESS2). Weaving all the parts together into a coherent whole is the job of the classroom narrative.

The set of units presented in Figure 15.3 addresses more than a single PE. Typically, a set of investigations that engages students in the practices of science will simultaneously

Figure 15.3

An example set of six units that addresses the big idea "planets have a history of change" and *NGSS* performance expectation MS-ESS1-4

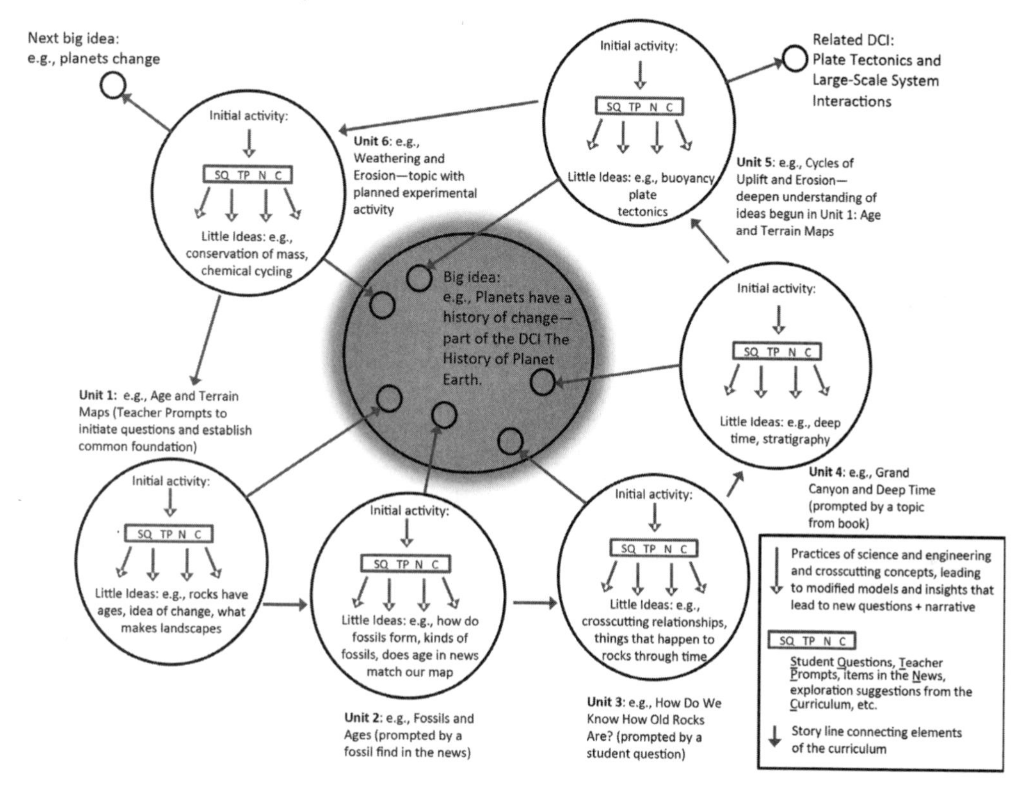

Notice that this set of units does not cover the entire scope of the big idea (or the *NGSS* DCI) as symbolized by the small circles within the larger circle, but it does address them in a substantial way. Elements of the story line are symbolized by the gray arrows and by the two-tone arrows. The practices of science and crosscutting concepts—the important *doing* of science that contribute to the narrative—are also present in the two-tone arrows in this figure, as in Figure 15.2.

address multiple PEs, allowing PEs to be "bundled" in the curriculum.[C43] In other words, any one unit can teach toward multiple PEs at the same time. For example, the set of activities shown in Figure 15.3 also addresses PE MS-ESS2-2, "Construct an explanation based on evidence for how geoscience processes have changed Earth's surface at varying time and spatial scales."

There isn't one "right" way to classify, organize, or sequence the ideas of earth science. In my (Russ's) first term in college, an honors biology professor made a comment about classification that has stuck with me. He arrived to class a bit late and one of the students asked, "So, what were you doing?" "I was working on my paper," he said. "What's the paper about?" the student asked. The professor laughed. "Classification. That's what you do when you don't have any real research to present!"

From his tone and following explanation, I knew he was only half joking. He explained that a new classification can open up new ways of approaching ideas or thinking about problems, but it does not change the nature of underlying discoveries. Classification is only a tool in communicating a scientific idea, not an end goal.

John Rodgers, former member of the National Academy of Science and expert on the geology of Appalachia, where we (Russ and Mary) did our graduate work, once said that the good classification is one that leads to good observations. It follows that the good classification will be different in different situations, depending on what kinds of observations are needed. Different individuals will select different classifications as being more helpful in characterizing a system under different circumstances, but there is no one "right" classification.

Unsurprisingly, then, there are many different ways for organizing and structuring—that is, classifying—curriculum. For example, at an EQuIP (Educators Evaluating the Quality of Instructional Products) training workshop in Florence, Kentucky, in June 2015, Sean Elkins, one of the workshop leaders, talked about the possibility of organizing curriculum around the *NGSS* crosscutting concepts. Rodger Bybee has promoted the BSCS 5E Instructional Model for learning science—providing a mnemonic that encourages use of the important instructional strategies of Engage, Explore, Explain, Elaborate, and Evaluate (Bybee 2015).

I (Russ) often base my college curricular sequence on concepts of "How We Know" (e.g., how we know about Earth's interior, how we read a planet's history, how we know the composition of stars) or "How Things Work" (e.g., how streams erode or deposit sediment, what causes geysers, how earth differentiates). I (Mary) tend to organize my curriculum around "big ideas," such as that shown in Figure 15.3. My big ideas form a structure within which driving questions arise organically and the practices of science and crosscutting concepts naturally fit.

The *NGSS* offer three different paradigms for how to organize the breadth of science: the three dimensions of the *NGSS*. Each of the dimensions provides overarching thematic concepts within which all areas of scientific investigation fit. There are many ways to pursue scientific investigation, but they all fit within the framework of the *NGSS* science and engineering practices and crosscutting concepts. There are many topics of study, but they all fit under the umbrella of the *NGSS* DCIs.

A reviewer of an early draft of the *NGSS* commented that key earth science topics—such as the rock cycle—were missing from the document. No. It's there as part of the PEs and DCIs. But the PEs and DCIs are not a list of topics to cover. They neither mandate nor omit specific topics of study.

Reading the Earth *Is* the Content Goal

The title of this book, *Learning to Read the Earth and Sky*, emphasizes the practices of science by which we read the stories rather than the stories themselves. We point out that if learning to read the earth and sky is the goal, it doesn't really matter all that much which "stories" students and teachers use as the vehicle for learning to read. The important curricular choice is not which facts, conceptual ideas, or theories to pursue in the classroom, but rather how to incorporate the practices by which they are pursued.

The *NGSS* DCIs offer the following thematic concepts for Earth and Space Sciences:

- Earth's Place in the Universe includes the following component ideas: The Universe and Its Stars; Earth and the Solar System; and The History of Planet Earth.
- Earth's Systems includes the following component ideas: Earth Materials and Systems; Plate Tectonics and Large-Scale System Interactions; The Roles of Water in Earth's Surface Processes; Weather and Climate; and Biogeology.
- Earth and Human Activity includes the following component ideas: Natural Resources; Natural Hazards; Human Impacts on Earth Systems; and Global Climate Change.

Few classification systems produce perfectly separate categories, meaning that there is always overlap among the categories. The DCIs are no different. In fact, the writers of *A Framework for K–12 Science Education: Practices, Crosscutting Concepts, and Core Ideas* (NRC 2012) chose to emphasize the overlaps. For example, there are few, if any, aspects of earth science that do not fit into both Earth Materials and Systems (ESS2) and The History of Planet Earth (ESS1C). The idea of global climate change, specifically listed under the DCI Earth and Human Activity, overlaps with the history of planet Earth because understanding how climate changed in the past is foundational to understanding present change. Climate change is also related to the ideas of water cycling, energy and matter cycling, and weather and climate, all of which fall under Earth's Systems.

Not only is there significant overlap among the branches of the *NGSS* DCIs, but the Earth and Space Sciences DCIs are completely nested; human activity takes place within the context of the Earth systems, and Earth systems take place within the context of Earth's relationship to the universe. This overlap and nesting means that the Earth and Space Sciences DCIs don't work particularly well as a topical guideline for curriculum development. Instead, they work as a conceptual framework that invites consideration of the interconnected and interacting nature of Earth systems.[C39]

Developing Big Ideas in Student-Friendly Language

If the objective of the curriculum narrator is for students to understand both the practices of science and how those practices inform our conceptual understanding of the universe, then the goal is not for the teacher to explain the information well. Nor is the goal to engage students in brainstorming sessions that start with a theory and ask students to come up with evidence or arguments without prior experimental or observational experience. Rather, the goal is to integrate the students' explorations into the fabric of larger models and ideas of science.

Getting students to see how their classroom-size investigations fit within planet- and universe-size models requires repeated opportunities to investigate, argue from evidence, model, and discuss. It requires multistep opportunities to apply observational and experimental evidence to construction of models and to fit expanding ideas into increasingly complex models.

The *NGSS* DCIs provide high-level links among intersecting branches of earth science, but in some cases they may be too abstract or large scale for students to connect to their own observations and small-scale experiments. Teachers might want to come up with some big ideas tailored to their instruction that form a bridge for students between classroom work and the bigger concepts. I (Mary) post my big ideas prominently in the classroom. Throughout the year, I provide opportunities for students to connect what they learn to those ideas. Here is one set of big ideas:

1. Planets change.
2. Planets have a history of change.
3. Energy and mass move through planetary systems.
4. Bodies move in space.
5. Earth is part of an immense and structured universe.

These big ideas are shown in Figure 15.4 as four overlapping circles enclosed by a fifth, larger circle. These five circles are superimposed on the *NGSS* DCIs, which are represented by the three concentric nested circles. Other sets of big ideas that serve as a bridge between small-scale classroom investigations and big-scale disciplinary core ideas are possible.

Teachers might be concerned that the all-encompassing DCIs of the *NGSS* cut the link to more familiar big ideas of earth science. For example, the DCIs do not specifically identify classics such as telling stories from rocks and strata (the wellspring of nearly everything we know about Earth's past), the movement of cyclones and fronts (the

Figure 15.4

Illustration of the nested relationship among the three *NGSS* earth science DCIs, with superimposed big ideas

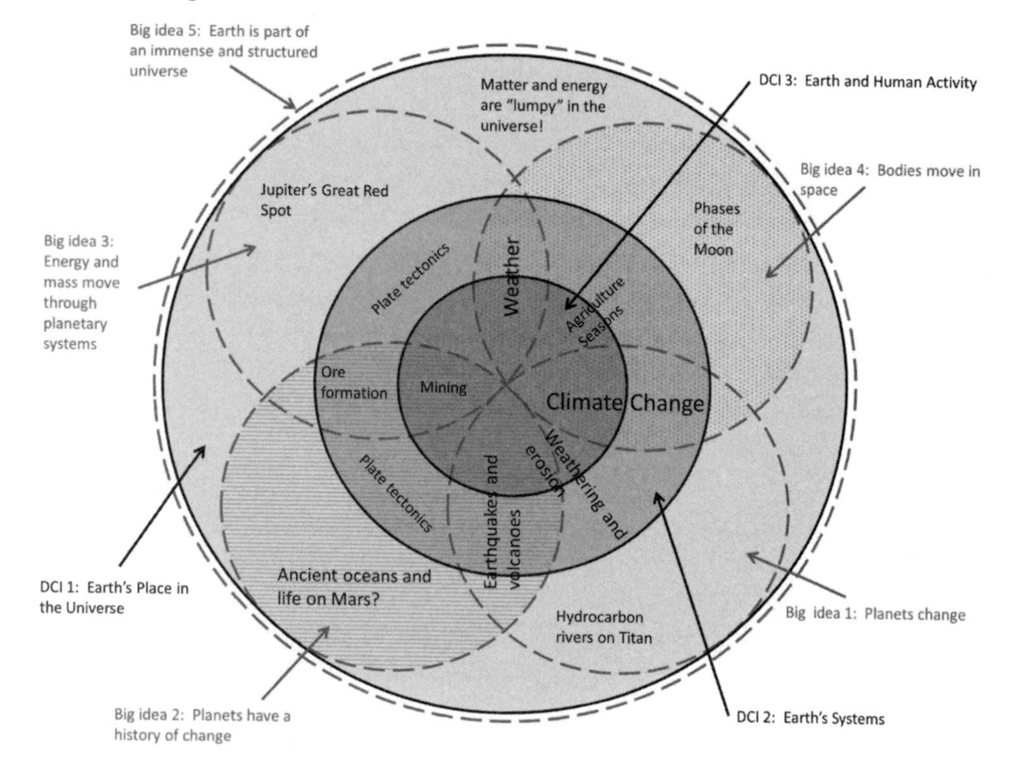

The three DCIs are shown as three nested and concentric circles bounded by solid black lines. The five big ideas are shown as five circles bounded by lighter dashed lines—four overlapping circles nested within a larger circle that represents big idea 5. The big idea circles occupy the same space as the DCI circles, meaning that they address the same material but organize it in a different way.

A few traditional earth science topics are shown to illustrate how these relate to classroom-friendly big ideas and the DCIs. For example, climate change and weather on Earth could be associated with several of the big ideas shown here. Depending on what aspect a teacher wants to emphasize, these studies could be associated with movements of bodies in space, cycles of matter and energy in Earth, or planets changing (and past climate change could be associated with the big idea that planets have a history of change, a relationship not shown on the diagram). Likewise, climate change and weather on Earth could be associated with either DCI 2, Earth's Systems, or DCI 3, Earth and Human Activity. Although not shown explicitly on the diagram, climate change or weather on other planets, which would fall outside the DCI 2 circle, could be associated with DCI 1, Earth's Place in the Universe.

traditional foundation for understanding weather), or the processes that shape planetary surfaces (one of the primary new discoveries of the last half century). Instead, the DCIs emphasize system interactions (energy and matter) and Earth's place in space and time. We argue that traditional earth science topics can fit into this scheme in a variety of ways, some examples of which are shown in the Figure 15.4.

A key takeaway point from Figure 15.4 is that many ideas of earth science are connected to each other in multiple ways. For teachers, trying to memorize complex educational classification schemes (including the *NGSS*, this book, or others) may be less useful in the classroom than starting with a driving question, recognizing that everything is linked to everything else, and connecting the investigation to appropriate big ideas, core ideas, and practices of science through curricular narrative. For example, a driving question such as "Why do cloud streamers rise above Minnesota lakes on a frosty fall morning?" is related to changes in heating patterns caused by movements of bodies in space (a core idea), to energy transfer across systems (a crosscutting concept), and to modeling of systems (a practice of science).

One key to successful classroom narration is to be aware of the many different ways that particular experimental investigations or topics can fit into the *NGSS* DCIs (like the examples shown in Figures 15.1–15.4) and to be flexible in pursuing activities that help students make those connections even when student-initiated investigations don't go in the exact direction prescribed by the planned curriculum. A second key is to remember that classroom narration is not achieved by simply telling students the connections but by helping students discover those connections through activities, discussion, and argument.

We summarize by noting that the DCIs for Earth and Space Sciences are not intended as "information to be taught" but rather as a conceptual framework for learning.[C39] Real research begins with small-scale experiments that are applied to understanding planet-size concepts. We suggest that classroom instruction do the same—engage students with experiments and practices that are at a scale they can investigate, and link their results to bigger ideas such as the DCIs.

STORY THEME: WHAT'S THE POINT?

Every good fictional story needs a purpose that gives the narrative more significance than merely a series of plot turns and challenges for the protagonist. Likewise, the classroom narrative needs a purpose that gives it greater significance than merely a series of activities and arguments to challenge the students.

Narrating a classroom story (that is, curriculum) that begins with student-initiated questions, meanders through a variety of investigations and activities, and ends with big ideas of science, requires an awareness of the key purpose of science education.

Although the *NGSS* can appear to be a complex tangle of ideas and directives with multiple agendas and goals, we argue that the core message is quite simple: First, students should learn to *do* science, and second, they should be able to fit what they do into a wider fabric of science.

In previous chapters, we argued that, as a nation, we have been reasonably good at teaching students the models and theories of science—even how to explain the universe in terms of the models or to apply theories to a variety of problems. We've done less well at exposing students to the basic practices of science on which those models and theories are based. Part of the reason is that we struggle to distinguish between basic practice and theoretical model, as we discussed in Chapters 8 and 9. For example, suppose the curriculum calls for learning about convection and students do an activity where they watch bits of oatmeal circulate in a beaker of warming water. Is this an investigative activity? Not really. It's a visual model, not different in kind from an illustration in the textbook or an animation online. There is no measurement of variables or figuring out a mechanism or cause for convection. No graphing. No arguing from evidence.

Many classroom activities in earth science are like the oatmeal model for convection—addressing the models and theories, and applying those to an understanding of large-scale processes, but not addressing the more basic practices of science such as experimentation, graphing, and data interpretation. It is through these basic practices of science that knowledge is constructed, both by scientists and by student-scientists in the classroom.

Some years back I (Russ) read an article in a newspaper about a poll taken to determine science knowledge among Americans. The article reported that somewhere in the range of 50% of people did not know that the Earth is billions of years old. As I remember it, the writer was astonished that so many people had not learned the age of the Earth. The article conveyed an expectation that if those people had heard how old the Earth is from authoritative sources, they would then "know" it. My interpretation of the poll results was quite different. The significance was not that 50% of those people had not *learned* that the Earth is old, but that they did not *agree*. The problem was not that they had not been told the Earth's age, but that they had not explored, experimented, and tested the observations of the Earth to make that understanding their own. They had not done science, they had only been told science, and they didn't believe it.

The important goal of science education, its theme and purpose, is that students understand the practices of science and gain ownership in investigations of their own making. We believe that a focus on teaching the facts and theories of science, even if they are applied to real problems, and even if based on the *NGSS* DCIs, will fall short of this goal. Cynthia Passmore wrote the following on the *NSTA Blog* in 2014:

CHAPTER 15

Part of the promise—and the challenge—inherent in the vision portrayed in the Framework *is for us as educators to move away from lists of discrete facts organized into separate units and toward a coherent set of ideas that can provide a foundation for further thought and exploration in the discipline. Indeed, the writers of the* Framework *and most science educators agree that it is impossible to cover all the scientific content that is relevant to modern science.*

Back in the 1990s, one educational buzz phrase was "less is more"—meaning that teachers shouldn't try to teach all the "material" but rather should leave time for exploration and discovery. We argue that this important concept applies to classroom application of the *NGSS* DCIs. All of earth science comes under the umbrella of the DCIs, which, as Passmore points out, cannot possibly be covered in its entirety. We suggest that the *NGSS* DCIs *can* be addressed without teaching the entire scope of earth science; they can be addressed by recognizing that their purpose is to provide a framework and destination for the practices of science.

The word *content* is often used to refer to factual ideas and theories of science. Passmore used the word in this way in the passage above. When secondary teachers worry about not covering all the content, they, too, mean the facts and theories. The title of an article by Nicki Monahan in the October 12, 2015, issue of *Faculty Focus* is "More Content Doesn't Equal More Learning." Again, the word *content* refers to facts and theories, more of which don't necessarily produce more learning.

We (Russ and Mary) suggest that we (the science education community) should instead use the term *content* to refer to the sum total of the practices of science, core ideas, and crosscutting concepts. *Doing* science is the content. And doing more of it does, in fact, produce more learning. Science—that is, the construction of an understanding of nature based on observation and testing—is something to be practiced, and more practice leads to deeper understanding.

We can't emphasize this enough: If we are to successfully implement the vision of the *NGSS,* we need to stop using the word *content* to refer to information and instead use the word *content* to refer to the entire sweep of skills and practices that make up the enterprise of science. This is the content that science classes should convey, not only in secondary schools but in college as well. Factual information can be part of that, but probably not the most important part.

Let's compare teaching science to another type of teaching. Suppose we wished to teach someone how to make a shirt. The fabric itself is not the goal of teaching, nor is the finished product, the shirt, the goal. Rather, it is the process—how to cut and sew the fabric—that is the goal (although we may on occasion try to impress our friends with our knowledge of sewing by pointing to the bolts of fabric with which we plan to make shirts someday!)

Often, in education, we focus on the topics, ideas, or theories because these show up as named items in the curriculum and are easily identified as elements to learn. What's more, they are more easily assessed. For example, the questions "Do you know this fact?" and "Can you explain a process in terms of this theory?" are much easier to test than "Can you carry out and interpret the results of an investigation?"

The key message of the *NGSS*, as we interpret it, is that the heart of science is found in the practices of science and crosscutting concepts—the arrows in Figures 15.2 and 15.3. The topics, ideas, and theories are the products of doing the practices and applying the crosscutting concepts. The scientifically literate person is the one who has done the arrows, not the one who accepts the theories.

In the end, we are left with Albert Einstein's famous statement on education: "The value of an education ... is not the learning of many facts, but the training of the mind to think something that cannot be learned from textbooks" (Frank 1947).

FINAL THOUGHTS

Researchers don't usually start out thinking "OK, I need to make sure I get crosscutting concepts into my study" or "Let's see, I need to first ask a question, and then plan and carry out an investigation, and then argue from evidence." Instead, researchers have a driving question they want to answer and an understanding of the big picture within which that question fits. They use the practices of science and crosscutting concepts as appropriate to answer the question.

Implementing an authentic science experience in the classroom requires a similar focus on the question at hand. Classroom narration allows that focus while still exposing students to the full sweep of the practices of science, crosscutting concepts, and core ideas. The classroom narration integrates and connects ideas and practices without simply dropping them into the class as learning objectives in their own right, disembodied from the questions and interests of the students.

Individual *NGSS* PEs are not student learning objectives.[C43] One way to think about a PE is that it provides an example story line linking a particular science and engineering practice, crosscutting concept, and DCI. The emphasis is that the PEs are *example* story lines. As curriculum narrator, the teacher is constrained less by any particular PE than by the need to connect student curiosity and initiative to the big ideas of the discipline through the practices of science. In other words, the *NGSS* PEs are not a curriculum, which the *NGSS* goes out of its way to affirm in Appendix A:

> *The* Next Generation Science Standards *are student performance expectations—not curriculum. Even though within each performance expectation Science and Engineering Practices (SEPs) are partnered with a particular Disciplinary Core*

Idea (DCI) and Crosscutting Concept (CC) in the NGSS*, these intersections do not predetermine how the three are linked in curriculum, units, or lessons.* (NGSS Lead States 2013)

REFERENCES

Bybee, R. W. 2015. *The BSCS 5E Instructional Model: Creating teachable moments.* Arlington, VA: NSTA Press.

Frank, P. 1947. *Einstein: His life and times.* Translated by G. Rosen, edited and revised by S. Kusaka. New York: Alfred A. Knopf.

Michaels, S., A. W. Shouse, and H. A. Schweingruber. 2008. *Ready, set, science! Putting research to work in K–8 classrooms.* Board on Science Education, Center for Education, Division of Behavioral and Social Sciences and Education. Washington, DC: National Academies Press.

Monahan, N. 2015. More content doesn't equal more learning. *Faculty Focus. www.facultyfocus.com/articles/curriculum-development/more-content-doesnt-equal-more-learning.*

National Research Council (NRC). 2012. *A framework for K–12 science education: Practices, crosscutting concepts, and core ideas.* Washington, DC: National Academies Press.

NGSS Lead States. 2013. *Next Generation Science Standards: For states, by states.* Washington, DC: National Academies Press. *www.nextgenscience.org/next-generation-science-standards.*

Passmore, C. *NSTA Blog*. 2014. Implementing the *Next Generation Science Standards*: How Your Classroom Is Framed Is as Important as What You Do in It. November 10. *http://nstacommunities.org/blog/2014/11/10/implementing-the-next-generation-science-standards-how-your-classroom-is-framed-is-as-important-as-what-you-do-in-it.*

Reiser, B. J. 2014. Designing coherent storylines aligned with *NGSS* for the K–12 classroom. Presentation made at the National Science Education Leadership Association Professional Development Institute, Boston, MA.

Strube, P. 1994. Narrative in the science classroom. *Research in Science Education* 24 (1): 313–321.

Wiggins, G. 2010. What is a big idea? Big Ideas. *www.authenticeducation.org/ae_bigideas/article.lasso?artid=99.*

16

EARTH SCIENCE WHERE YOU ARE

"Don't reinvent the wheel" might be good advice in the business world, but our purpose in teaching is to show students how to invent, not just how to use wheels. If we don't engage our class in the process of inventing, our students will grow to depend on someone else's wheel and never learn to invent.

With an emphasis on the invention inherent in science—that is, the practice of *doing* science—students can learn to make sense of observations. They can learn how scientists have worked for centuries to put together an understanding of the earth and universe. They can learn that even scientists don't know at first what to look for or how to look for it. They can experience the confusion that scientists experience with new and puzzling observations. They can sense the disappointment of committing years to figuring something out only to realize that embryo ideas were wrong and have to be changed. They can share the exuberance when a new idea makes sense not only of one scientist's observations, but also the observations and ideas of other scientists.

They can realize that discovery comes from their own efforts and not from a teacher with an answer key in their back pocket.

STEP BEYOND THE COMFORTABLE

I (Mary) remember when I first knew I wanted to be a geologist—when I realized that I could discover things about the earth *on my own.* It was all around me—in the sandstone and shale bedrock of the Allegheny Valley, in the sky that opened up between the flat-topped hills of the Allegheny Plateau, and in the processes I could see at work in the channel and banks of the Allegheny River—and *I could figure things out by observation and reasoning.* The earth revealed its secrets to me if I knew how to look and if I thought about what I saw.

The *Next Generation Science Standards* (*NGSS;* NGSS Lead States 2013) propose that the key "content" of a science class should not be the factual material that scientists have accumulated over the centuries. Rather, the key content passed on to the next generation of thinkers, voters, and inventors should be how we read the earth and sky. The key content should be a participant's understanding of the process of science, engaging students in learning for themselves how we know what we think we know.

This call to do science, not just learn it, is not a new one. The *NGSS* predecessor, the *National Science Education Standards* (NRC 1996), made a similar call 20 years ago. The language has been updated, and the emphasis on doing science in the classroom renewed, with the hope that even more teachers and students can participate in this most important part of science.

The Call to *Do* Science

Unless our educational system focuses more on teaching students how to think than on what to think, our populace will become increasingly credulous. Scientists and educators alike need to realize that the educated person is not the person who can answer the questions, but the person who can question the answers. In our age of rapidly changing information, knowing how to distinguish truth from falsity is more important than knowing what was once considered true or false.

—Theodore Schick Jr., "The End of Science?" *Skeptical Inquirer,* Vol. 21.2, March/April 1997

Let's suppose that you buy into the idea that learning how to do science is more important than knowing science facts and theories. How do you abandon the answer key and engage your students in authentic investigations—that is, investigations in which you and your students don't already know the outcome? Diving into investigative activities without an answer key is a scary prospect, and, with the next-period students already swarming through the door, when will you find time to plan it?

We suggest starting where you are. Start with the curriculum you know and the activities you understand. But don't stay there. Reach beyond the confines of your curriculum and premade activities and head off into the unknown.

Follow your students' momentum—what driving questions arise from the class? Follow your own curiosity—what's new for you? Follow your strengths—what particular insights can you offer your students? Are you particularly good at arguing from evidence? Interpreting graphs? Working in the field? Follow your passion. I (Mary) first learned geology within the context of the flat-lying Paleozoic rocks of the Appalachian Plateau. There are myriad other geological contexts, but what engaged my passion was learning earth science *where I was.*

You can invoke that same passion in the classroom, not only in your students but also in yourself. Start with where you are in location.[C44] Where you are in your existing activities. Where you are in your own understanding or the understanding of your students. Where you are with a puzzling problem you can't answer. Where you are with what's in the news or what's going on in your community.

Start where you are and explore earth science with your students.

START WHERE YOU ARE

There are many ways to start from where you are. In the following subsections, we explore options from various perspectives, including location, familiar activities, present understanding, discrepant events, and what's in the news.

Location

We have given examples of field activities in several previous chapters. Engaging students in activities specific to your region achieves two important educational goals. First, it gives students a sense of ownership in discovery—they are exploring material that comes from their own life, not simply from the lab manual. Second, it lets students see that you, the teacher, are engaged with science and are not simply a conduit of science information from the textbook. Their perception of whether *you* are *doing* science is perhaps the strongest message that you send them about the nature and importance of science.

Engagement with a region doesn't necessarily have to include the logistical complexity of a field excursion. Earth science happens all around us all the time, and if we're alert we can identify features or processes in our different regions that students can relate to.

There aren't any exposures of bedrock in the Red River Valley of northwestern Minnesota where we teach, nor is our region noted for its grand erosional features. However, in our farming region, most students have noticed erosional processes that happen in farm fields. In 2007, I (Mary) gave a talk at a Minnesota Earth Science Teachers Association conference about a question-asking game that I invented to encourage students to make and interpret observations of a small delta that formed in a local farm field after a big rain (Colson 2007). The game engaged students in trying to figure out how the feature formed, thinking through the processes that made it, why it had a flat top, and so on. A teacher at the conference asked me "How did you know how to do this?" The question caught me off guard and I answered, "I'm a geologist. Of course I know how to do it." My unprepared answer felt arrogant and unhelpful, and shortly after I regretted saying it.

So, with time to reflect on the question, how did I do it? First, I had to notice the feature and recognize its similarities to a river delta (the takeaway message: be looking for geological features in your region). Then, I had to think through the crosscutting relationships and their implications (the takeaway message: be engaged with your own scientific exploration of the world around you). I had to think about what the flat surface of the delta told me about how it formed. I had to reason through the implications of the various levels of wave-cut shorelines. Some of my reasoning is shown in Figure 16.1 (p. 330).

Figure 16.1 ✪

Delta feature in a farm field associated with a storm that hit northwestern Minnesota in June 2000, showing the relationships that were necessary to figure out before developing the question-asking game

Surfaces A, C, and E formed by wave action eroding into the farm field at a variety of water levels.

Gully B eroded downward rapidly because it was above the level of the standing water.

The flat-topped feature D is a delta formed as sediment filled in a low area up to the surface of the standing water.

Because these features are related to changing water level as the water evaporated and/or soaked into the ground, the first events took place at the highest water level (A and D) and the later events at lower water level (C and E).

The wave-cut shoreline of the little pond cuts across the trend of the planted rows—so, the planted rows had to come first.

There isn't much dead plant material poking up from the pond. Thus, little if any plant growth had occurred before the storm and the water suppressed any growth subsequent to the storm.

Finally, I had to think of a creative way to introduce the feature and its interpretation to my students. I chose the question-asking game to encourage my students to look critically at the photos and ask questions. I constrained them to the kinds of questions that could only be answered "yes" or "no," which kept us focused on the observational evidence. They couldn't ask *why* questions. This game made interpreting the feature and inferring geological processes fun (at least for me, and I think for my students). My activity is shown in the annex (pp. 340–343).

Although laboratory work plays an important part in earth science, earth science is by its nature an outdoor field science. The data for understanding the earth is found in the earth. Regardless of curriculum quality or how well you do your classroom activities, your students can never get a complete understanding of the discipline unless you take them out of the classroom in some way. Engage your students with the earth science in your region. Like a fine suit that needs to be tailor-made to the person and the situation, and not just picked off the rack, some of your best activities will be tailor-made to your location and for your students.

Get Your Students Outside

Be creative with your local opportunities to get students out of the classroom. When I (Russ) first started teaching in northwestern Minnesota, I faced five of the snowiest winters on record. Rocks or features in the region were covered by snow for entire terms. So I took my introductory college students outside to study the stratigraphy of snow layers and the sequence of storms and snowplow clearings that the layers recorded.

Ken Huff, one of the writers on the *NGSS*, reported doing a different snow-layer activity with his sixth graders in upstate New York. They measured the temperature of the different layers of snow, yielding the surprising (to students) result that the lower layers were warmer, a consequence of the flow of heat from the earth—probably heat stored over the summer in this case, rather than heat flow from Earth's mantle (Huff and Lange 2010).

Familiar Activities

In October 2015, a teacher posted the following request on the NSTA Listserv:

> *I am teaching Earth Science and preparing for my unit in Earth's Structure. This includes the systems, how Earth was formed, and the layers of Earth. It's been a while since I taught this subject so I'm struggling with finding appropriate resources and materials for 6th graders.*

Of the first six responses, five of them involved various labs with food—M&M's, Italian salad dressing, Milky Way bars, and graham crackers. Food labs are a favorite classroom activity in earth science. They create a fun experience that helps students remember terms or concepts by associating the terminology or concept with a physical experience outside the textbook, lecture, or worksheet. However, the pedagogical goal of edible models is usually to reinforce vocabulary or to create a memorable physical model, not to measure relationships or explain observations seen in the real world. Food labs usually don't engage students in the practices of science, which is a key theme of this book.[C32]

Safety Note

Remind students in any food lab that, given the risk of cross-contamination, no food used in the lab is to be eaten!

How can you start with a fun food activity, in which students learn terminology or a concept, and expand it to introduce some investigative science? Let's take the Milky Way bar activity cited by one of the responders. The version of the activity posted in the Listserv was prepared by Barbara Ferri in May 2009 and (to our knowledge) is not published, but it is adapted from a very similar activity by Caroline Singler and Alison Sanders-Fleming, which can be viewed at *www.lsrhs.net/departments/science/faculty/Sanders-Fleming/PDFfiles/CurrentPDF/PlateTectES2EdiblTctLab.pdf.* The caramel represents the Earth's plastic asthenosphere, and other layers represent Earth's crust and deeper mantle. Pulling the candy bar apart represents a divergent boundary. Pushing it together represents a convergent boundary. Your fingers represent the forces of convection in the Earth's interior tugging at the tectonic plates. And so on.

This activity, called "Edible Tectonics," engages students in filling out a worksheet that compares parts of the candy bar to parts of the Earth. In this way you can lead students through a significant vocabulary and concept lesson with a candy bar at its heart. But students are not taking any measurements or making any interpretations of the candy bar's behavior. They are not addressing questions that can be answered by the results of their activity. The terminology and concepts of plate tectonics must come from outside the activity itself and are not a product or discovery of the activity. Thus, this activity is not a scientific investigation. But it might be a great place from which to launch an investigation.

For example, suppose that you prompt your students to compress the candy bar from each end—the convergent boundary in the activity—and then measure the change in dimensions, as shown in Figure 16.2. How much does the candy bar shorten? How much does it thicken? Can you chart the relationship on a graph—amount of shortening versus amount of thickening? Does the total volume of the candy bar change or does only its shape change? Does the shortening occur because of plastic deformation (oozing of the material) or because of brittle deformation (breaking and faulting)? Does the thickening take place equally over the entire length of the candy bar, or is it focused on some particular part of the candy bar?

A similar set of experiments, measurements, and interpretations might be done based on pulling the candy bar apart. How much does it lengthen? How does the thickness change? Is there a net change in volume? Does the candy bar stretch equally along its length or is the stretching focused in a particular location? What might cause the

Figure 16.2

Compression experiments with a Milky Way bar at room temperature

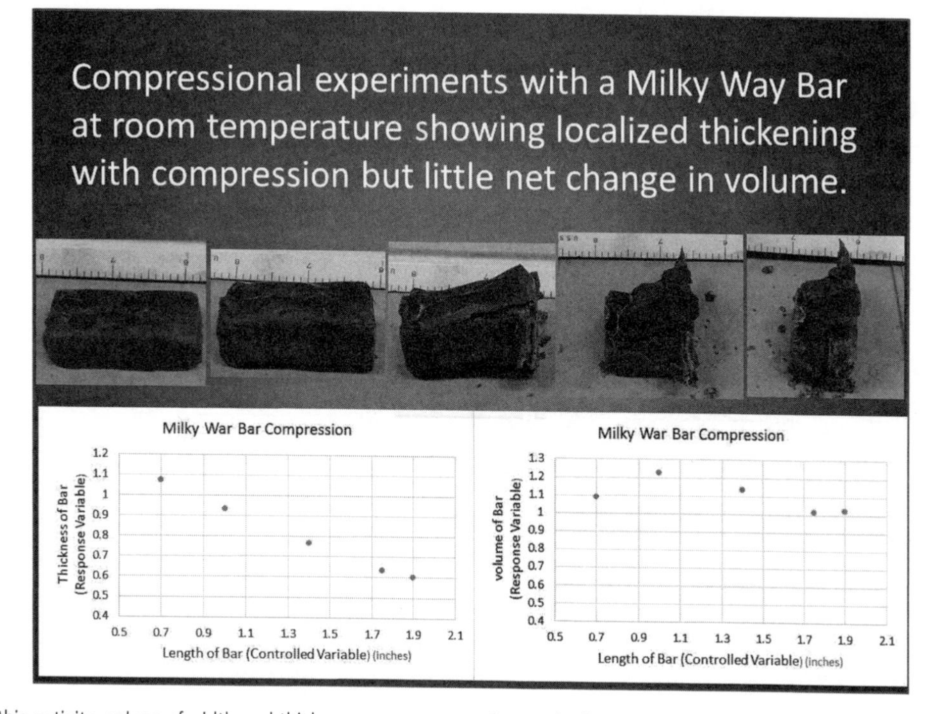

In this activity, values of width and thickness were measured at each of the ends of the candy bar as well as the middle. Thickness reported on the graph is the average of these three measurements. Volume is calculated by the average thickness times average width times length. The apparent increase in volume observed at intermediate compression is likely due to the formation of void spaces within the brittle fractures.

stretching to be focused in one location? Can your students come up with a possible explanation that they can test experimentally?

Doing compression experiments on real rock at high temperature is outside the reach of most classrooms, but analog materials such as a Milky Way bar can behave qualitatively like rock and provide for experimental investigation. After doing these experiments with Milky Way bars, students can apply their results to real rocks. For example, mountains form where tectonic plates are scrunched together and crust thickens. If the crust shortens by a given amount, can you predict the amount of thickening? What might cause the thickening to occur in one place rather being spread out equally over the entire tectonic plate?

A more elaborate set of experiments might test how changes in temperature affect the properties of the candy bar, as shown in Figure 16.3 (p. 334). Students might do a set of compression or extension experiments on three different candy bars, one cooled in the refrigerator (or on ice), one at room temperature, and one warmed by floating it in its wrapper in hot water. Can students observe or measure any differences in behavior?

Figure 16.3 ✪

Compression experiments with a Milky Way bar at three different temperatures

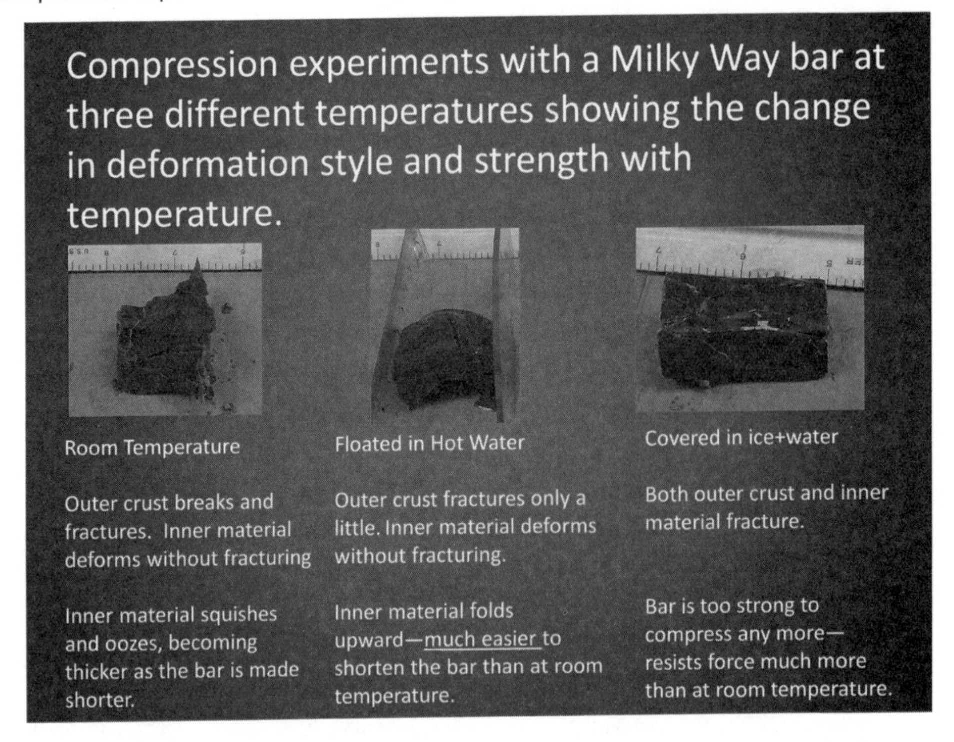

The results of these experiments yield insight into how temperature of rock influences the style of deformation. Hot water used in these experiments was from the faucet, warm enough to put your hand in but too warm to hold it there indefinitely.

What might those differences tell us about the effect of temperature on the behavior of rocks during plate tectonics events? Do materials (such as rocks and candy bars) have a greater tendency to deform plastically (ooze like Silly Putty) when warm or cold? Does the region of thickening or thinning expand or contract with increasing temperature?

In summary, starting with a familiar non-experimental learning activity and converting it into an experimental investigation is one way to start where you are. To generalize this idea, start with an activity that you already do and think of a way to add something to it. Ask how the activity can yield real information about the nature of our universe—not only act as a mnemonic for information that comes from the textbook or other authority.

Present Understanding

"Where you are" refers not only to your physical location, or the activities you are already doing in class, but also to where you and your students are in your journey as scientific explorers. What questions engage you and your students? Where are the frontiers of your understanding?

Spontaneous questions that don't come from the textbook are particularly important opportunities to step away from the planned curriculum and engage in science investigation. The diversions don't have to take a lot of time. In Chapter 4, we presented a study of weathering rates of tombstones in Australia. A student commented that the weathering rate didn't seem particularly fast. Mary asked, "Would the tombstone still be there in a million years?" This contribution from both teacher and student launched the class into an investigation of what a million years might do to the tombstones.

In Chapter 12, we introduced the idea of "authentic" questions and investigations—that is, questions whose answers you don't know ahead of time and investigations whose outcome isn't preordained by the teacher's answer key. Investigating a question that neither textbook nor teacher knows the answer to engages students far more thoroughly than an experiment whose primary purpose is to show students that yes, indeed, the textbook got it right.

Sometimes, we teachers fear that following the rabbit trail of student questions "into the weeds" will get away from the "core curriculum" or proceed in the wrong pedagogical sequence. We don't think so. The main goal of teaching science is to teach the enterprise of doing science. Getting "into the weeds" is the whole point. Most scientific discovery did not proceed in some imagined logical progression—for example, we measured the distance to the Sun thousands of years before we even knew that the Earth had a core, and we measured the speed of light before we measured the much slower speed of sound. Often, real science followed the interests of the scientists and the technologies available to them, not which "content area" was the next "step" in scientific learning. Likewise, classroom science can take a few authentic diversions into real investigation without disrupting the pedagogical sequence.

Starting with present understanding doesn't only mean letting students take the lead. The classroom investigative experience is a collaborative effort between students and teacher. Good teachers, as well as students, have puzzles they would like to solve.

There is a misconception that student-led investigation requires the teacher to stand aside and let students proceed unfettered. We think this idea is wrong for two reasons. First, without guidance, student investigations will likely produce poor or even meaningless results—it would take students at least hundreds of years to reproduce discoveries that took scientists hundreds of years to figure out. Second, an unengaged teacher sends a powerful negative message to the students about the importance of science.

Collaboration is an essential element of the mentoring relationship. Teacher and students pursue investigations together, from where they are, and the new questions they pursue are an outgrowth of that collaboration.

I (Mary) have found participating in discovery not too hard—because I have so much yet to learn. In fact, I've spent my entire teaching career understanding the concepts that I got As on in high school science. My own passions and interests always play a part in my students' classroom investigations.

An example of my own contributions is found in an investigation that we first introduced in Chapter 6, prompted by the following student comment: "Water evaporating into the air warms the air because it's the warm water that evaporates." To explore and test this idea, we measured how fast the temperature rose in different volumes of water being heated at the same rate. We explained some of the mathematics of these experiments in the "Implicit Math" section of Chapter 7 (p. 127).

This activity bears on many aspects of weather and climate. But the roots of my personal interest in the activity grew from another puzzling bit of science. In previous years, I had taught about the thermosphere, the very hot upper layer of Earth's atmosphere. Although the temperature there is very high, the textbook told us that it wouldn't feel hot if a person were in it because the total amount of heat (thermal energy) is low. This was a problem of heat capacity and density of molecules that I never quite understood. If the thermosphere was hot, didn't that mean it held lots of heat?

The experiments with heating different amounts of water in a beaker revealed the difference between heat and temperature to my students and me. In the process, the experiments helped me clarify my own thinking on the thermosphere. When I go back to my original notes and lesson plans for this activity, I read my final note, which reveals my own passion to understand an idea that I had been working on for years: "Conclusion is made based on experimental results and related to the thermosphere."

We've offered a number of examples of interactive student/teacher investigations throughout this book. Sometimes the initial question comes from students, sometimes from the teacher, but always the ensuing investigation involves many new questions and ideas arising from the interaction between students and teacher. For example, Chapter 1 suggests how the teacher can guide questions from students in a study of craters on the Moon. Chapter 2 offers an example of iterative improvements in classroom experiments over a number of years, with each class of students building on the results of previous classes in the same way that Isaac Newton "stood on the shoulders of giants." Chapter 9 gives examples of starting with either teacher questions or student ideas and then testing those ideas through interactive experiment and reasoning.

Discrepant Events

The unexpected or startling is another great place to start a class investigation. Students get engaged when the experiment doesn't work in a straightforward way to prove what the textbook says.

Wait! Why did that happen? What does it mean?

In Chapter 2, we talked about the importance of revising experimental design on the basis of unexpected results. In Chapter 6, we gave examples of how unexpected experimental or observational results bring us to revise our models. In Chapter 9, we discussed an investigation of the heat capacity of soil and water based primarily on results that differed from the textbook.

Often, it's better to start out with a lab activity that has *not* been thoroughly pretested and perfected, as we pointed out in in Chapter 2. And, as we noted in Chapter 14, failed experiments, such as Biosphere 2, are the best kind.

Most scientists prefer to explore the unknown rather than simply reproduce the discoveries of others using already-perfected methods. The activity on differential heating of soil and water mentioned earlier (from Chapters 2 and 9) actually has a bit of history. I (Mary) started using this activity years ago as a beginning teacher. Over several years, I refined the procedure to eliminate any variables that students might inadvertently introduce that would give them the wrong answer. I wanted to help students get meaningful data. But in perfecting the activity, I eliminated student participation in the practice of planning and carrying out an investigation. The perfected lab worked to illustrate and reinforce a concept, but not to give students an opportunity to figure something out.

What's in the News

Items in the news are a good place to start an investigation. Volcanoes, earthquakes, hurricanes, and tsunamis are obvious attention grabbers that can be places to start. Chapters 2 and 4 addressed activities related to climate change, a concept often in the news. Chapters 6, 8, and 10 offered activities inspired by the Curiosity rover that was in the news at the time. Chapter 14 offered two activities based on news items: Biosphere 2 and the weathering away of Hawaii.

Students can be engaged with a broader range of cultural items than stories in the news. One time I (Mary) was studying rivers with my students when one student said, "Mrs. Colson, guess what I learned from Minecraft [the computer game]? I learned that water flows faster down steeper slopes." And so we were off on the influence of slope on velocity, an idea that we addressed in Chapters 7 and 10.

Engaging fiction stories can be a starting place for a science investigation. In the spirit of the old TV show *Mythbusters,* you might explore whether a claim in the story makes sense. Or you might use some event or science in the story as a launching place for an experiment or an exercise in arguing from evidence. We used the idea of a murder mystery in Chapter 13 to engage students with making predictions of Moon phases based on their conceptual model. In Chapter 14, we used the story of Grandma's well to engage students with geochemical partitioning of pollutants. Stories that use earth science concepts to solve mysteries or escape plot dilemmas can be found at the Issues in Earth Science website, *http://earthscienceissues.net/home,* along with teacher activity suggestions.

FINAL THOUGHTS

When I (Russ) was in high school, a Native American artist visited my school to do a sand painting. He didn't talk much, but he let us watch as, with elaborate ceremony, he poured different-color sands onto the floor of the gymnasium where we were all gathered. At first the colored sand seemed to form random patches and lines. But over the span of about an hour, a picture emerged, a wonderful, beautiful picture.

When he finished, he let us enjoy his painting for only a brief moment. In seconds, he stirred his picture into oblivion, destroying all the patterns and colors and lines that had captured my eye. "Given to God," he said, a spiritual commentary on the nature of human endeavor.

Perhaps the scene, and my sense of horror, has remained with me because the act was so contrary to my personality. My inclination is to perfect and preserve—I would have been perfectly happy as a museum curator! The artist's destruction of his lovely painting seemed pointless to me.

As a teacher, I've had to fight my tendency toward perfection and preservation. True perfection in teaching requires that we stir our paintings into oblivion and start over every year. The value of what we give our students is not in the painting itself, but in the process of making it. Students need to see us making the painting—practicing science—and paint with us. It's not enough for them to merely receive someone else's painting from the textbook or last year's curriculum.

When students see teachers launch from classroom questions into new and unplanned investigations, they know that the teacher's passion for discovery remains alive. When they see teachers create activities that are unique to their own community, the student's own creativity emerges. Students learn to take risks and make mistakes in the process of learning. When they practice science with the teacher, they can understand science as a scientist understands it.

REFERENCES

Colson, M. C. 2007. Earth Stories: A game for learning about what was by examining what is. Presentation made at the Minnesota Earth Science Teachers Association conference, St. Paul.

Huff, K., and C. Lange. 2010. SSSNOW project: Helping make science cool for students. *Science Scope* 33 (5): 36–41.

National Research Council (NRC). 1996. *National science education standards.* Washington, DC: National Academies Press.

NGSS Lead States. 2013. *Next Generation Science Standards: For states, by states.* Washington, DC: National Academies Press. *www.nextgenscience.org/next-generation-science-standards.*

ANNEX
A GAME OF YES/NO QUESTIONS

I (Mary) used this question-asking game to introduce a geological feature. The game was based on information about a small delta that formed in a local farm field after a heavy rain. When using this game, all the students get copies of the pictures and the questions must be ones that can be answered by "yes" or "no."

TEACHER'S PLAN

DAY ONE

First divide the class into groups of 4 students each and pass out student worksheets. The end goal for the students is to write a paragraph that tells a plausible story of the events that happened to form all the features visible in the photograph.

Round One Directions – for the students to do

1. Discuss, in your small group, what you see in the photo. Ask questions of each other about the things in the photo that puzzle you.
2. Decide on two questions that the group would like to ask about the landscape and the features you see. Write them down on your worksheet.
3. Each group will ask one of the questions in turn.
4. As each group asks their question, listen to each question and to the response. Jot down the information you learn about what has happened to shape the landscape in the photo.
5. Round One is complete when each group has had the chance to ask one question.

Round Two Directions – for the students to do

6. Discuss all the new information you have. Consider things like what are the features, how did they form, in what order did they form and how can you know what happened in the past if you didn't actually see what happened.
7. Generate two more questions that you might ask during Round Two. These questions may include your unasked question from Round One. Be sure to refer to the photos often as you try to understand what your observations tell you about what happened.
8. Each group will ask one of the questions in turn.
9. As each group asks their question, listen to each question and to the response. Jot down the information you learn about what has happened to shape the landscape in the photo. Remember, your goal is to write a story about what happened.
10. Round Two is complete when each group has had the chance to ask one question.

End Game – for the students to do

11. Write a paragraph that tells a plausible story for how this landscape developed.
12. Your story should describe, from first to last, the events and the results that occurred. Your story should be include the observations of the features in the photograph and your explanations for how these feature formed.
13. I will collect your paragraphs and judge the tonight
14. The winning story is the one that is most correct (not the most outlandish or imaginative). The winning team gets jolly ranchers tomorrow.

DAY TWO

A self/formative assessment for students

1. Give each group of 4 students the event cards (the need to be pre-mixed up) and photos of the landscape again.
2. Tell the students to put the event cards in order, beginning with what had to happen first (the oldest event) and ending with the last event to happen (the youngest event). Students should not just declare an opinion, but explain their reasoning and cite evidence for why they think what they think.
3. Lead a class discussion on the order of the events; have the groups keep their ordered event cards out in front of the group. Allow students to describe the evidence for each of the events and their reasoning for the order of the events. Facilitate the scientific argumentation that will pop up as students discuss with one another.
4. Announce the winning group, give reward and read the winning essay out loud. All groups can refer to their event cards.
5. Use the theme of *geology is learning about what was by examining what is* during your lessons on rocks, fossils, geologic time, relative dating, stratigraphy and cross-cutting relationships.

Pictures of the delta and a related gully uphill from the delta

STUDENT WORKSHEET FOR DAY 1

Names ______________________ **Date** _____ **Period** _____

Earth Stories: Learning about what was, by examining what is.

Information about the photograph and the feature

1. I took the picture about ½ km from my house; my house is surrounded by farmland.
2. The land there is not flat, but gently rolling. To take the picture I stood on the gravel road, about 20 m from the pond.
3. The blue is water; the brown is sand and gravel (sediment).
4. The relief in the photograph is on the order of 5 m.
5. The width of the fan-shaped feature is about 3 m.
6. Two weeks before I snapped this picture, we had had some rain.

Your Questions

Round One

1. ______________________

2. ______________________

Round Two

1. ______________________

2. ______________________

List of Observations

Be sure to listen to and write down other groups' answers.

Names ______________________ **Date** _____ **Period** _____

Earth Stories: Learning about what was, by examining what is.

Directions for your geologic story

In the space below, write a paragraph that tells a plausible story for how this landscape developed. Your story should describe, from first to last, the events and the results that occurred from those events. Your story should be based on the observations of the features in the photograph and your explanations for how those features formed. I will collect your paragraphs and judge them tonight. The winning story is the one that is most correct (not the most outlandish or imaginative). The winning team gets jolly ranchers tomorrow.

STUDENT EVENT CARDS FOR DAY 2

The farmer cultivated the field and planted soybeans.	Some of the rainwater soaked into the ground.	A tiny delta formed at the mouth of the gully where water and sediment flowed into the new pond.	The water rose in the pond to be level with the still-living soybeans. The soybeans, that spent time underwater, drowned.
The soybean seeds germinated and the bean plants grew.	Some rainwater ran off the water-soaked ground and flowed down slope toward the low spot.	Water evaporated from the pond causing the water level to go down.	A windy day made waves on the pond that carved the highest "step-like" feature.
A large amount of rain fell in a short period of time.	In one portion of the field, runoff eroded the soil and underlying sediment to form a gully.	Water continued to evaporate and periodically waves carved lower steps.	The water in the pond reached its current level.

TEACHER AS MENTOR, PRACTITIONER, AND SCHOLAR

We have a neighbor, Joyce Grier, who lives on a farm here in northwestern Minnesota. She has no formal training in geology and is not part of the academic ivory tower. Yet Joyce is a coauthor on two research articles in the *Journal of Paleontology* and got to name a new species of ammonite (an extinct creature a bit like the chambered nautilus). She has learned to read maps, decipher clues in fossils, and interpret layers of rock by doing it for fun. Her excitement about science is infectious. She says, "When I'm out there, my spirit is just free. It's just me, and God, and the snakes, and the rocks. It's total play!"

When we were at her house once, looking at her wonderful collection of fossils, Joyce picked up a coiled ammonite shell and said, "See, it had an injury back here"—she pointed to one of the fossil chambers formed in the creature's youth—"but it recovered from the injury. There's all kinds of clues about its life!"

Joyce can do real science, and do it with passion, despite having no formal training in it. We suggest that science teachers, trained in the discipline, can do it as well. We can do science, do science with our students, and do it with passion!

WITH FRESH THOUGHT AND GENUINE EXCITEMENT

Stephen Jay Gould commented in *Bully for Brontosaurus* that "it is essential to my notion of scholarship that good teaching requires fresh thought and genuine excitement" (1991, p. 166). He was concerned that all textbooks are alike, covering the same information and often even using the same anecdotal and pictorial illustrations. To Dr. Gould, this uniformity reflected a stagnation or laziness of thinking. As teachers—and curriculum creators—we need to do better than simply spew out the same ideas, offering the same illustrations, and using the same activities. We need to show students not only our excitement but also our fresh thought.

Our excitement and fresh thought are important not only because they invigorate learning but also because they invigorate the *desire* to learn. By participating in discovery and doing it with passion, teachers act as role models for their students. The realization that teachers care enough about the subject to invest themselves in it can have a greater impact on students than learning the material.

In a 1995 lecture, Dr. James Comer said, "No significant learning occurs without a significant relationship." Learning is a social activity. Students are learning some facts and (we hope) learning to think, but first they are asking the questions "What does the teacher think is important? Does the teacher think science is important enough to be engaged in it her or himself? Does the teacher think it is interesting, fun, and exciting?"

TEACHER AS MENTOR AND COACH

In *Earth: An Intimate History*, Richard Fortey writes that "understanding is always a journey, never a destination" (2005, p. 24). This is true on at least two levels: (1) understanding is an ongoing, lifelong, process, and (2) the important aspects of understanding, at least in science, are found in the process of discovery, not in the conclusions at the end.

To authentically implement the vision of the Next Generation Science Standards (NGSS), *the teacher's main job is to mentor and facilitate the practices of science, not to convey information.* Our goal in writing this book has been to provide examples, context, and illustrations for how this mentoring might happen. For example, in Chapter 1 we talked about the importance of mentoring as students begin to ask questions of observations. In Chapter 2, we considered how to mentor students through the practice of experimental investigation. In Chapter 3, we addressed the particularly difficult challenge that teachers face as they mentor their students through field investigations.

The mentor does not give students assignments and then step back and let students sink or swim. The mentor guides, advises, nudges, and—most important—*shows* students how science can be done. The mentor is an active participant in exploration.

Mentoring and coaching are closely aligned, as we've noted in some previous chapters. A coach not only teaches the rules (the facts and theories of science) but also shows by example how to play the game (the practices of science). The coach not only assigns challenges but also encourages students to believe in their ability to solve those challenges and to believe in the value of the hard work needed to solve them. A coach sees students not as static learners but as players in the great game of discovery.

Mentoring and coaching require a balance between too little guidance and too much. How can the teacher show students how to investigate without preempting student initiative in the investigation? This difficult goal sums up our main purpose in writing this book—to inspire teachers to do science with their students, an interaction in which student and teacher are engaged together in the complex enterprise of science.[C45]

In starting your curricular sequence, you might first show your students your own investigative thinking, perhaps pursuing some of your own questions, before sending students off on their own investigative adventures. Short practice exercises of this sort offer ways to *show* (not tell) your students how to reason through problems and give them an opportunity to work through smaller problems in preparation for larger classroom activities. Many of the short activities in this book—such as our graph-reading, math-reading, and evidence-conclusion challenges in Chapters 4, 7, and 8, and the short activities with spectroscopic data in Chapter 13—are offered as this kind of activity.

However, the biggest contribution to mentoring comes during the fog of the students' investigation itself, through small prompts, questions, and suggestions as teacher and students puzzle over results together. Many of the classroom anecdotes offered throughout this book are intended to provide insight into how you can do this.

These interactions require pursuing investigative activities that don't have an inevitable "right" answer or outcome. I (Mary) used to do paper timelines with my students—physical models to show how small the period of human occupation on Earth is compared with the age of the Earth. This activity did a fine job of informing students of geological time in a memorable way. Of the long strip of adding machine paper stretched the entire length of the whiteboard, only the width of a pencil line at the end corresponded to human presence and the last of the dinosaurs was only 5 inches before that. However, the activity was designed to have a clear "right answer" and provided little opportunity for students to interact with me in a science investigation.

More recently, I have dropped the paper timeline in favor of exploring the evidence and processes for geological change. This requires much more student interaction with me as we explore a multitude of observations and experiments. Students experience a much more visceral reaction to the vastness of time. For example, while studying the Wilson cycles—seafloor spreading, subduction, mountain building, and so on—one student commented with a sense of awe, "Oh wow! That would take a long time."

Dr. Claudia Alexander, one of the scientists on the Rosetta mission to Comet 67P/Churyumov-Gerasimenko, was interviewed in 2014 by Anna North of the *New York Times*. The interview addressed the issue of being a woman of color in a field dominated by white males. Her take was that we don't have trouble attracting women to the geosciences, but we have trouble keeping them there. She pointed to the need for mentoring:

> *People tend to learn science from somebody that takes them out and shows them cool things. Father to son used to be the way that it was done: Your father would take you out and show you how to connect up a battery. And I think it's been shown that the more we have real mentoring going on, that is also part of retaining women and women of color in science.*

TEACHER AS PRACTITIONER OF SCIENCE

As we suggested in Chapter 1, engaging students in open-ended explorations requires that the teacher be able and willing to depart from a lesson plan and fly out into the unknown. Engaging students in doing authentic science investigations—that is, investigations in which the outcome is not preordained by the textbook and where students make real choices in what questions to pursue and real contributions to the experiments and interpretations—requires that the teacher be a *practitioner of science.*

Practitioners Know Their Craft

No one can mentor what they don't understand or show how to do something that they do not do themselves.[C46] If we want to buy a quilt, we can go to a store and buy it from anyone who has a quilt for sale. But if we want to learn how to make a quilt, we apprentice ourselves to a master quilt maker. Teachers need to be master quilt makers, showing students how to piece together all the practices, crosscutting concepts, and core ideas of science and stitch them into a coherent whole, a goal consistent with the three-dimensional philosophy of the *NGSS* (NGSS Lead States 2013).

Which one of us would feel comfortable being wheeled into an operating room run by a surgeon who learned from a teacher who had never performed a surgery? We would be uneasy regardless of the pedagogical skill of the surgeon's teacher.

The practitioner continues to learn. For the teacher as practitioner of science, this entails not only learning better methods for teaching (as we ourselves discovered in writing Chapters 8 and 9) but also deepening understanding of the discipline through the continued practice of science. In writing this book, we learned that a glass marble's impact into sand does not yield crater diameters that match the theoretical size expected for hypersonic impacts (Chapter 7). Richard Oldham's and Inge Lehmann's discoveries of Earth's inner and outer core in Chapter 12 gave us new understanding of how these discoveries were made and changed how we will teach about the Earth's interior in the future.

So, if we continue to learn, doesn't that mean that we don't really know our discipline? As we asked in Chapter 8, should we really dive into things we don't fully understand and let our students see that we don't have all the answers?

You bet. That's what it means to be a practitioner of science.

Practitioners Contribute Creatively to Their Craft

As busy teachers, most of us won't develop all of our own activities. But developing *some* sends the significant message that we are engaged, that we are practitioners of science, and that science is something we all can do. It also keeps our activities fresh and "uncanned." As we pointed out in Chapter 9, if an activity is "teacher-proofed" and "student-proofed" so that it yields the right answer with no problems to solve, then that activity has very little resemblance to real science and provides little opportunity to learn the practices of science.

Because of this philosophy, only a few of the activities suggested in this book are fully fleshed out. Our purpose is not to provide ready-to-go activities, but rather to provide inspiration for teachers to develop or modify activities for our own classes. We offered example activities that range from providing only starting guidelines for an earthquake and seismic activity in Chapter 12 to offering a fairly extensively developed impact lab in Chapters 2, 4, and 7.

Providing inspiration for the teacher as practitioner of science is also the main purpose of the historical anecdotes sprinkled throughout the book. Chapter 11 developed a detailed stratigraphy activity based on the work of John Wesley Powell at the Grand Canyon. Chapter 13 proposed a starter activity based on Galileo's study of the phases of Venus. Chapter 3 considered the classic work of Desmarest with volcanoes but left development of an activity to the teacher. These stories show people doing science in context, and we hope they've inspired you to develop some activities based on historical or classic investigations.

We've argued the importance of investigative exploration arising from the interaction between the teacher and students rather than from a set of textbook activities. We've argued that teachers need to engage in doing activities with their students, including some activities of their own design. In so engaging, teachers showcase their own curiosity and passion for exploration. In exploring and reasoning through problems with students, teachers can be mentors, *showing* students their own methods of exploration.

Practitioners Can Mentor Other People in Their Craft

If the teacher is not able to take the students along a spontaneous trail of exploration, then the perfect *NGSS*-aligned curriculum will not engage students in doing science or make the practices of science come alive in the classroom.

In discussing the *NGSS,* many teachers have expressed concern that letting students do what they want in an open-ended investigative activity is a terrifying prospect. They foresee a classroom dissolving into chaos in which student speculation runs wild and no one learns anything.

Yes … an authentic investigation without a knowledgeable teacher to guide it creates a terrifying image in our minds as well. Having students make sense of phenomena through investigation is not an invitation to "anything goes." Students need a participating expert mentor to help focus their questions, improve their experiments, and facilitate their arguments. To be that expert mentor, the teacher needs to be a practitioner of science.

TEACHER AS SCHOLAR

Trust and respect are important in the classroom. Students need to believe that the teacher has more to offer than what they can get by reading the textbook on their own. You can't mentor students if they don't believe that you truly know what you're talking about. Having a basket of teaching tricks and methodologies is fine. But to be most effective, you need to know your discipline and the practices of science, and your students need to know you know. You need to be a scholar.

The Scholar Is the Person Who Is Curious

When interviewing scientists, I (Mary) am always struck by the bone-deep curiosity reflected in their language. "Right now I'm thinking about …" and "I'm trying to figure out the relationship between …."

As teachers, what are we trying to figure out? When students see us pursuing problems with them, curious and asking questions, their respect grows. They see that we are not simply classroom managers who shuffle pieces of curriculum that someone else gave us.

The Scholar Is the Person Who Learns

One year, I (Mary) had been looking at energy transfer with my class as part of a lesson on energy flow through Earth's systems. The concepts were hard for my eighth graders and, after a few days of intense activities, I decided that we needed to take a fun break. So I brought marshmallows to class and we toasted our marshmallows over the classroom hot plates!

We toasted, and toasted, and toasted, but the marshmallows never got toasted. Why not? I realized that the reason (which I should have anticipated but didn't) was in itself a great lesson in energy transfer. My classroom hot plates are conduction hot plates, in which the heat is conveyed by conduction, not radiation. Holding the marshmallows near but not on the hot plates was never going to get them toasted.

When I figured out why our efforts failed, my students' respect for me as a scholar grew. The fact that I hadn't anticipated the problem beforehand mattered not at all (to them—it still bugged me).

The Scholar Is the Person Who Isn't Too Quick With an Answer

Scholars have the ability to hold multiple working hypotheses in mind while working forward on a puzzling problem. This is harder than we might think. In the classroom, we get used to the idea of having the answer for every student question. Sometimes, we feel pressured to have the right answer. More than one student has said to me (Mary),

"What do you mean you don't know the answer? You're the teacher!" Implication: you aren't doing your job.

Often, the students wouldn't know if we made the answer up. This can lull us into a complacency of thinking we have the answer when perhaps we only have the beginning of an answer.

A year or so ago, I (Russ) found myself caught in this trap. I'm on a lunar Listserv—an e-mail forum for a community of planetary researchers and others with particular interest in and ties to the Moon. This question came through on the Listserv: "I was asked a question by a kid today, I hope someone will know the answer. Do Moon rocks and soil have any smell? If so, what kind of smell?"

I immediately imagined how I would answer were I asked that question. My answer was a short, simplistic response about how there wouldn't be many volatiles in the vacuum of the Moon's surface so we couldn't expect much of a smell to emanate from the materials, and the Moon is in vacuum anyway so there wouldn't be any way to smell anything on the surface of the Moon.

Then came the flood of responses from the Listserv community. Stories from old-timers who remembered that the astronauts reported the smell of gunpowder. Suggestions for how to give students insight into what that smell was like—light a match and blow it out, or break a rock with a sledgehammer and give it a quick sniff. Others offered explanations for the cause—unsatisfied broken bonds in the minerals which on Earth would be quickly satisfied by the water vapor abundant in our atmosphere. More detailed explanations included considering how sulfur, common in the lunar regolith, reacts with hydrogen in the solar wind and how metallic iron common in lunar regolith reacted with oxygen in the lunar module. Either might produce unsatisfied bonds and a burnt gunpowder odor. From there, discussion continued into the molecular processes for generating the unsatisfied bonds on the Moon.

The scholar realizes that any answer is only a starting place and that answers can go deep indeed. They keep their thinking in flux, ready to learn and not too quick to declare a question "answered." This type of thinking encourages further study—the starting place for scientific investigation.

Seeing Ourselves as Scholars

At a National Science Teachers Association meeting in Boston in April 2014, Stephen Pruitt, then a vice president at Achieve Inc., commented that he no longer wanted to hear anyone say "I'm just a teacher." We're with him on that. To be effective, teachers need the respect of students, and that means they need to respect themselves. They also need the respect of their communities. They need to be scholars, to view themselves as scholars, and to be seen as scholars by their community and students. But, despite the common idea that the teacher's job is one of the most important in our culture, there remain substantial cultural hurdles to the idea of teacher as scholar.

In fall 2013, a teaching colleague was selected to be the assistant superintendent and I (Mary) offered him my congratulations. I realized that we always congratulate people when they "move up" from teaching into administration. We never congratulate anyone for staying in teaching. The only status promotion available to most teachers is to leave teaching.

A 2014 report on NPR addressing the *Common Core State Standards* in English language arts and mathematics (NGAC and CCSSO 2010; Turner 2014) noted that there is a pressing need for a curriculum to meet these standards—implying that teachers, supposedly experts in their field, are not up to the task of meeting the standards of their own disciplines without a curriculum prepared for them by someone else. Even the idea that teachers need answer keys—as though they can't figure out the answers to questions in their own discipline—undermines our concept of a teacher as scholar.

We propose that the teacher as scholar is not straightjacketed by the curricular materials, by state or national standards, or even by a book purporting to interpret the standards for the classroom, like this one. The teacher as scholar can create, discover, pursue, and explore in all the ways that are open to the human mind.

KEEPING YOUR EYE ON THE BALL

As we look back on the chapters in this book, we see that learning to teach science as something that we do rather than something that we know is not a trivial challenge. Approaching science teaching as a coach and mentor, practitioner and scholar, is a journey that is not traveled in a single day. We suggest that you take that journey one day, one lesson, one chapter at a time. Scientists don't discover everything at once. They study one thing in depth. Then another thing in depth. And pretty soon the pieces of the big picture start to come together.

You can do that in the classroom. But keep your eye on the ball. The goal is to practice science, not to know facts and theories. The goal is to engage interactively with your students, not to either tell students the information or leave them to fend for themselves in impossible open-ended activities. In short, remember to *do* science *with your students.*

Dive into the details of doing science, developing some of your own activities and engaging in explorations where the outcome is not preordained by the textbook. But don't get lost in the details. Remind yourself of your wider purpose. Mary has a mantra to help her remember that wider purpose: Have an important message. Know what you're talking about. Love those you're talking to.

And don't forget to be passionate and engaged in your discipline. We end with our neighbor Joyce Grier's passion as a reminder to us all:

> *[When] I saw these concretions … and the sunlight … sparkles off of these, I mean it's just fire off the ammonite shells! It's just brilliant and you open that up and you, my gosh you have a glimpse into God's creation 70 million years ago right here in your hands! … I was hooked! There was an instant passion there that you could crack this [rock] open and find these, these wonderful creatures inside!*

REFERENCES

Comer, J. P. 2001. Schools that develop children. *The American Prospect* 12 (7): 30–35.

Fortey, R. 2005. *Earth: An intimate history*. New York: Vintage Books.

Gould, S. J. 1991. *Bully for brontosaurus.* New York: W.W. Norton.

National Governors Association Center for Best Practices and Council of Chief State School Officers (NGAC and CCSSO). 2010. *Common core state standards.* Washington, DC: NGAC and CCSSO.

NGSS Lead States. 2013. *Next Generation Science Standards: For states, by states.* Washington, DC: National Academies Press. *www.nextgenscience.org/next-generation-science-standards.*

North, A. *New York Times.* 2014. How It Feels to Land a Spacecraft on a Comet. November 17. *http://op-talk.blogs.nytimes.com/2014/11/17/how-it-feels-to-land-a-spacecraft-on-a-comet/?_r=0.*

Turner, C. 2014. The *Common Core* curriculum void. *www.npr.org/sections/ed/2014/06/03/318228023/the-common-core-curriculum-void.*

AFTERWORD

SOME THINGS WE'VE LEFT OUT

In *Learning to Read the Earth and Sky,* we emphasize the *doing* of science. We have given less attention to other important aspects of the teacher's job, such as how to link classroom-size investigation to big issues like cost/benefit of extracting natural resources and the human effects on our environment, how to adapt specific activities to different ages or abilities, and how the teacher can engage students in doing science and still give them sufficient factual knowledge to do well on required standardized tests.

Earth science has a long tradition of practical application, such as to problems in natural resource extraction, engineering of landscapes and rivers, and mitigation of human impacts on environment and climate. However, large-scale applied problems might not be the best starting point for classroom investigation, and so they received less attention in this book. Questions that are "too big" tend to generate hypotheses that are not easily testable in the classroom and lead to sociopolitical discussions, which—although an important educational endeavor—do not engage students in doing many of the basic aspects of science. Even when simplified, an investigative foray into the big picture quickly becomes complex enough to require many weeks of class time—as seen in the resources extraction exercise in the "Creating and Using Algorithms" section in Chapter 7 (pp. 130–134).

Nevertheless, students need to put the concepts and practices they learn from doing science into the context of real-world, practical problems to understand the significance of their learning. We hope that teachers, as mentoring scholars, can help students connect the small-scale investigations of this book to the bigger-picture ideas. We have provided some example applications to practical problems in engineering, the environment, and cycles of matter and energy in the earth. For example, when considering flow volume and flow viscosity in Chapter 7, we applied the reasoning challenges to real problems in river management—a story of how humans can have an impact on natural processes. In Chapter 14, we considered how to take big-picture ideas of pollutant migration or ore formation and break them down into classroom-size investigations. In Chapter 15, we considered a variety of ways to connect classroom experience to big ideas.

Likewise, we provide example activities appropriate for a variety of age and ability levels but don't offer multiple versions of each activity tailored to different ages. In keeping with our philosophy of teacher as practitioner of science, we believe that teachers can recognize students' varied abilities and adapt these activities to the appropriate ability level.

We've offered some insight into formative assessment but largely avoided summative assessment.[C47] We've also avoided the challenge of how the teacher can engage students in the kinds of investigations that address the most important learning objectives of the *Next Generation Science Standards* and still have enough class time to "cover all the material" that will be on standardized tests. We note that the kind of learning that comes from practicing science is more difficult to assess than knowledge of facts or theoretical models. For example, how do you assess whether a student has learned to ask insightful questions about a data set, turn that insight into the invention of a new experimental approach, and apply the result to construction of an improved conceptual model?

Creative leaps take time—something that isn't available on a typical assessment "test." As much as we are able, we should base our assessments on what we want to teach rather than basing what we teach on what is most easy to assess. Consequently, we hope that standardized tests move toward testing true science abilities.[C48] We believe that *Learning to Read the Earth and Sky* can provide insight into achieving that goal.

APPENDIX A

Chapter Activities and Anecdotes Related to *NGSS* Performance Expectations

This appendix lists *Next Generation Science Standards* (*NGSS;* NGSS Lead States 2013) middle and high school Earth and Space Sciences (ESS) performance expectations (PEs) with example chapter activities and anecdotes that address each. Most of the activities, anecdotes, and other content listed in this appendix address part, but not all, of the corresponding PEs. The content shown in bold addresses the corresponding PEs more substantially than that shown in regular type.

Recognize that the philosophy of the PEs is that they are not "met" by a single activity, or even a group of activities, but rather they are an ongoing objective of learning at which students become more skilled with practice. One might argue that no one completely "reaches" the PE goals—research at the frontiers of science can be viewed as the ongoing search for a better grasp of the PE.

Often, the PEs invoke modeling, argumentation, and prediction—advanced practices that can't be reached before students first have a firm understanding of how the models were derived and the experimental and observational evidence on which they are based. Thus, many of the activities in this book address the experimental and observational foundation needed to reach the advanced practices in the PEs.

All of the PEs are much broader than one might expect from simply reading the PE without its accompanying clarification statement. We have included the PE clarification statement where the breadth of the PE is particularly obscure and the connection to our example activities might be unclear.

REFERENCE

NGSS Lead States. 2013. *Next Generation Science Standards: For states, by states.* Washington, DC: National Academies Press. *www.nextgenscience.org/next-generation-science-standards.*

NGSS Performance Expectations in Earth and Space Sciences	Relevant activities, anecdotes, and other content in this book (by chapter)
MS-ESS1-1. Develop and use a model of the Earth-sun-moon system to describe the cyclic patterns of lunar phases, eclipses of the sun and moon, and seasons.	**13. Enlightened: Reading the Sky With Our Naked Eye (phases of the Moon); SDA: Constructing a Moon Phase Model From Observation**
MS-ESS1-2. Develop and use a model to describe the role of gravity in the motions within galaxies and the solar system.	**13. Enlightened: Reading the Sky With a Telescope (phases of Venus)** Fully addressing this PE might involve an investigation of the role of gravity, which we don't address in this book.
MS-ESS1-3. Analyze and interpret data to determine scale properties of objects in the solar system.	**2. EAD: The Controlled Experiment: Impact Cratering (quantitative experiments)** **4. EAD: The Impact-Cratering Experiment (graphing data and making predictions)** 6. Building on a Seed Model (example using Gale Crater on Mars) 10. Reading the Stories (rocks on Mars and environments of deposition) **13. Aristarchus and the distance to the Moon; Dangers of Pictorial and Physical Models (box) (includes the problem of scale in making models for Moon phases)**
MS-ESS1-4. Construct a scientific explanation based on evidence from rock strata for how the geologic time scale is used to organize Earth's 4.6-billion-year-old history.	1. Observations of the surface of Earth and Moon, preliminary to impact-cratering experiments in Chapter 2; EAD: Go and See: Geology in the Field **5. A Tapestry of Time and Terrain; EAD: Maps and Cross-Sections** **10. Reading the Stories (dinosaurs and environments of deposition);** Reading the Stories (rocks on Mars and environments of deposition) **11. Stratigraphy: Earth's Storybook; Classroom Lessons and Challenges From the Grand Canyon (crosscutting relationships in Powell's units A, B, and C); SDA: Reading Stories of Deep Time** 15. Units and sequences that bundle PEs, including Figures 15.1–15.3

NGSS Performance Expectations in Earth and Space Sciences	Relevant Activities, Anecdotes, and Other Content in This Book (by Chapter)
MS-ESS2-1. Develop a model to describe the cycling of Earth's materials and the flow of energy that drives this process.	1. EAD: Go and See: Geology in the Field 4. Graph-reading challenges (tombstone weathering) 5. A Case Study in Spatial Reasoning **9. Arguing Through the Fog; Wrapping Up the Argument** 10. Leaf inset into the snow; **SDA: Experimental Sedimentary Petrology With Downspouts; SDA: Experimental Igneous Petrology With Thymol** 11. SDA: Reading Stories of Deep Time **14. Mass balance (lakes and magma); SDA: Experimental Test of the Mathematical Model for Geochemical Differentiation** 15. Units and sequences that bundle PEs, including Figures 15.1–15.3 16. A game of yes/no questions (Figure 16.1 and annex)
MS-ESS2-2. Construct an explanation based on evidence for how geoscience processes have changed Earth's surface at varying time and spatial scales.	**1. EAD: Go and See: Geology in the Field** **3. Test a hypothesis with field observations; Focused Field Discovery** 5. A Tapestry of Time and Terrain; A Case Study in Spatial Reasoning; EAD: Maps and Cross-Sections **6. EAD: Modeling Isostatic Equilibrium** **9. EAD: The Columns of Pozzuoli** 10. Reading the Stories (rocks on Mars and environments of deposition) **12. When You Can't Look Up the Answer** 15. Units and sequences that bundle PEs, including Figures 15.1–15.3 **16. A game of yes/no questions (Figure 16.1 and annex); edible models to launch an investigation (Figures 16.2 and 16.3)**

NGSS Performance Expectations in Earth and Space Sciences	Relevant Activities, Anecdotes, and Other Content in This Book (by Chapter)
MS-ESS2-3. Analyze and interpret data on the distribution of fossils and rocks, continental shapes, and seafloor structures to provide evidence of the past plate motions.	3. Test a hypothesis with field observations; Focused Field Discovery; EAD: Field Observation: Layer-Cake Rocks (geological mapping) 5. A Tapestry of Time and Terrain; **A Case Study in Spatial Reasoning** **8. Difficulties in identifying evidence; Solutions for learning to identify evidence; EAD: Wegener's Evidence for Continental Drift (arguing from evidence)** 16. Edible models to launch an investigation (Figures 16.2 and 16.3)
MS-ESS2-4. Develop a model to describe the cycling of water through Earth's systems driven by energy from the sun and the force of gravity.	**2. Solubility of water vapor as a function of temperature** 4. Graph-Reading Challenges (solubility of water in air) 5. DD: Contour Mapping With Weather; Figure 5.7 (global circulation in map and cross-sectional views [as part of the water cycle]) 6. Starting With the "Wrong" Model (thermal energy and evaporation, isostasy) 14. SDA: Experimental Test of the Mathematical Model for Geochemical Differentiation 15. Units and sequences that bundle PEs, including Figures 15.1–15.3 16. A game of yes/no questions (Figure 16.1 and annex)
MS-ESS2-5. Collect data to provide evidence for how the motions and complex interactions of air masses results in changes in weather conditions.	2. Is water naturally cooler than air?; **Playing with cups, water, and air** **5. DD: Contour Mapping With Weather; Figure 5.7 (global circulation in map and cross-sectional views [as part of the water cycle])** **6. Modeling air pressure with a classroom barometer** 9. Finding a Place to Start 15. Units and sequences that bundle PEs, including Figures 15.1–15.3

NGSS PERFORMANCE EXPECTATIONS IN EARTH AND SPACE SCIENCES	RELEVANT ACTIVITIES, ANECDOTES, AND OTHER CONTENT IN THIS BOOK (BY CHAPTER)
MS-ESS2-6. Develop and use a model to describe how unequal heating and rotation of the Earth cause patterns of atmospheric and oceanic circulation that determine regional climates.	1. Conversation, questions, and argumentation among students about the moderating influence of oceans on coastal climates **2. Is water naturally cooler than air?** **6. Modeling air pressure with a classroom barometer** 9. Arguing Through the Fog; Wrapping Up the Argument 15. Units and sequences that bundle PEs, including Figures 15.1–15.3
MS-ESS3-1. Construct a scientific explanation based on evidence for how the uneven distributions of Earth's mineral, energy, and groundwater resources are the result of past and current geoscience processes.	3. Focused Field Discovery; EAD: Field Observation: Layer-Cake Rocks (geological mapping) 14. Mass balance (lakes and magma); **DD: Gold Ore in South Africa;** SDA: Experimental Test of the Mathematical Model for Geochemical Differentiation
MS-ESS3-2. Analyze and interpret data on natural hazards to forecast future catastrophic events and inform the development of technologies to mitigate their effects.	7. Math-Reading Challenges: Connecting Math to Nature (Reynolds number); DD: Mathematical Modeling Related to Flooding; Math-Reading Challenges: Stream Velocity and Discharge **12. SDA: Authentic Questions and Real Seismic Data (frequency of big earthquakes and magnitude of deep earthquakes)**
MS-ESS3-3. Apply scientific principles to design a method for monitoring and minimizing a human impact on the environment.*	This PE involves multi-element algorithmic thinking to establish the relationship between input variables and the design objective, including concepts from both inside and outside earth science. It might be addressed in a fashion analogous to the reasoning presented in the Chapter 7 section "Creating and Using Algorithms."

NGSS PERFORMANCE EXPECTATIONS IN EARTH AND SPACE SCIENCES	RELEVANT ACTIVITIES, ANECDOTES, AND OTHER CONTENT IN THIS BOOK (BY CHAPTER)
MS-ESS3-4. Construct an argument supported by evidence for how increases in human population and per-capita consumption of natural resources impact Earth's systems.	This PE asks students to connect experimental and observational evidence from classroom learning to system-level applications of that evidence. The evidence might come from small-scale studies related to climate change, erosion, chemical alterations of surface water, changes in air quality, and so on or to how ancient changes on Earth give us indications for how future changes might occur. Many activities in this book could thus play a part in this PE, if tied through story lines (analogous to those shown in Figures 15.1–15.3) on the effects of human activity on Earth's systems. Once students have an investigative base for understanding the underpinning processes that affect Earth's systems, they might address this PE by considering human-induced factors affecting those systems, analogous to the algorithmic thinking in the Chapter 7 section "Creating and Using Algorithms."
MS-ESS3-5. Ask questions to clarify evidence of the factors that have caused the rise in global temperatures over the past century.	**4. A Case Study in Graph Reading: The Keeling Curve** **9. The Tyranny of the Answer Key**
HS-ESS1-1. Develop a model based on evidence to illustrate the life span of the sun and the role of nuclear fusion in the sun's core to release energy that eventually reaches Earth in the form of radiation.	**13. Enlightened: Reading the Sky With a Spectroscope (composition of stars and atmospheres of exoplanets)** Fully addressing this PE might involve consideration of the Hertzsprung-Russell (H-R) diagram and how it reveals stellar evolution, which we don't address in this book.
HS-ESS1-2. Construct an explanation of the Big Bang theory based on astronomical evidence of light spectra, motion of distant galaxies, and composition of matter in the universe.	**13. Enlightened: Reading the Sky With a Spectroscope (composition of stars and atmospheres of exoplanets)** Fully addressing this PE might include an examination of redshift and its significance and an examination of particle physics, which we don't address in this book.

NGSS PERFORMANCE EXPECTATIONS IN EARTH AND SPACE SCIENCES	RELEVANT ACTIVITIES, ANECDOTES, AND OTHER CONTENT IN THIS BOOK (BY CHAPTER)
HS-ESS1-3. Communicate scientific ideas about the way stars, over their life cycle, produce elements.	**13. Enlightened: Reading the Sky With a Spectroscope (composition of stars and atmospheres of exoplanets)** Fully addressing this PE might involve a lesson on nucleosynthesis, which we don't address in this book.
HS-ESS1-4. Use mathematical or computational representations to predict the motion of orbiting objects in the solar system.	9. Arguing Through the Fog; Wrapping Up the Argument This PE might also be addressed with appropriate math-reading challenges analogous to those in several sections of Chapter 7.
HS-ESS1-5. Evaluate evidence of the past and current movements of continental and oceanic crust and the theory of plate tectonics to explain the ages of crustal rocks.	**3. Test a hypothesis with field observations; EAD: Field Observation: Layer-Cake Rocks (geological mapping)** **5. A Tapestry of Time and Terrain; A Case Study in Spatial Reasoning; EAD: Maps and Cross-Sections** **8. Difficulties in identifying evidence; Solutions for learning to identify evidence; EAD: Wegener's Evidence for Continental Drift (arguing from evidence)** **11. Stratigraphy: Earth's Storybook; Classroom Lessons and Challenges From the Grand Canyon (crosscutting relationships in Powell's units A, B, and C); SDA: Reading Stories of Deep Time**

NGSS Performance Expectations in Earth and Space Sciences	Relevant Activities, Anecdotes, and Other Content in This Book (by Chapter)
HS-ESS1-6. Apply scientific reasoning and evidence from ancient Earth materials, meteorites, and other planetary surfaces to construct an account of Earth's formation and early history.	1. Observations of the surface of Earth and Moon, preliminary to impact-cratering experiments in Chapter 2 **2. EAD: The Controlled Experiment: Impact Cratering (quantitative experiments)** 3. Age of Earth, heat flow, and Lord Kelvin **4. EAD: The Impact-Cratering Experiment (graphing data and making predictions)** 5. EAD: Maps and Cross-Sections **7. EAD. The Impact-Cratering Experiment: Identifying and Testing a Mathematical Model** **10. SDA: Experimental Igneous Petrology With Thymol** Addressing this PE could include many other aspects of solar system history and evidence, such as examination of meteoritic chondrules, or condensation of materials from a high-temperature vapor, which are not addressed in this book.
HS-ESS2-1. Develop a model to illustrate how Earth's internal and surface processes operate at different spatial and temporal scales to form continental and ocean-floor features.	**1. EAD: Go and See: Geology in the Field** 3. Focused Field Discovery 5. A Tapestry of Time and Terrain **6. EAD: Modeling Isostatic Equilibrium** 8. Difficulties in identifying evidence; Solutions for learning to identify evidence; EAD: Wegener's Evidence for Continental Drift (arguing from evidence) 10. Reading the Stories (dinosaurs and environments of deposition); Reading the Stories (rocks on Mars and environments of deposition) 14. Mass balance (lakes and magma) 16. Edible models to launch an investigation (Figures 16.2 and 16.3)

NGSS PERFORMANCE EXPECTATIONS IN EARTH AND SPACE SCIENCES	RELEVANT ACTIVITIES, ANECDOTES, AND OTHER CONTENT IN THIS BOOK (BY CHAPTER)
HS-ESS2-2. Analyze geoscience data to make the claim that one change to Earth's surface can create feedbacks that cause changes to other Earth systems.	2. Solubility of water vapor as a function of temperature **4. A Case Study in Graph Reading: The Keeling Curve** **6. EAD: Modeling Isostatic Equilibrium** 10. Leaf inset into the snow **14. Mass Balance in the News (Biosphere 2)** 15. Units and sequences that bundle PEs, including Figures 15.1–15.3
HS-ESS2-3. Develop a model based on evidence of Earth's interior to describe the cycling of matter by thermal convection. [Clarification Statement: Emphasis is on both a one-dimensional model of Earth, with radial layers determined by density, and a three-dimensional model, which is controlled by mantle convection and the resulting plate tectonics. Examples of evidence include maps of Earth's three-dimensional structure obtained from seismic waves, records of the rate of change of Earth's magnetic field (as constraints on convection in the outer core), and identification of the composition of Earth's layers from high-pressure laboratory experiments.]	3. Age of Earth, heat flow, and Lord Kelvin 4. Graph-Reading Challenges (geothermal gradient) 5. A Case Study in Spatial Reasoning 6. Starting With the "Wrong" Model (thermal energy and evaporation, isostasy); EAD: Modeling Isostatic Equilibrium **12. Science in the Real World: Witness to the Unseen; A Lesson in Measuring What We Can't See; Richard Oldham and the outer core (developing and revising models [Figure 12.4] and arguing from evidence); Reading Scientific Literature: Inge Lehmanns's Pesky P Waves and the Inner Core** 16. Edible models to launch an investigation (Figures 16.2 and 16.3)
HS-ESS2-4. Use a model to describe how variations in the flow of energy into and out of Earth's systems result in changes in climate.	7. Math-Reading Challenges: Connecting Math to Nature (Reynolds number); DD: Mathematical Modeling Related to Flooding (Math-Reading Challenges: Stream Velocity and Discharge) This activity does not directly address this PE, rather, a similar "challenges" approach could be applied—for example, by considering a mathematical relationship between energy in and out of the Earth system and how changes in reflectivity, absorption of outgoing radiation, and other factors might affect the balance. 9. Arguing Through the Fog; Wrapping Up the Argument; **The Tyranny of the Answer Key**

NGSS Performance Expectations in Earth and Space Sciences	Relevant activities, anecdotes, and other content in this book (by chapter)
HS-ESS2-5. Plan and conduct an investigation of the properties of water and its effects on Earth materials and surface processes. [Clarification Statement: Emphasis is on mechanical and chemical investigations with water and a variety of solid materials to provide the evidence for connections between the hydrologic cycle and system interactions commonly known as the rock cycle. Examples of mechanical investigations include stream transportation and deposition using a stream table, erosion using variations in soil moisture content, or frost wedging by the expansion of water as it freezes. Examples of chemical investigations include chemical weathering and recrystallization (by testing the solubility of different materials) or melt generation (by examining how water lowers the melting temperature of most solids).]	1. EAD: Go and See: Geology in the Field **2. Is water naturally cooler than air?; Solubility of water vapor as a function of temperature** 4. Graph-Reading Challenges (solubility of water in air) 6. Starting With the "Wrong" Model (thermal energy and evaporation, isostasy); **7. Math-Reading Challenges: Connecting Math to Nature (Reynolds number); DD: Mathematical Modeling Related to Flooding (Math-Reading Challenges: Stream Velocity and Discharge)** **10. SDA: Experimental Sedimentary Petrology With Downspouts** **14. Mass balance (lakes and magma); Mass Balance in the News (erosion on Oahu); SDA: Experimental Test of the Mathematical Model for Geochemical Differentiation** 15. Units and sequences that bundle PEs, including Figures 15.1–15.3
HS-ESS2-6. Develop a quantitative model to describe the cycling of carbon among the hydrosphere, atmosphere, geosphere, and biosphere.	4. A Case Study in Graph Reading: The Keeling Curve 14. Mass Balance in the News (Biosphere 2) The quantitative modeling in this PE might be approached in a similar fashion to the quantitative modeling we suggest in the Chapter 7 section "Creating and Using Algorithms."
HS-ESS2-7. Construct an argument based on evidence about the simultaneous coevolution of Earth's systems and life on Earth.	5. A Tapestry of Time and Terrain; EAD: Maps and Cross-Sections **10. Reading the Stories (dinosaurs and environments of deposition)** 11. SDA: Reading Stories of Deep Time 14. Mass Balance in the News (Biosphere 2) To address this PE more substantially, one might consider the impact of oxygen in the atmosphere on formation of ore deposits (iron) or weathering (oxidation of iron), or the oxygen cycle. The oxygen cycle is linked to the carbon cycle through the formation of carbon dioxide during respiration.

NGSS PERFORMANCE EXPECTATIONS IN EARTH AND SPACE SCIENCES	RELEVANT ACTIVITIES, ANECDOTES, AND OTHER CONTENT IN THIS BOOK (BY CHAPTER)
HS-ESS3-1. Construct an explanation based on evidence for how the availability of natural resources, occurrence of natural hazards, and changes in climate have influenced human activity.	6. Modeling air pressure with a classroom barometer Weather and weather modeling did not make it into any high school PEs, but air pressure is a key element in predicting severe weather, part of HS-ESS3-1. **12. SDA: Authentic Questions and Real Seismic Data (frequency of big earthquakes and magnitude of deep earthquakes)** 15. Units and sequences that bundle PEs, including Figures 15.1–15.3
HS-ESS3-2. Evaluate competing design solutions for developing, managing, and utilizing energy and mineral resources based on cost-benefit ratios.*	**7. Creating and Using Algorithms**
HS-ESS3-3. Create a computational simulation to illustrate the relationships among management of natural resources, the sustainability of human populations, and biodiversity.	**7. Creating and Using Algorithms**
HS-ESS3-4. Evaluate or refine a technological solution that reduces impacts of human activities on natural systems.*	**14. DEATH-X: An experimental partitioning activity** The migration of pollutants can be minimized by selection of containment materials that adsorb the pollutants, based on experimental measurement of partition coefficients.
HS-ESS3-5. Analyze geoscience data and the results from global climate models to make an evidence-based forecast of the current rate of global or regional climate change and associated future impacts to Earth systems.	**5. DD: Contour Mapping With Weather; Figure 5.7 (global circulation in map and cross-sectional views [as part of the water cycle])** Although this activity does not use data from climate models, output from climate models will likely involve contoured information or profiles.

NGSS PERFORMANCE EXPECTATIONS IN EARTH AND SPACE SCIENCES	RELEVANT ACTIVITIES, ANECDOTES, AND OTHER CONTENT IN THIS BOOK (BY CHAPTER)
HS-ESS3-6. Use a computational representation to illustrate the relationships among Earth systems and how those relationships are being modified due to human activity.	**7. DD: Mathematical Modeling Related to Flooding (A Computational Simulation for Rainfall Runoff)** **14. DEATH-X: An experimental partitioning activity (experimental data are a necessary foundation for computer modeling)** 15. Units and sequences that bundle PEs, including Figures 15.1–15.3 16. A game of yes/no questions (Figure 16.1 and annex) The changes to Earth's surface in this activity could be the foundation for a computational representation of interactions among Earth's systems.

Note: DD = "Digging Deeper" section; EAD = "Example Activity Design" section; SDA = "Suggestions for Designing an Activity" section.

* This performance expectation integrates traditional science content with engineering through a practice or disciplinary core idea.

APPENDIX B

Chapter Activities and Anecdotes Related to the Three Dimensions and Performance Expectations of the *NGSS*

Appendix B lists some of the activities and anecdotes of each chapter as they relate to the *Next Generation Science Standards* (*NGSS*; NGSS Lead States 2013): science and engineering practices (SEPs), crosscutting concepts (CCs), disciplinary core ideas (DCIs), and performance expectations (PEs). Rather than think of each SEP, CC, DCI, or PE as the outcome of a particular activity or lesson, we recommend thinking of them as outcomes of many activities that together provide insight into the practice of science. In this spirit, many of the SEPs that we list for a particular activity or anecdote are in addition to the SEPs embedded in the listed PEs. Likewise, you shouldn't think of a particular lesson as addressing only a single PE, but rather as addressing many PEs that can be bundled into particular lessons or sequences of lessons, as illustrated in Chapter 15. In the following table, PEs shown in bold are addressed more substantially in the associated chapter activities. Some connections to *A Framework for K–12 Science Education: Practices, Crosscutting Concepts, and Core Ideas* (*Framework;* NRC 2012) are also included.

REFERENCES

National Research Council (NRC). 2012. *A framework for K–12 science education: Practices, crosscutting concepts, and core ideas*. Washington, DC: National Academies Press.

NGSS Lead States. 2013. *Next Generation Science Standards: For states, by states.* Washington, DC: National Academies Press. *www.nextgenscience.org/next-generation-science-standards.*

Activities and anecdotes	The three dimensions and performance expectations of the *NGSS*			
	Science and engineering practices*	Disciplinary core ideas (DCIs)	Crosscutting concepts	Performance expectations (PEs)
Chapter 1: Go and See				
Observations of the surface of Earth and Moon, preliminary to impact-cratering experiments in Chapter 2 (pp. 6–8)	• Ask questions that can be investigated within the scope of the classroom and frame a hypothesis based on observations and scientific principles. (Practice 1, grades 6–8; see also grades 9–12)	• ESS1.C: The History of Planet Earth	• Patterns • Stability and Change	• MS-ESS1-4 • HS-ESS1-6
Conversation, questions, and argumentation about the moderating influence of oceans on coastal climates (pp. 9–10)	• Ask questions that arise from careful observation of phenomena to clarify and/or seek additional information. (Practice 1, grades 6–8 and 9–12) • Ask questions that challenge the premise(s) of an argument or the interpretation of a data set. (Practice 1, grades 6–8; see also grades 9–12)	• ESS2.D: Weather and Climate	• Systems and System Models	• Builds on elementary grade 5-ESS2-1 • MS-ESS2-6

Note: DD = "Digging Deeper" section; EAD = "Example Activity Design" section; SDA = "Suggestions for Designing an Activity" section.

*Unless otherwise stated, the practices in this column are taken from Appendix F of the *NGSS*. Some of the practices from Appendix F have been shortened to focus on the ideas addressed in the activity or anecdote.

Activities and Anecdotes	The Three Dimensions and Performance Expectations of the *NGSS*			
	Science and Engineering Practices*	Disciplinary Core Ideas (DCIs)	Crosscutting Concepts	Performance Expectations (PEs)
Chapter 1: Go and See (*continued*)				
EAD: Go and See: Geology in the Field (pp. 12–15)	• Ask questions that arise from careful observation of phenomena to clarify and/or seek additional information. (Practice 1, grades 6–8 and 9–12) • Builds on: Make observations (firsthand) to construct an evidence-based account for natural phenomena. (Practice 6, grades K–2) • Apply scientific ideas, principles, and/or evidence to construct, revise, and/or use an explanation for real-world phenomena. (Practice 6, grades 6–8) • Apply scientific ideas, principles, and/or evidence to provide an explanation of phenomena, taking into account possible unanticipated effects. (Practice 6, grades 9–12) • Respectfully provide and receive critiques about one's explanations and questions by citing relevant evidence and posing and responding to questions that elicit pertinent elaboration and detail. (Practice 7, grades 6–8; see also grades 9–12)	• ESS1.C: The History of Planet Earth • ESS2.C: The Roles of Water in Earth's Surface Processes [specifically, the role of water in rock-forming processes]	• Patterns • Scale, Proportion, and Quantity • Structure and Function • Stability and Change	• Builds on elementary grade 4-ESS1-1 • MS-ESS1-4 • MS-ESS2-1 • **MS-ESS2-2** • **HS-ESS2-1** • HS-ESS2-5
Chapter 2: The Controlled Experiment				
Is water naturally cooler than air? (pp. 18–19)	• Ask questions that can be investigated within the scope of the classroom and frame a hypothesis based on observations and scientific principles. (Practice 1, grades 6–8; see also grades 9–12)	• ESS2.C: The Roles of Water in Earth's Surface Processes • ESS2.D: Weather and Climate	• Cause and Effect: Mechanism and Explanation	• MS-ESS2-5 • **MS-ESS2-6** • **HS-ESS2-5**

Note: DD = "Digging Deeper" section; EAD = "Example Activity Design" section; SDA = "Suggestions for Designing an Activity" section.

*Unless otherwise stated, the practices in this column are taken from Appendix F of the *NGSS*. Some of the practices from Appendix F have been shortened to focus on the ideas addressed in the activity or anecdote.

Activities and anecdotes	The three dimensions and performance expectations of the *NGSS*			
	Science and engineering practices*	Disciplinary core ideas (DCIs)	Crosscutting concepts	Performance expectations (PEs)
Chapter 2: The Controlled Experiment (*continued*)				
Playing with cups, water, and air (pp. 19–20)	• Develop or modify a model—based on evidence—to match what happens if a variable or component of a system is changed. (Practice 2, grades 6–8)	• ESS2.D: Weather and Climate	• Cause and Effect: Mechanism and Explanation	• **MS-ESS2-5**
Solubility of water vapor as a function of temperature (pp. 21–23)	• Make directional hypotheses that specify what happens to a dependent variable when an independent variable is manipulated. (Practice 3, grades 9–12) • Use multiple types of models to provide mechanistic accounts and/or predict phenomena and move flexibly between model types based on merits and limitations. (Practice 2, grades 9–12)	• ESS2.C: The Roles of Water in Earth's Surface Processes	• Cause and Effect: Mechanism and Explanation • Scale, Proportion, and Quantity	• **MS-ESS2-4** • HS-ESS2-2 • **HS-ESS2-5**
EAD: The Controlled Experiment: Impact Cratering (quantitative experiments) (pp. 28–34)	• Plan an investigation individually and collaboratively, and in the design identify independent and dependent variables and controls, what tools are needed to do the gathering, how measurements will be recorded, and how many data are needed to support a claim. (Practice 3, grades 6–8; see also grades 9–12)	• ESS1.C: The History of Planet Earth (*Framework*, p. 177): Study of other planets and their moons, many of which exhibit such features as volcanism and meteor impacts similar to those found on Earth, also help illuminate aspects of Earth's history and changes.	• Cause and Effect: Mechanism and Explanation	• **MS-ESS1-3** • **HS-ESS1-6**

Note: DD = "Digging Deeper" section; EAD = "Example Activity Design" section; SDA = "Suggestions for Designing an Activity" section.

*Unless otherwise stated, the practices in this column are taken from Appendix F of the *NGSS*. Some of the practices from Appendix F have been shortened to focus on the ideas addressed in the activity or anecdote.

Activities and Anecdotes	The Three Dimensions and Performance Expectations of the *NGSS*			
	Science and Engineering Practices*	Disciplinary Core Ideas (DCIs)	Crosscutting Concepts	Performance Expectations (PEs)
Chapter 3: Field Observations				
Age of Earth, heat flow, and Lord Kelvin (pp. 35–36)	• Make and defend a claim based on evidence about the natural world. (Practice 7, grades 9–12)	• ESS2.A: Earth Materials and Systems	• Systems and System Models • Energy and Matter: Flows, Cycles, and Conservation	• HS-ESS1-6 • HS-ESS2-3
Test a hypothesis with field observations (pp. 39–41)	• Develop a model to show the relationships among variables, including those that are not observable but predict observable phenomena. (Practice 2, grades 6–8) • Collect data to produce data to serve as the basis for evidence to answer scientific questions. (Practice 3, grades 6–8)	• ESS2.C: The Roles of Water in Earth's Surface Processes	• Systems and System Models	• **MS-ESS2-2** • MS-ESS2-3 • **HS-ESS1-5**
Focused Field Discovery (pp. 41–42)	• Ask questions based on observations to find more information about the natural world. (Practice 1, grades K–2)	• ESS2.C: The Roles of Water in Earth's Surface Processes	• Systems and System Models	• MS-ESS2-3 • **MS-ESS2-2** • MS-ESS3-1 • HS-ESS2-1
EAD: Field Observation: Layer-Cake Rocks (geological mapping) (pp. 47–51)	• Develop and use a model to generate data to support explanations and predict phenomena. (Practice 2, grades 9–12)	• ESS1.C: The History of Planet Earth • ESS2.A. Earth Materials and Systems	• Patterns • Scale, Proportion, and Quantity	• MS-ESS2-3 • MS-ESS3-1 • **HS-ESS1-5**

Note: DD = "Digging Deeper" section; EAD = "Example Activity Design" section; SDA = "Suggestions for Designing an Activity" section.

*Unless otherwise stated, the practices in this column are taken from Appendix F of the *NGSS*. Some of the practices from Appendix F have been shortened to focus on the ideas addressed in the activity or anecdote.

Activities and Anecdotes	The Three Dimensions and Performance Expectations of the *NGSS*			
	Science and Engineering Practices*	Disciplinary Core Ideas (DCIs)	Crosscutting Concepts	Performance Expectations (PEs)
Chapter 4: Analyzing and Interpreting Data, Part 1: Graphing				
A Case Study in Graph Reading: The Keeling Curve (pp. 55–58)	• Distinguish between causal and correlational relationships in data. (Practice 4, grades 6–8) • Analyze data to make valid and reliable scientific claims. (Practice 4, grades 9–12) • Consider limitations of data analysis when analyzing and interpreting data. (Practice 4, grades 9–12)	• ESS2.D: Weather and Climate	• Cause and Effect: Mechanism and Explanation	• **MS-ESS3-5** • **HS-ESS2-2** • HS-ESS2-6
Graph-Reading Challenges (tombstone weathering, solubility of water in air, and geothermal gradient) (pp. 60–62)	• Construct, analyze, and/or interpret graphical displays of data. (Practice 4, grades 6–8) • Analyze and interpret data to determine similarities and differences in findings. (Practice 4, grades 6–8)	• ESS2.A: Earth's Materials and Systems • ESS2.C: The Role of Water in Earth's Surface Processes	• Scale, Proportion, and Quantity	• MS-ESS2-1 • MS-ESS2-4 • HS-ESS2-3 • HS-ESS2-5 (also applicable to a variety of PEs when graphed data are used as evidence to support arguments or explanations)
EAD: The Impact-Cratering Experiment (graphing data and making predictions) (pp. 67–71)	• Construct, analyze, and interpret graphical displays of data to identify linear and nonlinear relationships. (Practice 4, grades 6–8) • Compare and contrast various types of data sets to examine consistency of measurements and observations. (Practice 4, grades 9–12)	• ESS1.C: The History of Planet Earth	• Patterns • Cause and Effect: Mechanism and Explanation	• **MS-ESS1-3** • **HS-ESS1-6**

Note: DD = "Digging Deeper" section; EAD = "Example Activity Design" section; SDA = "Suggestions for Designing an Activity" section.

*Unless otherwise stated, the practices in this column are taken from Appendix F of the *NGSS*. Some of the practices from Appendix F have been shortened to focus on the ideas addressed in the activity or anecdote.

Activities and Anecdotes	The Three Dimensions and Performance Expectations of the *NGSS*			
	Science and Engineering Practices*	Disciplinary Core Ideas (DCIs)	Crosscutting Concepts	Performance Expectations (PEs)
Chapter 5: Analyzing and Interpreting Data, Part 2: Maps and Cross-Sections				
An Introduction to Spatial Reasoning (pp. 75–77)	• Use graphical displays (e.g., maps, charts, graphs, and/or tables) of large data sets to identify temporal and spatial relationships. (Practice 4, grades 6–8)	• ESS2.A: Earth Materials and Systems	• Scale, Proportion, and Quantity	• **MS-ESS1-4** • MS-ESS2-2 • MS-ESS2-3 • **HS-ESS1-5** • HS-ESS2-1 • HS-ESS2-7
A Case Study in Spatial Reasoning (GIS, big data, and plate tectonics) (pp. 77–83)	• Analyze and interpret data to provide evidence for phenomena. (Practice 4, grades 6–8) • Analyze and interpret data to find similarities and differences in findings. (Practice 4, grades 6–8) • Develop and/or revise a model to show the relationships among variables, including those that are not observable but predict observable phenomena. (Practice 2, grades 6–8)	• ESS2.B: Plate Tectonics and Large-Scale System Interactions	• Patterns • Cause and Effect: Mechanism and Explanation • Energy and Matter: Flows, Cycles, and Conservation	• MS-ESS2-1 • MS-ESS2-2 • **MS-ESS2-3** • **HS-ESS1-5** • HS-ESS2-3
EAD: Maps and Cross-Sections (pp. 89–92)	• Develop and/or use a model to generate data to support explanations, predict phenomena, analyze systems, and/or solve problems. (Practice 2, grades 9–12)	• ESS2.A: Earth Materials and Systems • ESS2.B: Plate Tectonics and Large-Scale System Interactions	• Patterns • Scale, Proportion, and Quantity	• **MS-ESS1-4** • MS-ESS2-2 • **HS-ESS1-5** • HS-ESS1-6 • HS-ESS2-7
DD: Contour Mapping With Weather (pp. 93–94); Figure 5.7 illustrating global circulation in map and cross-sectional views (p. 85)	• Develop, revise, and/or use a model based on evidence to illustrate and/or predict the relationships between systems or between components of a system. (Practice 2, grades 9–12)	• ESS2.D: Weather and Climate	• ESS2.D: Weather and Climate	• MS-ESS2-4 • **MS-ESS2-5** • **HS-ESS3-5**

Note: DD = "Digging Deeper" section; EAD = "Example Activity Design" section; SDA = "Suggestions for Designing an Activity" section.

*Unless otherwise stated, the practices in this column are taken from Appendix F of the *NGSS*. Some of the practices from Appendix F have been shortened to focus on the ideas addressed in the activity or anecdote.

Activities and Anecdotes	The Three Dimensions and Performance Expectations of the *NGSS*			
	Science and Engineering Practices*	Disciplinary Core Ideas (DCIs)	Crosscutting Concepts	Performance Expectations (PEs)
Chapter 6: Modeling, Part 1: Conceptual Modeling				
Starting With the "Wrong" Model (thermal energy and evaporation, isostasy) (pp. 98–100, 127)	• Develop or modify a model—based on evidence—to match what happens if a variable or component of the system is changed. (Practice 2, grades 6–8) • Develop, revise, and/or use a model based on evidence to illustrate and/or predict the relationships between systems or between components of a system. (Practice 2, grades 9–12)	• ESS2.C: The Roles of Water in Earth's Surface Processes • ESS2.B: Plate Tectonics and Large-Scale System Interactions	• Energy and Matter: Flows, Cycles, and Conservation • Stability and Change	• MS-ESS2-4 • HS-ESS2-3 • HS-ESS2-5
Building on a Seed Model (example using Gale Crater on Mars) (pp. 100–104)	• Construct an explanation using models or representations. (Practice 6, grades 6–8; see also grades 9–12)	• ESS2.C: The Roles of Water in Earth's Surface Processes	• Systems and System Models	• MS-ESS1-3
Modeling air pressure with a classroom barometer (pp. 105–108)	• Develop and/or use a model to generate data to test ideas about phenomena in natural systems, including those representing inputs and outputs and those at unobservable scales. (Practice 2, grades 6–8; see also grades 9–12)	• ESS2.D: Weather and Climate	• Systems and System Models • Stability and Change	• **MS-ESS2-5** • **MS-ESS2-6** • HS-ESS3-1 (Weather and weather modeling did not make it into any high school PEs, but air pressure is a key element in predicting severe weather.)

Note: DD = "Digging Deeper" section; EAD = "Example Activity Design" section; SDA = "Suggestions for Designing an Activity" section.

*Unless otherwise stated, the practices in this column are taken from Appendix F of the *NGSS*. Some of the practices from Appendix F have been shortened to focus on the ideas addressed in the activity or anecdote.

Activities and Anecdotes	The three dimensions and performance expectations of the *NGSS*			
	Science and engineering practices*	Disciplinary core ideas (DCIs)	Crosscutting concepts	Performance expectations (PEs)
Chapter 6: Modeling, Part 1: Conceptual Modeling *(continued)*				
EAD: Modeling Isostatic Equilibrium (pp. 111–118)	• Develop or modify a model—based on evidence—to match what happens if a variable or component of the system is changed. (Practice 2, grades 6–8) • Develop and/or use multiple types of models to provide mechanistic accounts and predict phenomena and move flexibly between model types based on merits and limitations. (Practice 2, grades 9–12)	• ESS2.B: Plate Tectonics and Large-Scale System Interactions • PS2.A: Forces and Motion • PS2.B: Types of Interactions	• Systems and System Models • Stability and Change	• **MS-ESS2-2** • **HS-ESS2-1** • **HS-ESS2-2** • HS-ESS2-3
Chapter 7: Modeling, Part 2: Mathematical Modeling				
Math-Reading Challenges: Connecting Math to Nature (Reynolds number) (pp. 121–123); DD: Mathematical Modeling Related to Flooding (Math-Reading Challenges: Stream Velocity and Discharge) (pp. 143–145)	• Apply mathematical concepts and/or processes to scientific and engineering questions and problems. (Practice 5, grades 6–8) • Use simple limit cases to test mathematical expressions to see if a model "makes sense" by comparing the outcomes with what is known about the real world. (Practice 5, grades 9–12)	• ESS2.C: The Roles of Water in Earth Surface Processes	• Systems and System Models • Scale, Proportion, and Quantity	• MS-ESS3-2 (flood and erosion mitigation) • **HS-ESS2-5** (Analogous math-reading challenges using equations for planetary motion or energy balance could be applied to HS-ESS1-4 and HS-ESS2-4.)

Note: DD = "Digging Deeper" section; EAD = "Example Activity Design" section; SDA = "Suggestions for Designing an Activity" section.

*Unless otherwise stated, the practices in this column are taken from Appendix F of the *NGSS*. Some of the practices from Appendix F have been shortened to focus on the ideas addressed in the activity or anecdote.

Activities and Anecdotes	The Three Dimensions and Performance Expectations of the *NGSS*			
	Science and Engineering Practices*	Disciplinary Core Ideas (DCIs)	Crosscutting Concepts	Performance Expectations (PEs)
Chapter 7: Modeling, Part 2: Mathematical Modeling (*continued*)				
Deriving Mathematical Expressions (pp. 124–126)	• Use mathematical representations to describe scientific conclusions. (Practice 5, grades 6–8) • Apply techniques of algebra and functions to represent and solve scientific problems. (Practice 5, grades 9–12)	• ESS2.C: The Roles of Water in Earth Surface Processes • ESS2.A: Earth Materials and Systems	• Scale, Proportion, and Quantity • Systems and System Models	• HS-ESS1-4 (also supports mathematical thinking implicit in many other PEs, such as HS-ESS2-6)
Creating and Using Algorithms (pp. 130–134)	• Create algorithms (a series of ordered steps) to solve a problem. (Practice 5, grades 6–8) • Use mathematical, computational, and/or algorithmic representations of phenomena to support claims and explanations. (Practice 5, grades 9–12)	• ESS3.C: Human Impacts on Earth Systems	• Scale, Proportion, and Quantity • Systems and System Models	• **HS-ESS3-2** • **HS-ESS3-3** • (The quantitative modeling in this activity could be applied analogously to other PEs, such as MS-ESS3-3, MS-ESS3-4, and HS-ESS2-6.)
DD: Mathematical Modeling Related to Flooding (A Computational Simulation for Rainfall Runoff) (pp. 145–148)	• Create and/or revise a computational model or simulation of a phenomenon or system. (Practice 5, grades 9–12)	• ESS3.A: Natural Resources • ESS3.C: Human Impacts on Earth systems	• Scale, Proportion, and Quantity • Systems and System Models	• **HS-ESS3-6**
EAD: The Impact-Cratering Experiment: Identifying and Testing a Mathematical Model (pp. 136–142)	• Use simple limit cases to test mathematical expressions to see if a model "makes sense" by comparing the outcomes with what is known about the real word. (Practice 5, grades 9–12)	• ESS1.C: The History of Planet Earth • PS3.C: Relationship between Energy and Forces	• Systems and System Models • Energy and Matter: Flows, Cycles, and Conservation	• **HS-ESS1-6**

Note: DD = "Digging Deeper" section; EAD = "Example Activity Design" section; SDA = "Suggestions for Designing an Activity" section.

*Unless otherwise stated, the practices in this column are taken from Appendix F of the *NGSS*. Some of the practices from Appendix F have been shortened to focus on the ideas addressed in the activity or anecdote.

Activities and Anecdotes	The Three Dimensions and Performance Expectations of the *NGSS*			
	Science and Engineering Practices*	Disciplinary Core Ideas (DCIs)	Crosscutting Concepts	Performance Expectations (PEs)
Chapter 8: Arguing From Evidence, Part 1: Focus on Evidence				
Difficulties in identifying evidence (pp. 152–156)	• Distinguish between opinions and evidence in one's own explanations. (Practice 7, grades K–2) • Compare and critique two arguments on the same topic and analyze whether they emphasize similar or different evidence and/or interpretations of facts. (Practice 7, grades 6–8) • Evaluate the claims, evidence, and/or reasoning behind currently accepted explanations to determine the merits of arguments. (Practice 7, grades 9–12)	• ESS2.B: Plate Tectonics and Large-Scale Interactions	• Patterns • Cause and Effect: Mechanism and Explanation	• **MS-ESS2-3** • **HS-ESS1-5** • HS-ESS 2-1
Solutions for learning to identify evidence (pp. 156–162)	• Apply scientific reasoning to show why the data or evidence is adequate for the explanation or conclusion. (Practice 6, grades 6–8) • Apply scientific reasoning, theory, and/or models to link evidence to the claims to assess the extent to which the reasoning and data support the explanation or conclusion. (Practice 6, grades 9–12) • Make and defend a claim based on evidence about the natural world that reflects scientific knowledge and student-generated evidence. (Practice 7, grades 9–12)	• ESS2.B: Plate Tectonics and Large-Scale Interactions	• Patterns • Cause and Effect: Mechanism and Explanation	• **MS-ESS2-3** • **HS-ESS1-5** • HS-ESS 2-1

Note: DD = "Digging Deeper" section; EAD = "Example Activity Design" section; SDA = "Suggestions for Designing an Activity" section.

*Unless otherwise stated, the practices in this column are taken from Appendix F of the *NGSS*. Some of the practices from Appendix F have been shortened to focus on the ideas addressed in the activity or anecdote.

Activities and anecdotes	The three dimensions and performance expectations of the *NGSS*			
	Science and engineering practices*	Disciplinary core ideas (DCIs)	Crosscutting concepts	Performance expectations (PEs)
Chapter 8: Arguing From Evidence, Part 1: Focus on Evidence (*continued*)				
EAD: Wegener's Evidence for Continental Drift (arguing from evidence) (pp. 165–168)	• Respectfully provide and receive critiques about one's explanations, procedures, models, and questions by citing relevant evidence and posing and responding to questions that elicit pertinent elaboration and detail. (Practice 7, grades 6–8) • Respectfully provide and/or receive critiques on scientific arguments by probing reasoning and evidence, challenging ideas and conclusions, responding thoughtfully to diverse perspectives, and determining additional information required to resolve contradictions. (Practice 7, grades 9–12)	• ESS1.C: The History of Planet Earth	• Patterns • Scale, Proportion, and Quantity	• **MS-ESS2-3** • **HS-ESS1-5** • HS-ESS 2-1
Chapter 9: Arguing From Evidence, Part 2: Claim and Argument				
Finding a Place to Start (pp. 172–175)	• Apply scientific ideas, principles, and/or evidence to construct, revise, and/or use an explanation for real-world phenomena, examples, or events. (Practice 6, grades 6–8)	• ESS2.A: Earth Materials and Systems • ESS2.C: The Roles of Water in Earth's Surface Processes	• Cause and Effect: Mechanism and Explanation	• MS-ESS2-5

Note: DD = "Digging Deeper" section; EAD = "Example Activity Design" section; SDA = "Suggestions for Designing an Activity" section.

*Unless otherwise stated, the practices in this column are taken from Appendix F of the *NGSS*. Some of the practices from Appendix F have been shortened to focus on the ideas addressed in the activity or anecdote.

Activities and Anecdotes	The Three Dimensions and Performance Expectations of the *NGSS*			
	Science and Engineering Practices*	Disciplinary Core Ideas (DCIs)	Crosscutting Concepts	Performance Expectations (PEs)
Chapter 9: Arguing From Evidence, Part 2: Claim and Argument *(continued)*				
Arguing Through the Fog; Wrapping Up the Argument (pp. 175–182)	• Apply scientific reasoning, theory, and/or models to link evidence to the claims to assess the extent to which the reasoning and data support the explanation or conclusion. (Practice 6, grades 9–12) • Construct, use, and/or present an oral and written argument supported by empirical evidence and scientific reasoning to support or refute an explanation or model for a phenomenon. (Practice 7, grades 6–8) • Construct, use, and /or present an oral and written argument or counter-arguments based on data and evidence. (Practice 7, grades 9–12)	• ESS2.A: Earth Materials and Systems • ESS2.D: Weather and Climate	• Patterns • Cause and Effect: Mechanism and Explanation • Systems and System Models	• **MS-ESS2-1** • MS-ESS2-6 • HS-ESS1-4 • HS-ESS2-4
The Tyranny of the Answer Key (pp. 183–185)	• Plan an investigation individually and collaboratively to produce data to serve as the basis for evidence as part of building and revising models and supporting explanations for phenomena. Consider possible confounding variables or effects and evaluate the investigation's design to ensure variables are controlled. (Practice 3, grades 9–12)	• ESS2.D: Weather and Climate	• Cause and Effect: Mechanism and Explanation	• **MS-ESS3-5** • **HS-ESS2-4**

Note: DD = "Digging Deeper" section; EAD = "Example Activity Design" section; SDA = "Suggestions for Designing an Activity" section.

*Unless otherwise stated, the practices in this column are taken from Appendix F of the *NGSS*. Some of the practices from Appendix F have been shortened to focus on the ideas addressed in the activity or anecdote.

Activities and Anecdotes	The Three Dimensions and Performance Expectations of the *NGSS*			
	Science and Engineering Practices*	Disciplinary Core Ideas (DCIs)	Crosscutting Concepts	Performance Expectations (PEs)
Chapter 9: Arguing From Evidence, Part 2: Claim and Argument *(continued)*				
EAD: The Columns of Pozzuoli (pp. 187–193)	• Construct an explanation based on valid and reliable evidence obtained from sources and the assumption that theories and laws that describe the natural world operate today as they did in the past and will continue to do so in the future. (Practice 6, grades 6–8 and grades 9–12; see also Practice 7, grades 6–8 and 9–12)	• ESS1.C: The History of Planet Earth	• Cause and Effect: Mechanism and Explanation	• **MS-ESS2-2**
Chapter 10: Stories in Rocks				
Leaf inset into the snow (pp. 197–198)	• Ask questions based on observations to find more information about the natural or designed world. (Practice 1, grades K–2; see grades 3–5, 6–8, and 9–12 for an expansion on this practice) • The goal of science is to construct explanations for the causes of phenomena. (Practice 6, introductory text)	• PS3.B: Conservation of Energy and Energy Transfer • ESS2.A: Earth Materials and Systems	• Energy and Matter: Flows, Cycles, and Conservation • Stability and Change	• MS-ESS2-1 • HS-ESS2-2
Reading the Stories (dinosaurs and environments of deposition) (pp. 199–203)	• Apply scientific ideas, principles, and evidence to construct revise, and/or use an explanation for real-world phenomena, examples, or events. (Practice 6, grades 6–8 and 9–12)	• ESS1.C: The History of Planet Earth • ESS2.A: Earth Materials and Systems	• Scale, Proportion, and Quantity • Stability and Change	• **MS-ESS1-4** • HS-ESS2-1 • **HS-ESS2-7**

Note: DD = "Digging Deeper" section; EAD = "Example Activity Design" section; SDA = "Suggestions for Designing an Activity" section.

*Unless otherwise stated, the practices in this column are taken from Appendix F of the *NGSS*. Some of the practices from Appendix F have been shortened to focus on the ideas addressed in the activity or anecdote.

Activities and Anecdotes	The Three Dimensions and Performance Expectations of the *NGSS*			
	Science and Engineering Practices*	Disciplinary Core Ideas (DCIs)	Crosscutting Concepts	Performance Expectations (PEs)
Chapter 10: Stories in Rocks *(continued)*				
Reading the Stories (rocks on Mars and environments of deposition) (pp. 202–203)	• Apply scientific ideas, principles, and evidence to construct and revise an explanation for real-world phenomena, examples, or events. (Practice 6, grades 6–8) • Apply scientific ideas, principles, and/or evidence to provide an explanation of phenomena, taking into account possible unanticipated effects. (Practice 6, grades 9–12)	• ESS1.C: The History of Planet Earth • ESS2.A: Earth Materials and Systems	• Patterns • Cause and Effect: Mechanism and Explanation • Energy and Matter: Flows, Cycles, and Conservation	• MS-ESS1-3 • MS-ESS1-4 • MS-ESS2-2 • HS-ESS2-1
SDA: Experimental Sedimentary Petrology With Downspouts (pp. 205–208)	• Plan and conduct an investigation individually and collaboratively, and in the design identify independent and dependent variables and controls, what tools are needed to do the gathering, how measurements will be recorded, and how many data are needed to support a claim. (Practice 3, grades 6–8; see grades 9–12 for an expansion on this practice)	• ESS2.A: Earth Materials and Systems • ESS2.C: The Roles of Water in Earth's Surface Processes	• Systems and System Models • Energy and Matter: Flows, Cycles, and Conservation	• **MS-ESS2-1** • **HS-ESS2-5**
SDA: Experimental Igneous Petrology With Thymol (pp. 209–212)	• Plan and conduct an investigation individually and collaboratively, and in the design identify independent and dependent variables and controls, what tools are needed to do the gathering, how measurements will be recorded, and how many data are needed to support a claim. (Practice 3, grades 6–8; see grades 9–12 for an expansion on this practice)	• ESS2.A: Earth Materials and Systems • PS1.A: Structure and Properties of Matter	• Energy and Matter: Flows, Cycles, and Conservation • Structure and Function	• **MS-ESS2-1** • **HS-ESS1-6** (Petrology, although not specifically called out in the PE, applies to Earth's entire history.) • **MS-PS1-4** • HS-PS3-2

Note: DD = "Digging Deeper" section; EAD = "Example Activity Design" section; SDA = "Suggestions for Designing an Activity" section.

*Unless otherwise stated, the practices in this column are taken from Appendix F of the *NGSS*. Some of the practices from Appendix F have been shortened to focus on the ideas addressed in the activity or anecdote.

Activities and Anecdotes	The Three Dimensions and Performance Expectations of the *NGSS*			
	Science and Engineering Practices*	Disciplinary Core Ideas (DCIs)	Crosscutting Concepts	Performance Expectations (PEs)
Chapter 11: Stories in Rock Layers				
Stratigraphy: Earth's Storybook (pp. 213–214)	• Construct a scientific explanation based on valid and reliable evidence obtained from sources (including the students' own experiments) and the assumption that theories and laws that describe the natural world operated today as they did in the past and will continue to do so in the future. (Practice 6, grades 6–8 and 9–12)	• ESS1.C: The History of Planet Earth • ESS2.E: Biogeology • LS4.A: Evidence of Common Ancestry and Diversity [rationale: the fossil record is embedded in the rock record, not separate from, the rock record]	• Scale, Proportion, and Quantity	• **MS-ESS1-4** • **HS-ESS1-5** (Both PEs showcase the use of evidence.)
Classroom Lessons and Challenges From the Grand Canyon (crosscutting relationships in Powell's units A, B, and C) (pp. 218–228)	• Distinguish between explanations that account for all gathered evidence and those that do not. (Practice 7, grades K–2) • Analyze why some evidence is relevant to a scientific question and some is not. (Practice 7, grades K–2) • Use data to evaluate claims about cause and effect. (Practice 7, grades 3–5) • Construct and use an argument supported by empirical evidence and scientific reasoning to support or refute an explanation or a model for a phenomenon. (Practice 7, grades 6–8 and 9–12) • Use graphical displays of large data sets [the simplified geological cross-section of the Grand Canyon] to identify temporal and spatial relationships. (Practice 4, grades 6–8) • Evaluate the impact of new data on a working explanation and/or model of a proposed process or system. (Practice 4, grades 9–12)	• ESS1.C: The History of Planet Earth	• Patterns • Scale, Proportion and Quantity • Systems and System Models • Stability and Change	• **MS-ESS1-4** • **HS-ESS1-5**

Note: DD = "Digging Deeper" section; EAD = "Example Activity Design" section; SDA = "Suggestions for Designing an Activity" section.

*Unless otherwise stated, the practices in this column are taken from Appendix F of the *NGSS*. Some of the practices from Appendix F have been shortened to focus on the ideas addressed in the activity or anecdote.

Activities and Anecdotes	The Three Dimensions and Performance Expectations of the *NGSS*			
	Science and Engineering Practices*	Disciplinary Core Ideas (DCIs)	Crosscutting Concepts	Performance Expectations (PEs)
Chapter 11: Stories in Rock Layers *(continued)*				
SDA: Reading Stories of Deep Time (pp. 229–235)	• Critically read scientific texts adapted for classroom use to determine the central ideas and obtain scientific information to describe patterns in and evidence about the natural world. (Practice 8, grades 6–8 and 9–12) • Apply scientific ideas, principles, and/or evidence to construct, revise, and/or use an explanation for real-world phenomena, examples, or events. (Practice 6, grades 6–8) • Apply scientific ideas, principles, and/or evidence to provide an explanation of phenomena, taking into account possible unanticipated effects. (Practice 6, grades 9–12) (*Note:* This project asks students to argue from evidence as they develop their explanation. The practices listed for the Grand Canyon challenges above apply here as well.)	• ESS1.C: The History of Planet Earth • ESS2.E: Biogeology • LS4.A: Evidence of Common Ancestry and Diversity [rationale: the fossil record is embedded in, not separate from, the rock record]	• Patterns • Scale, Proportion, and Quantity • Systems and System Models • Stability and Change	• **MS-ESS1-4** • MS-ESS2-1 • **HS-ESS1-5** • HS-ESS2-7

Note: DD = "Digging Deeper" section; EAD = "Example Activity Design" section; SDA = "Suggestions for Designing an Activity" section.

*Unless otherwise stated, the practices in this column are taken from Appendix F of the *NGSS*. Some of the practices from Appendix F have been shortened to focus on the ideas addressed in the activity or anecdote.

Activities and Anecdotes	The Three Dimensions and Performance Expectations of the *NGSS*			
	Science and Engineering Practices*	Disciplinary Core Ideas (DCIs)	Crosscutting Concepts	Performance Expectations (PEs)
Chapter 12: Stories of Places We Can't See but Can Hear				
When You Can't Look Up the Answer (pp. 237–239)	• Gather, read and evaluate scientific and/or technical information from multiple authoritative sources, assessing the evidence and usefulness of each source. (Practice 8, grades 9–12; see also grades 6–8)	• ESS2.A: Earth Materials and Systems	• Patterns	• **MS-ESS2-2** (Constructing explanations about Earth's changing surface rests on an understanding of present topography.)
Science in the Real World: Witness to the Unseen (authentic science in the *Kursk* investigation) (pp. 239–241)	• Ask questions that can be answered by an investigation. (Practice 1, grades K–2) • Ask questions to clarify and/or refine a model, an explanation, or an engineering problem. (Practice 1, grades 6–8 and 9–12)	• ESS2.B: Plate Tectonics and Large-Scale System Interactions	• Energy and Matter: Flows, Cycles, and Conservation	• **HS-ESS2-3** (The practice of Asking Questions is implicit in all the PEs even though it doesn't appear as a stand-alone practice for the ESS PEs for grades 6–12.)
A Lesson in Measuring What We Can't See (pp. 242–245)	• Conduct an investigation and evaluate/revise the experimental design to produce data to serve as the basis for evidence that meet the goals of the investigation. (Practice 3, grades 6–8; see grades 9–12 for an expansion on this practice)	• ESS2.A: Earth Materials and Systems	• Energy and Matter: Flows, Cycles, and Conservation	• **HS-ESS2-3** • HS-PS4-1

Note: DD = "Digging Deeper" section; EAD = "Example Activity Design" section; SDA = "Suggestions for Designing an Activity" section.

*Unless otherwise stated, the practices in this column are taken from Appendix F of the *NGSS*. Some of the practices from Appendix F have been shortened to focus on the ideas addressed in the activity or anecdote.

Activities and Anecdotes	The Three Dimensions and Performance Expectations of the *NGSS*			
	Science and Engineering Practices*	Disciplinary Core Ideas (DCIs)	Crosscutting Concepts	Performance Expectations (PEs)
Chapter 12: Stories of Places We Can't See but Can Hear (*continued*)				
Richard Oldham and the outer core (developing and revising models [Figure 12.4, p. 247] and arguing from evidence) (pp. 247–249)	• Analyze data using tools, technologies, and/or models in order to make valid and reliable scientific claims. (Practice 4, grades 9–12) • Develop, revise, and use a model based on evidence to illustrate and predict the relationships between systems or between components of a system. (Practice 2, grades 9–12) • Evaluate the claims, evidence, and/or reasoning behind currently accepted explanations to determine the merits of the arguments. (Practice 7, grades 9–12)	• Systems and System Models • Energy and Matter: Flows, Cycles, and Conservation	• Systems and System Models • Energy and Matter: Flows, Cycles, and Conservation	• **HS-ESS2-3** • HS-PS4-1
Reading Scientific Literature: Inge Lehmanns's Pesky P Waves and the Inner Core (p. 249)	• Critically read scientific literature adapted for classroom use to determine the central ideas and to obtain scientific information to summarize complex evidence and concepts. (Practice 8, grades 9–12; see also the practices listed for Richard Oldham and the Outer Core, above).	• ESS2.A: Earth Materials and Systems	• Systems and System Models • Energy and Matter: Flows, Cycles, and Conservation	• **HS-ESS2-3** • HS-PS4-1

Note: DD = "Digging Deeper" section; EAD = "Example Activity Design" section; SDA = "Suggestions for Designing an Activity" section.

*Unless otherwise stated, the practices in this column are taken from Appendix F of the *NGSS*. Some of the practices from Appendix F have been shortened to focus on the ideas addressed in the activity or anecdote.

Activities and Anecdotes	The Three Dimensions and Performance Expectations of the *NGSS*			
	Science and Engineering Practices*	Disciplinary Core Ideas (DCIs)	Crosscutting Concepts	Performance Expectations (PEs)
Chapter 12: Stories of Places We Can't See but Can Hear (*continued*)				
SDA: Authentic Questions and Real Seismic Data (frequency of big earthquakes and magnitude of deep earthquakes) (pp. 253–255)	• Ask questions that require sufficient and appropriate empirical evidence to answer. (Practice 1, grades 6–8) • Ask questions that arise from careful observation of phenomena, or unexpected results, to clarify and seek additional information. (Practice 1, grades 9–12) • Analyze and interpret data to determine similarities and differences in findings. (Practice 4, grades 6–8) • Apply concepts of statistics and probability to analyze and characterize data, using digital tools when feasible. (Practice 4, grades 6–8 and 9–12)	• ESS2.A: Earth Materials and Systems • ESS3.B: Natural Hazards	• Patterns • Cause and Effect: Mechanism and Explanation • System and System Models • Energy and Matter: Flows, Cycles, and Conservation	• **MS-ESS3-2** • **HS-ESS3-1**
Chapter 13: Stories of Places We Can't Go but Can See				
Aristarchus and the distance to the Moon (pp. 257–258); Dangers of Pictorial and Physical Models (box) (pp. 261–262)	• Construct an explanation using models or representations. (Practice 6, grades 6–8)	• ESS1.B: Earth and the Solar System	• Scale, Proportion, and Quantity	• **MS-ESS1-3** (Consideration of problems in scaling is important for this PE.)
Enlightened: Reading the Sky With Our Naked Eye (phases of the Moon) (pp. 258–264)	• Develop and use a model to predict and describe phenomena. (Practice 2, grades 6-8)	• ESS1.B: Earth and the Solar System	• Patterns • Scale, Proportion, and Quantity	• **M-ESS1-1**

Note: DD = "Digging Deeper" section; EAD = "Example Activity Design" section; SDA = "Suggestions for Designing an Activity" section.

*Unless otherwise stated, the practices in this column are taken from Appendix F of the *NGSS*. Some of the practices from Appendix F have been shortened to focus on the ideas addressed in the activity or anecdote.

Activities and Anecdotes	The Three Dimensions and Performance Expectations of the *NGSS*			
	Science and Engineering Practices*	Disciplinary Core Ideas (DCIs)	Crosscutting Concepts	Performance Expectations (PEs)
Chapter 13: Stories of Places We Can't Go but Can See (*continued*)				
Enlightened: Reading the Sky With a Telescope (phases of Venus) (pp. 264–267)	• Scientific explanations are subject to revision and improvement in light of new evidence. (Appendix H [nature of science], p. 99) • Compare and critique two arguments on the same topic and analyze whether they emphasize similar or different evidence and/or interpretation of facts. (Practice 7, grades 6–8)	• ESS1.B: Earth and the Solar System	• Systems and System Models • Technological advances have influenced the progress of science and science has influenced advances in technology. (Appendix H [nature of science], p. 100)	• **MS-ESS1-2**
Enlightened: Reading the Sky With a Spectroscope (composition of stars and atmospheres of exoplanets) (pp. 267–271)	• Analyze and interpret data to provide evidence for phenomena. (Practice 4, grades 6–8) • Analyze data using tools, technologies, and/or models in order to make valid and reliable scientific claims. (Practice 4, grades 9–12)	• ESS1.A: The Universe and Its Stars	• Patterns	• **HS-ESS1-1** • **HS-ESS1-2** • **HS-ESS1-3**
SDA: Constructing a Moon Phase Model From Observations (pp. 272–274)	• Develop and use a model to predict and describe phenomena. (Practice 2, grades 6–8) • Conduct an investigation to produce data to serve as the basis for evidence that meet the goals of the investigation. (Practice 3, grades 6–8)	• ESS1.B: Earth and the Solar System	• Patterns • Cause and Effect: Mechanism and Explanation	• **MS-ESS1-1**

Note: DD = "Digging Deeper" section; EAD = "Example Activity Design" section; SDA = "Suggestions for Designing an Activity" section.

*Unless otherwise stated, the practices in this column are taken from Appendix F of the *NGSS*. Some of the practices from Appendix F have been shortened to focus on the ideas addressed in the activity or anecdote.

Activities and Anecdotes	The Three Dimensions and Performance Expectations of the *NGSS*			
	Science and Engineering Practices*	Disciplinary Core Ideas (DCIs)	Crosscutting Concepts	Performance Expectations (PEs)
Chapter 14: Stories Told by Atoms				
Mass balance (lakes [pp. 275, 280, 287] and magma [pp. 281–284])	• Apply mathematical concepts and processes (e.g., ratios and simple algebra) to scientific questions. (Practice 5, grades 6–8) • Use mathematical representations to describe and support scientific conclusions. (Practice 5, grades 6–8) • Use mathematical representations of phenomena to describe and support claims and/or explanations. (Practice 5, grades 9–12)	• ESS2.A: Earth Materials and Systems • ESS3.A: Natural Resources	• Scale, Proportion, and Quantity	• **MS-ESS2-1** • MS-ESS3-1 • MS-PS1-5 • HS-ESS2-1 • **HS-ESS2-5**
Mass Balance in the News (Biosphere 2) (pp. 284–286)	• Apply mathematical concepts and processes (e.g., ratios and simple algebra) to scientific questions. (Practice 5, grades 6–8) • Use mathematical representations to describe and support scientific conclusions. (Practice 5, grades 6–8) • Use mathematical representations of phenomena to describe and support claims and/or explanations. (Practice 5, grades 9–12)	• ESS2.D: Weather and Climate • ESS2.E: Biogeology • PS1.B: Chemical Reactions	• Systems and System Models • Energy and Matter: Flows, Cycles, and Conservation • Stability and Change	• **HS-ESS2-2** • HS-ESS2-6 • HS-ESS2-7
Mass Balance in the News (erosion on Oahu) (pp. 287–288)	• Apply ratios, rates, percentages, and unit conversions in the context of complicated measurement problems involving quantities with derived or compound units. (Practice 5, grades 9–12)	• ESS2.A: Earth Materials and Systems	• Energy and Matter: Flows, Cycles, and Conservation • Stability and Change	• **HS-ESS2-5**

Note: DD = "Digging Deeper" section; EAD = "Example Activity Design" section; SDA = "Suggestions for Designing an Activity" section.

*Unless otherwise stated, the practices in this column are taken from Appendix F of the *NGSS*. Some of the practices from Appendix F have been shortened to focus on the ideas addressed in the activity or anecdote.

Activities and Anecdotes	The Three Dimensions and Performance Expectations of the *NGSS*			
	Science and Engineering Practices*	Disciplinary Core Ideas (DCIs)	Crosscutting Concepts	Performance Expectations (PEs)
Chapter 14: Stories Told by Atoms (*continued*)				
DEATH-X: an experimental partitioning activity (pp. 290–295)	• Select appropriate tools to collect, record, analyze, and evaluate data. (Practice 3, grades 9–12) • Design, evaluate, and/or refine a solution to a complex real-world problem, based on scientific knowledge, student-generated sources of evidence, prioritized criteria, and tradeoff considerations. (Practice 6, grades 9–12)	• ESS2.A: Earth Materials and Systems • ESS2.C: The Roles of Water in Earth's Surface Processes	• Systems and System Models • Energy and Matter: Flows, Cycles, and Conservation	• **HS-ESS3-4** • **HS-ESS3-6** (Creating a computer model rests on an understanding of partition coefficients.)
DD: Gold Ore in South Africa (pp. 301–303)	• Apply scientific ideas, principles, and evidence to construct an explanation for real-world phenomena. (Practice 6, grades 6–8) • Develop a model to show relationships among variables, including those that are not observable but predict observable phenomena. (Practice 2, grades 6–8)	• ESS3.A: Natural Resources	• Systems and System Models • Energy and Matter: Flows, Cycles, and Conservation	• **MS-ESS3-1**

Note: DD = "Digging Deeper" section; EAD = "Example Activity Design" section; SDA = "Suggestions for Designing an Activity" section.

*Unless otherwise stated, the practices in this column are taken from Appendix F of the *NGSS*. Some of the practices from Appendix F have been shortened to focus on the ideas addressed in the activity or anecdote.

Activities and Anecdotes	The Three Dimensions and Performance Expectations of the *NGSS*			
	Science and Engineering Practices*	Disciplinary Core Ideas (DCIs)	Crosscutting Concepts	Performance Expectations (PEs)
Chapter 14: Stories Told by Atoms (*continued*)				
SDA: Experimental Test of the Mathematical Model for Geochemical Differentiation (pp. 297–300)	• Develop or modify a model—based on evidence—to match what happens if a variable or component of a system is changed. (Practice 2, grades 6–8) • Design a test of a model to ascertain its reliability. (Practice 2, grades 9–12) • Conduct an investigation to produce data to serve as the basis for evidence that meet the goals of the investigation. (Practice 3, grades 6–8) • Plan an investigation to produce data to serve as the basis for evidence as part of building and revising models, supporting explanations for phenomena, or testing solutions to problems. (Practice 3, grades 9–12) • Apply mathematical concepts (ratios) to scientific and engineering questions and problems. (Practice 5, grades 6–8) • Apply techniques of algebra and functions to represent and solve scientific and engineering problems. (Practice 5, grades 9–12)	• ESS2.C: The Roles of Water in Earth's Surface Processes	• Cause and Effect: Mechanism and Explanation • Systems and System Models • Energy and Matter: Flows, Cycles, and Conservation	• **MS-ESS2-1** • MS-ESS2-4 • MS-ESS3-1 • **HS-ESS2-5**

Note: DD = "Digging Deeper" section; EAD = "Example Activity Design" section; SDA = "Suggestions for Designing an Activity" section.

*Unless otherwise stated, the practices in this column are taken from Appendix F of the *NGSS*. Some of the practices from Appendix F have been shortened to focus on the ideas addressed in the activity or anecdote.

Activities and Anecdotes	The Three Dimensions and Performance Expectations of the *NGSS*			
	Science and Engineering Practices*	Disciplinary Core Ideas (DCIs)	Crosscutting Concepts	Performance Expectations (PEs)
Chapter 15: Teacher as Curriculum Narrator				
Does convection hold up hot-air balloons? (Figure 15.1 illustrating curricular narrative/ story line, p. 309)	• This figure includes investigative activities from other chapters in this book and will meet multiple practices, but the exact mix of practices will depend on which direction you take the work in the classroom.	• These sequences could support a variety of DCIs, depending on which activities and directions you and your students decide to pursue.	• Cause and Effect: Mechanism and Explanation • Systems and System Models • Energy and Matter: Flows, Cycles, and Conservation	• A coherent sequence that examines the mechanisms of convection and buoyancy could support part of each PE in a bundle of PEs, such as ○ MS-ESS2-1 ○ MS-ESS2-4 ○ MS-ESS2-5 ○ MS-ESS2-6
Story Hook: Awakening the Inner Scientist (pp. 310–313)	• Ask questions ("Asking questions is essential to developing scientific habits of mind." *Framework, p. 54)*	• Asking questions applies to all DCIs.	• All crosscutting concepts require asking questions—for example, without asking questions, one can't pursue Cause and Effect: Mechanism and Explanation.	• All PEs require asking questions.

Note: DD = "Digging Deeper" section; EAD = "Example Activity Design" section; SDA = "Suggestions for Designing an Activity" section.

*Unless otherwise stated, the practices in this column are taken from Appendix F of the *NGSS*. Some of the practices from Appendix F have been shortened to focus on the ideas addressed in the activity or anecdote.

Activities and Anecdotes	The Three Dimensions and Performance Expectations of the *NGSS*			
	Science and Engineering Practices*	Disciplinary Core Ideas (DCIs)	Crosscutting Concepts	Performance Expectations (PEs)
Chapter 15: Teacher as Curriculum Narrator (*continued*)				
Figure 15.2 illustrating example unit on weathering and erosion (p. 315)	• Planning and Carrying Out Investigations (Practice 3, all grades) • Constructing Explanations and Designing Solutions (Practice 6, all grades)	• ESS2.A: Earth Materials and Systems • ESS2.C: The Roles of Water in Earth's Surface Processes • ESS2.D: Weather and Climate • ESS3: Earth and Human Activity	• Cause and Effect: Mechanism and Explanation • Systems and System Models • Energy and Matter: Flows, Cycles, and Conservation	• Example PE bundle supported by the story lines in Figure 15.2: o HS-ESS2-2 o HS-ESS2-5 o HS-ESS3-1 o HS-ESS3-6
Figure 15.3 illustrating example set of units that address the big idea "Planets have a history of change" and *NGSS* ESS1.C (p. 317)	• Multiple practices are used in this sequence of units to investigate student-initiated questions and explore new avenues prompted by the teacher or science in the news.	• ESS1.C: History of Planet Earth • ESS2: Earth's Systems (Coherence emerges when teacher and students relate "first ideas" to "later ideas" and to big ideas and core ideas.)	• Patterns • Cause and Effect: Mechanism and Explanation • Systems and System models • Energy and Matter: Flows, Cycles, and Conservation	• Example PE bundle supported by the story lines in Figure 15.3: o MS-ESS1-4 o MS-ESS2-1 o MS-ESS2-2 o MS-ESS2-4 o MS-LS4-1 o MS-PS1-2

Note: DD = "Digging Deeper" section; EAD = "Example Activity Design" section; SDA = "Suggestions for Designing an Activity" section.

*Unless otherwise stated, the practices in this column are taken from Appendix F of the *NGSS*. Some of the practices from Appendix F have been shortened to focus on the ideas addressed in the activity or anecdote.

Activities and Anecdotes	The Three Dimensions and Performance Expectations of the *NGSS*			
	Science and Engineering Practices*	Disciplinary Core Ideas (DCIs)	Crosscutting Concepts	Performance Expectations (PEs)
Chapter 16: Earth Science Where You Are				
A game of yes/no questions (pp. 329–331, Figure 16.1 [p. 330], and annex [pp. 340–343])	• Ask questions (Practice 1, grades 6–8) o that arise from careful observation of phenomena to clarify or seek additional information, o to identify and clarify evidence, o to clarify and refine a model, o that require sufficient and appropriate empirical evidence to answer, and o that challenge the premise of an interpretation of a data set. • Ask questions (Practice 1, grades 9–12) o that arise from careful observation of phenomena to clarify or seek additional information and o to clarify and refine an explanation.	• ESS2.A: Earth Materials and Systems • ESS2.B: The Roles of Water in Earth's Surface Processes • ESS3.C: Human Impacts on Earth Systems	• Patterns • Cause and Effect: Mechanism and Explanation • Stability and Change	• MS-ESS2-1 • **MS-ESS2-2** • MS-ESS2-4 • HS-ESS3-6 (The changes to Earth's surface in this activity could be the foundation for a computational representation of interactions among Earth's systems.)

Note: DD = "Digging Deeper" section; EAD = "Example Activity Design" section; SDA = "Suggestions for Designing an Activity" section.

*Unless otherwise stated, the practices in this column are taken from Appendix F of the *NGSS*. Some of the practices from Appendix F have been shortened to focus on the ideas addressed in the activity or anecdote.

Activities and anecdotes	The three dimensions and performance expectations of the *NGSS*			
	Science and engineering practices*	Disciplinary core ideas (DCIs)	Crosscutting concepts	Performance expectations (PEs)
Chapter 16: Earth Science Where You Are *(continued)*				
Edible models to launch an investigation (pp. 331–334), Figures 16.2 and 16.3 [pp. 333–334])	• Develop and use a model to predict and describe phenomena. (Practice 2, grades 6–8) • Develop a model to describe unobservable mechanisms. (Practice 2, grades 6–8) • Develop a model to generate data to test ideas about phenomena in natural or designed systems, including those representing inputs and outputs and those at unobservable scales. (Practice 2, grades 6–8) • Develop and revise a model based on evidence to illustrate and predict the relationships between systems or between components of a system. (Practice 2, grades 9–12) • Develop and use multiple types of models to provide mechanistic accounts and predict phenomena, and move flexibly between model types based on merits and limitations. (Practice 2, grades 9–12)	• ESS2.A: Earth Materials and Systems • ESS2.B: Plate Tectonics and Large-Scale System Interactions	• Cause and Effect: Mechanism and Explanation • Scale, Proportion, and Quantity • System and System Models	• **MS-ESS2-2** • MS-ESS2-3 • HS-ESS2-1 • HS-ESS2-3

Note: DD = "Digging Deeper" section; EAD = "Example Activity Design" section; SDA = "Suggestions for Designing an Activity" section.

*Unless otherwise stated, the practices in this column are taken from Appendix F of the *NGSS*. Some of the practices from Appendix F have been shortened to focus on the ideas addressed in the activity or anecdote.

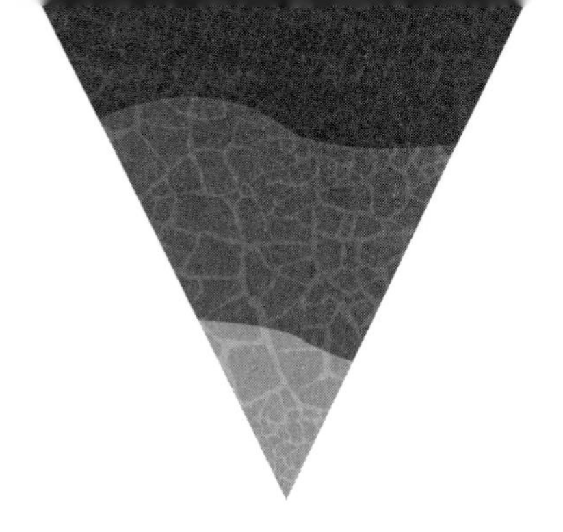

APPENDIX C

Resources for Further Reading

The citations in this appendix, organized by chapter, highlight some of the important ideas in science education and complement ideas that we present in *Learning to Read the Earth and Sky.* We provide a short excerpt from each resource, but we hope that the citations encourage further reading. Collectively, the citations give a sense of the breadth of the science teaching enterprise and offer a launching pad for further thinking.

INTRODUCTION

C1. Teaching to do, not just to know

> *Analogies for three-dimensional learning [integration of practices, core ideas and crosscutting concepts] can be drawn from extracurricular or community-based activities. Take, for example, the way youngsters learn to play basketball in a formal or informal setting. Young basketball players are not taught basketball skills in isolation, nor do they memorize the rules of the game or focus on ideas of sportsmanship. They slowly master how to pass, dribble, and shoot, the rules of basketball, and sportsmanship—a concept that cuts across all sports—within the context of a real game.*

Miller, E., and J. Krajcik. 2015. Reflecting on instruction to promote equity and alignment to the *NGSS*. In *NGSS for All Students,* ed. O. Lee, E. Miller, and R. Januszyk, 181. Arlington, VA: NSTA Press.

CHAPTER 1

C2. Asking questions of the observations

> *My colleagues had everything they needed to develop this curriculum [Elementary Science Study], except kids. So I served as the sample kid for my first 6 months on this job. And this was a brand new experience for me. My colleagues knew that science was of the world. It wasn't words about the world. … The basis of their curriculum was to give the kids the stuff. So I was given stuff that they were working*

on …. I learned about balances, and pendulums, and growing seeds, and the Moon and ice cubes and all kinds of stuff …. You don't need some other person in between you and the things. So it was me, my ideas, and the things. And it was exhilarating.

Duckworth, E. R. 2012. When teachers listen and learners explain [video]. TedxTalk. *www.youtube.com/watch?v=1sfgenKusQk.*

C3. Inquiry involves more than lab work, and more than simple observation.

Earth science researchers have used multiple modes of inquiry to cultivate concepts such as plate tectonics, geological time, the hydrologic cycle, and global climate change. Students who understand the process of science as comprising only laboratory experimentation are at risk of developing a disconnect between the content and process aspects of their Earth science education (Tsai 1999) … Six modes of inquiry widely used by practicing geoscientists include: 1) the classic laboratory experiment, 2) observation of change over time, 3) comparison of ancient artifacts with products of active processes, 4) observation of variations across space, 5) use of physical models, and 6) application of computer models.

Kastens, K., and A. Rivet. 2008. Multiple modes of earth science inquiry. *The Science Teacher* 75 (1): 26–31.

Tsai, C. C. 1999. "Laboratory exercises help me memorize scientific truths": A study of eighth graders' scientific epistemological views and learning in laboratory activities. *Science Education* 83 (6): 654–674.

CHAPTER 2

C4. Student investigation involves more than doing experiments from the textbook.

The emerging evidence suggests that learning how to design, set up, and carry out experiments and other kinds of scientific investigations can help students understand key scientific concepts, provide a context for understanding why science needs empirical evidence, and how tests can distinguish between explanations.

National Research Council (NRC) and Committee on Science Learning, Kindergarten Through Eighth Grade. 2007. *Taking science to school: Learning and teaching science in grades K–8.* R. A. Duschl, H. A. Schweingruber, and A. W. Shouse, eds. Washington, DC: National Academies Press, p. 257.

C5. Sometimes we practice science to practice science, not to reinforce factual content.

> *If students and teachers only encounter preplanned confirmatory investigations based on tried and true step-by-step procedures always ensuring the anticipated outcome(s), then an undesirable outcome for students is that important and relevant cognitive and materials struggles of doing science get stripped away. A negative outcome for teachers is that important formative assessment and feedback on learning opportunities get omitted, too.*

Duschl, R. A., and R. W. Bybee. 2014. Planning and carrying out investigations: an entry to learning and to teacher professional development around NGSS science and engineering practices. *International Journal of STEM Education* 1 (12): 1–9.

> *At the end of our week of exploration, students had not arrived at a full-blown modern understanding of cratering, energy exchange, or comets. However, they had successfully designed and carried out an experiment, graphically represented their data, used graphs to make predictions, and discussed the strengths and limitations of their classroom model. … As one student wrote in his final report: "I found out there is just one thing that separates us from the real scientists. That's degree of experience, experimental apparatus, and resources."*

Walker, G. 2002. Excavating cratering. *The Science Teacher* 69 (7): 44–47. *http://static.nsta.org/files/tst0210_44.pdf.*

CHAPTER 3

C6. Fieldwork is essential to both earth science research and teaching.

> *A hallmark of the geosciences is that theoretical advances are usually grounded in direct observations of the Earth, oceans, atmosphere, or planets … many geoscientists report that fieldwork was a central, formative experience, whether at geology field camp, on a research vessel, or during an atmospheric science field experiment … Two lines of reasoning may shed light on why field experiences are so fundamental. First, field experiences provide a concentrated opportunity to develop what anthropologists call "professional vision," the ability to see features that are important to professional practice. Like a criminal investigator at a crime scene, a geoscientist in the field sees differently than a novice at the same scene. Second, field experiences provide practice in transforming the raw material of nature into the words, signs, and symbols that geoscientists use to capture and communicate their observations.*

Kastens, K., C. A. Manduca, C. Cervato, R. Frodeman, C. Goodwin, L.S. Liben, D.W. Mogk, T.C. Spangler, N.A. Stillings, and S. Titus. 2009. How geoscientists think and learn. SERC [Science Education Resource Center at Carleton College]. *http://serc.carleton.edu/serc/EOS-90-31-2009.html.*

C7. Map and lab notations as inscriptions of real-world features

Science, as much as map-making, is a sort of semiotic practice, which lets you deal with signs and symbols instead of touching directly the messy physical nature. You cannot just go out and deal with the "real" nature every time you want to make a statement about nature. The nature has to be scaled down and inscribed in order for you to examine it.

Anonymous student. 2005. In conversation with Bruno Latour: Historiography of "Science in Action." Paper written for MIT OpenCourseWare History of Science course taught by David Jones. Available at *http://dspace.mit.edu/bitstream/handle/1721.1/103818/sts-310-fall-2005/contents/assignments/paper2.pdf* (accessed August 5, 2016).

C8. Field work requires identifying important observations among a blizzard of less relevant ones.

To investigate the strategies children use when recording their observations during field-based inquiries, fourth graders were asked to indicate the location of colored flags by placing similarly colored stickers on a map and, after each placement, to write down what clues they used to decide where to place each sticker. … For inaccurately placed stickers, the accompanying verbal reports suggest that, most commonly, students attended to irrelevant or insufficient information in the environment. The hypothesized interpretation is that students have not fully mastered the skill of selectively attending to task-relevant information in a visually complex landscape. Findings imply that instructors should anticipate students' difficulty …

Kastens, K., and L. Liben. 2010. Children's strategies and difficulties while using a map to record locations in an outdoor environment. *International Research in Geographical and Environmental Education* 19 (4): 315–340.

C9. Using classic field studies in the classroom

[In this lesson,] students pretend to visit four of the original study sites [of Louis Agassiz] by examining original images, maps and notes…then discuss and summarize their findings from their 'expedition' … The idea that glaciers alone could produce these results [erratics, scratches on the bedrock, boulders strewn about the valleys of the Swiss Alps] seemed preposterous because [glaciers] weren't liquid, seemingly couldn't flow, and didn't currently exist in many European valleys that

were studied (Hallam, 1989; Rudwick, 2010). It was into this context that Louis Agassiz introduced glacial theory as an alternative idea to explain the same geologic surface features.... Agassiz supported these claims by comparing glacial evidence, from existing glaciers in the Alps to similar evidence found in places such as Scotland in which glaciers were absent [today].

Biddy, Q. 2015. A cool controversy: Exploring the nature of science using a historical debate about glaciers. *The Science Teacher* 82 (7): 52–62.

Hallam, A. 1989. *Great geological controversies.* Vol. 2. Oxford, U.K.: Oxford University Press.

Rudwick, M. J. 2010. *Worlds before Adam: The reconstruction of geohistory in the age of reform.* Chicago: University of Chicago Press.

CHAPTER 4

C10. Graphs require interpretation: Reading the graph.

Student laboratory reports are often premised on the assumption that the graph itself has meaning embedded in it. ... However, meaning doesn't exist in the graphs themselves, but rather lies in the interpretation ... [When writing research reports,] scientists (1) have detailed captions explaining what the graphics are about and (2) provide detailed written interpretations of the graphs in the data section so that the reader reaches the same conclusions from the graphs that the authors themselves did. ... [W]hat we are arguing is that students need to spend more time interpreting the graphs (and/or tables) and providing more detailed written interpretations in their laboratory reports of the patterns or trends they see.

Bowen, M., and A. Bartley. 2013. *The basics of data literacy: Helping your students (and you) make sense of data.* Arlington, VA: NSTA Press, pp. 47–48.

C11. Graphical data represent real phenomena and need to be understood that way.

Current instruction often underestimates the difficulty of connecting representations with reasoning about the scientific phenomena they represent. Students need support in both interpreting and creating data representations that carry meaning. Students learn to use representations that are progressively more symbolic and mathematically powerful.

Michaels, S., A. W. Shouse, and H. A. Schweingruber. 2008. *Ready, Set, Science! Putting research to work in K–8 classrooms.* Board on Science Education, Center for Education, Division of Behavioral and Social Sciences and Education. Washington, DC: National Academies Press, p. 118.

C12. Understanding graphs is benefited when students generate their own data.

> *Students are better able to understand data if as much attention is devoted to how they are generated as to their analysis. First and foremost, students need to understand that data are constructed to answer questions, not provided in a finished form by nature. Questions are what determine the types of information that will be gathered, and many aspects of data coding and structuring also depend on the questions asked.*

Michaels, S., A. W. Shouse, and H. A. Schweingruber. 2008. *Ready, Set, Science! Putting research to work in K–8 classrooms.* Board on Science Education, Center for Education, Division of Behavioral and Social Sciences and Education. Washington, DC: National Academies Press, p. 111.

CHAPTER 5

C13. Tools for visualizing in four dimensions: Maps and cross-sections

> *In a science in which experimental manipulations are difficult or impossible for many important questions, geoscientists rely heavily on natural experiments in which a causal factor varies across space (Kastens and Rivet, 2008), and on the technique of trading space for time. Maps, profiles, and cross-sections are the tools by which such causally significant spatial patterns are conveyed, discussed, and reflected upon.*

Kastens, K., L. Pistolesi, and M. J. Passow. 2014. Analysis of spatial concepts, spatial skills and spatial representations in New York State Regents earth science examinations. *Journal of Geoscience Education* 62 (2): 278–289.

Kastens, K. A., and A. Rivet. 2008. Multiple modes of inquiry in earth science. *The Science Teacher* 75 (1):26–31.

C14. Spatial thinking is rarely taught in the United States.

> *Because spatial thinking is rarely taught explicitly in the U.S. education system, improving spatial thinking may be "low-hanging fruit" as far as improving science education … we find that students on average scored lower on items [in the New York high school earth science exam] that we had coded as spatial than on items we had coded as nonspatial. In the short run, these findings should motivate Earth Science teachers to attend more deliberately to fostering spatial thinking in their instruction.*

Kastens, K., L. Pistolesi, and M. J. Passow. 2014. Analysis of spatial concepts, spatial skills and spatial representations in New York State Regents earth science examinations. *Journal of Geoscience Education* 62 (2): 278–289.

C15. Gathering data is not the same as thinking and learning.

How and where do actions and expressed thoughts differ between novice and expert geologists when solving a geologic mapping problem? ... For the experts, instances of synthesis occur at nearly every site of collecting and recording data. For the novices, synthesis occurs only sporadically and at relatively fewer collecting and recording sites. An implication of these findings is that novices could become more expert-like if they make a greater effort to link their observations and data with their ideas about the large-scale geologic setting of the field area.

Callahan, C., K. M. Baker, and H. Petcovic. 2015. Rock, paper, hammer: Where do thoughts and actions count in making a geologic map? Poster presented at the Earth Educators' Rendezvous, Boulder, CO. Also available at *http://serc.carleton.edu/earth_rendezvous/2015/program_table/abstracts/100888.html.*

CHAPTER 6

C16. Models are not static.

Students moved from illustrative to explanatory models, and developed increasingly sophisticated views of the explanatory nature of models, shifting from models as correct or incorrect to models as encompassing explanations for multiple aspects of a target phenomenon. They also developed more nuanced reasons to revise models.

Schwarz, C., B. Reiser, E. Davis, L. Kenyon, A. Acher, D. Fortus, Y. Shwartz, B. Hug, and J. S. Krajcik. 2009. Developing a learning progression for scientific modeling: Making scientific modeling accessible and meaningful for learners. *Journal of Research in Science Teaching* 46 (1): 232–254.

C17. Student articulation of models through pictures and words

A key component of this approach [Modeling Instruction] is that it moves the teacher from the role of authority figure who provides the knowledge to that of a coach/facilitator who helps the students construct their own understanding. Since students systematically misunderstand most of what we tell them (due to the fact that what they hear is filtered through their existing mental structures), the emphasis is placed on student articulation of the concepts.

Jackson, J., L. Dukerich, and D. Hestenes. 2008. Modeling Instruction: An effective model for science education. *Science Educator* 17 (1): 10–17.

C18. Modeling can also serve as focus, organizer, and motivation for smaller-scale investigations.

> *This approach [developing explanations/models] provides a specific explanandum that can serve as the focus of students' investigative activities and learning (Windschitl et al. 2009). For example, in our curriculum, we use photographs and narrative to introduce students to the Atacama Desert, a region in South America, as presenting a puzzle. It is literally the driest place on Earth, receiving no annual rainfall, but is not far from the Amazon jungle, one of the world's wettest places. How is it that the two regions can be so close to one another, yet have such drastically different climates?*

Falk, A., and L. Brodsky. 2013. Scientific argumentation as a foundation for the design of inquiry-based science instruction. *Journal of Mathematics and Science: Collaborative Explorations* 13:27–55.

Windschitl, M., J. Thompson, and M. Braaten. 2009. The beginner's repertoire: Proposing a core set of instructional practices for teacher preparation. Paper presented at the Discovery Research K–12 Meeting, National Science Foundation, Washington, DC.

> *Modeling Instruction, under development since 1990 under the leadership of David Hestenes (Emeritus Professor of Physics, Arizona State University) ... organizes the course around a small number of scientific models, thus making the course coherent. It applies structured inquiry techniques to the teaching of basic skills and practices in mathematical modeling, proportional reasoning, quantitative estimation and technology-enabled data collection and analysis.*

American Modeling Teachers Association. n.d. Welcome. *http://modelinginstruction.org.*

CHAPTER 7

C19. Connecting math to the real world

> *In eighth grade, students usually learn about forces in science class and linear relationships in math class, crucial topics that form the foundation for further study in science and engineering … One of the important outcomes of this experiment [linear model for a spring] is that students see how math is used in the "real world" of science and engineering.*

Darling, G. 2012. Springing into linear models. *Science Scope* 36 (1): 18–25.

> *Philosophy [science] is written in this grand book—I mean the universe—which stands continually open to our gaze, but it cannot be understood unless one first learns to comprehend the language in which it is written. It is written in the language of mathematics, and its characters are triangles, circles, and other geometric figures,*

without which it is humanly impossible to understand a single word of it; without these, one is wandering about in a dark labyrinth.

Galilei, G. 1623. Il saggiatore. Translated by Richard Popkin. In *The philosophy of the sixteenth and seventeenth centuries*, 1967, p. 65

C20. High school students often need additional experience with and understanding of nonlinear functions.

This study examined Advanced Placement Calculus students' mathematical understanding of rate of change ... Student errors on the three instruments revealed a lack of understanding of the interpretation or meaning of rate of change regardless of the curricular path. Students successfully calculated the rate of change of linear functions; however, when the function was not linear, students struggled to calculate it, model it on a graph, or interpret it in a real-world context.

Teuscher, D., and R. Reys. 2012. Rate of change: AP calculus students' understandings and misconceptions after completing different curricular paths. *School Science and Mathematics* 112 (6): 359–376.

CHAPTER 8

C21. It's important to be able to distinguish evidence from theory.

Epistemic knowledge is knowledge of the constructs and values that are intrinsic to science. Students need to understand what is meant, for example, by an observation, a hypothesis, an inference, a model, a theory, or a claim and be able to readily distinguish between them. An education in science should show that new scientific ideas are acts of imagination ... [that] often survive because they are coherent with what is already known, and they either explain the unexplained, explain more observations, or explain in a simpler and more elegant manner.

National Research Council (NRC). 2012. *A framework for K–12 science education: Practices, crosscutting concepts, and core ideas.* Washington, DC: National Academies Press. *www.nap.edu/catalog/13165/a-framework-for-k-12-science-education-practices-crosscutting-concepts*, p. 79.

C22. Evidence is the foundation for scientific argumentation.

By shifting their classroom discussions toward scientific argumentation, middle school science teachers can enhance their students' ability to consider the strength of claims they encounter in any part of their lives so that students always say, "Show me the evidence!"

Mesa, J. C., R. M. Pringle, and L. Hayes. 2013. Show me the evidence! Scientific argumentation in the middle school classroom. *Science Scope* 36 (9): 60–64.

> *Students are helped by teachers who encourage questions such as: "How do you know?" or "What evidence do you have to back up your claim?"*

Konicek-Moran, R., and P. Keeley. 2015. *Teaching for conceptual understanding in science.* Arlington, VA: NSTA Press, p. 131.

C23. Explanation versus argumentation

> *Key to the distinction between explanation and argument is that an explanation should make sense of a phenomenon based on other scientific facts. Thus, explanations begin with ...the feature or phenomenon to be explained that is often phrased as a question, for example, why did the dinosaur die out or why do we have seasons? In an argument, however, there is not so much a feature or behavior to be explained but a claim to be justified ... there is always a substantial degree of tentativeness associated with any argument and, without this element, there would be no necessity for the argument itself.*

Osborne, J. F., and A Patterson. 2011. Scientific argument and explanation: A necessary distinction? *Science Education* 95 (4): 627–638.

> *["Explanation" aligns with constructing knowledge to develop understanding, whereas "argumentation" aligns with persuading peers of the quality of an explanation based on evidence.] We see the two practices of explanation and argumentation as having a complementary and synergistic relationship ... we referred to this combined argumentation and explanation product as a "scientific explanation" ... a term meant to communicate that the students' explanatory products would entail an explanation as well as the evidence and reasoning to justify that explanation.*

Berland, L., and K. McNeil. 2012. For whom is argument and explanation a necessary distinction? A response to Osborne and Patterson. *Science Education* 96 (5): 808–813.

C24. Classroom discourse is needed to realize that the goal is not to learn facts.

> *Discourse and classroom discussion are key to supporting learning in science. Students need encouragement and guidance to articulate their ideas and recognize that explanation rather than facts is the goal of the scientific enterprise.*

National Research Council (NRC) and Committee on Science Learning, Kindergarten Through Eighth Grade. 2007. *Taking science to school: Learning and teaching science in grades K–8.* R. A. Duschl, H. A. Schweingruber, and A. W. Shouse, eds. Washington, DC: National Academies Press, p. 251.

C25. An example activity identifying claim, evidence, and reasoning

> *Jeff Rohr, a fifth grade teacher in Beaver Dam, Wisconsin, suggests using the following Audi commercial [shown in the blog entry] to introduce students to the components of an explanation by asking them to identify the claim, the evidence, and the reasoning—or rule—that connects the evidence to the little girl's claim that her dad is a space alien.*

Brunsell, E. 2012. Designing science inquiry: claim + evidence + reasoning = explanation. Edutopia. *http://www.edutopia.org/blog/science-inquiry-claim-evidence-reasoning-eric-brunsell.*

C26. Identifying language of the argument

> *If student skills are to develop not only must there be explicit teaching of how to reason but also students need a knowledge of the meta-linguistic features of argumentation (claims, reasons, evidence, and counterargument) to identify the essential elements of their own and others' arguments (Herrenkohl and Guerra, 1998). Younger students, particularly, need to be desensitized to the negative connotation of conflict surrounding these words and to see argument as a fundamental process in constructing knowledge.*

Osborne, J. 2010. Arguing to learn in science: The role of collaborative critical discourse. *Science* 328 (5977): 463–466. Also available at *www.sciencemag.org/content/328/5977/463.full.*

Herrenkohl, L. R., and M. R. Guerra. 1998. Participant structures, scientific discourse, and student engagement in fourth grade. *Cognition and Instruction* 16 (4): 431–473.

CHAPTER 9

C27. More than an educational scaffolding

> *The Claim-Evidence-Reasoning format to writing explanations is not a trivial thing for your students. You will need to explicitly introduce and model it for them. They will need support throughout the year as they get better at writing explanations.*

Brunsell, E. 2012. Designing science inquiry: claim + evidence + reasoning = explanation. Edutopia. *http://www.edutopia.org/blog/science-inquiry-claim-evidence-reasoning-eric-brunsell.*

C28. Doing science in community is key to understanding science.

> *Strand 4: Participating Productively in Science … Proficiency in science entails skillful participation in a scientific community in the classroom and mastery of productive ways of representing ideas, using scientific tools, and interacting with peers about science … Like scientists, science students benefit from sharing ideas with peers, building interpretive accounts of data, and working together to discern which accounts are most persuasive … Strand 4 is often completely overlooked by educators, yet research indicates that it is a critical component of science learning, particularly for students from populations that are underrepresented in science.*

Michaels, S., A. W. Shouse, and H. A. Schweingruber. 2008. *Ready, Set, Science! Putting research to work in K–8 classrooms.* Board on Science Education, Center for Education, Division of Behavioral and Social Sciences and Education. Washington, DC: National Academies Press, p. 21.

C29. Cognitive conflict generates understanding.

> *The theoretical position adopted in this research is that developing student understanding requires the juxtaposition of competing alternatives—often between the students' pre-existing conception and the newly presented scientific idea. Argumentation is then a vital means of generating cognitive conflict forcing the student to identify their current conception and engage in the cognitive act of comparison, contrast, and evaluation of evidence.*

Osborne, J., S. Simon, A. Christodoulou, C. Howell-Richardson, and K. Richardson. 2014. Learning to argue: A study of four schools and their attempt to develop the use of argumentation as a common instructional practice and its impact on students. *Journal of Research in Science Teaching* 50 (3): 315–347.

CHAPTER 10

C30. Students should practice geological thinking, not learn names.

> *Students need ... apprentice-like activities in which they can practice geological ways of thinking. Geologists develop deep understanding by solving geological problems, especially in the field. Students likewise need authentic geological tasks to practice thinking like a geologist.*

Kusnick, J. 2002. Growing pebbles and conceptual prisms: Understanding the source of student misconceptions about rock formation. *Journal of Geoscience Education* 50 (1): 31–39.

Helping learners to connect what they know about the general rock category—be it igneous, sedimentary or metamorphic—with the specific conditions and location of formation for a particular sample, can help them develop more expert-like understandings of rock identification. In many elementary and middle school earth science curricula, these connections are not made, as earth history, landforms, and rock and mineral identification are treated separately.

Ford, D. 2003. Sixth graders' conceptions of rocks in their local environments. *Journal of Geoscience Education* 51 (4): 373–377.

C31. Rock stories involve more than the rock cycle.

Student responses indicate they perceive the rock cycle as the cause of rock formation, rather than a generalized model representing relationships among rock categories and their formation. For example, when asked how a rock formed, one student responded "It went through the rock cycle." … The model is also the source of misconceptions surrounding a perceived sequence or order to rock formation …

Ford, D. 2003. Sixth graders' conceptions of rocks in their local environments. *Journal of Geoscience Education* 51 (4): 373–377.

C32. True models explain or predict.

There are traditional "models" in classrooms across the country of which I imagine about 80% are edible. Models that students construct and use for the NGSS classroom are quite different. Students need to use models to explain or predict phenomena using evidence. Most "edible" models do not allow for that experience. Scientific and engineering practices are what students do, not teaching strategies. Students should be able, for example, to identify the components of a model, articulate the relationship of those components, and explain or predict future phenomena based on the model.

Pruitt, S. 2015. The *Next Generation Science Standards:* Where are we now and what have we learned. *Science Scope* 38 (9): 17–19.

CHAPTER 11

C33. Thinking in the four dimensions of time and space is unfamiliar to most of us.

Conceptualizing geologic space and time is particularly troublesome for children and their teachers, yet necessary for grasping the formation of rocks over millions of years.

Trend (1998) points out the challenges inherent in conceptualizing the vast amounts of time necessary for rock formation. According to his study, children ages 10–11 divide time into two general levels ("extremely" and "less" ancient), and make few connections between major Earth events and their place in geological time.

Ford, D. 2003. Sixth graders' conceptions of rocks in their local environments. *Journal of Geoscience Education* 51 (4): 373–377.

Trend, R. 1998. An investigation into understanding of geological time among 10- and 11-year old children. *International Journal of Science Education* 20 (8): 973–988.

C34. Reading the earth: Images of deep time

As [John] McPhee describes it, geology is virtually a literary exercise, a form of close-reading, devised by a retired Scottish farmer named James Hutton, who in 1785 developed an amazingly simple idea into several opaque but world-changing tomes. … You had only to keep your wits about you, give free rein to your interpretive faculty, and every cliffside, stream bed or mountain range became a text of sorts. … Mr. McPhee conveys the excitement of the geologist's interpretive jags, his struggle to divert the cozy prospects of human language to a new task: the description of events so glued in the slowness of mineral time that they annihilate the scale of our lives and furnish convincing images of eternity.

Zweig, P. *New York Times.* 1981. Rhapsodist of Deep Time. May 17. Also available online at *www.nytimes.com/books/98/07/05/specials/mcphee-basin.html.*

McPhee, J. 1981. *Basin and range.* New York: Farrar, Straus and Giroux.

C35. Knowing Earth's age is not the same as understanding it.

Humans view the world through human scales of space and time. All geology instructors recognize the difficulty in understanding geological scales of time. Thus we devote much attention to activities which illustrate the immensity of deep time (Zen, 2001). The difficulty we have is in recognizing the futility of those lessons. Students can quote the age of the earth, but they still cite surprisingly short time scales in describing rock formation.

Kusnick, J. 2002. Growing pebbles and conceptual prisms: Understanding the source of student misconceptions about rock formation. *Journal of Geoscience Education* 50 (1): 31–39.

Zen, E. 2001. What is deep time and why should anyone care? *Journal of Geoscience Education* 49 (1): 5–9.

CHAPTER 12

C36. Classroom science should resemble research science.

Throughout this book we examine different science classrooms in which educators strive to structure students' scientific practice so that it resembles that of scientists. In these classrooms, students engage in a process of logical reasoning about evidence. They work cooperatively to explore ideas. They use mathematical or mechanical models, develop representations of phenomena, and work with various technological and intellectual tools. Students participate in active and rigorous discussion—of predictions, of evidence, of explanations, and of the relationships between hypotheses and data. They examine, review, and evaluate their own knowledge.

Michaels, S., A. W. Shouse, and H. A. Schweingruber. 2008. *Ready, Set, Science! Putting research to work in K–8 classrooms.* Board on Science Education, Center for Education, Division of Behavioral and Social Sciences and Education. Washington, DC: National Academies Press, p. 6.

CHAPTER 13

C37. Students should develop and revise models based on evidence.

What is important to realize in these examples [explored in the article] is that these student models account for all the evidence they have regarding the properties of gases. The student was not told the features of the particle model but rather developed the particle model through carefully supported modeling activities in which students built models based upon evidence. This is the major feature of the modeling practice: developing and revising models.

Krajcik, J., and J. Merritt. 2012. Engaging students in scientific practices: What does constructing and revising models look like in the science classroom? *Science Scope* 35 (7): 6–10.

CHAPTER 14

C38. Most students, even in college, have no basic understanding of mass balance.

With recent U.S. government efforts to develop policy procedures for addressing climate change, public understanding of basic aspects of climate change is imperative in order for people to understand such policy ... Results indicated that most [college geoscience] students did not have a basic understanding of mass-balance problems and that their misunderstanding varied according to gender and their interest in science but not according to factors, such as students' opinions of the seriousness of climate change. Students also tended to exhibit poor graphical interpretation skills when examining mass-balance graphs.

Reichert, C. P., C. Cervato, M. Larsen, and D. Niederhauser. 2014. Conceptions of atmospheric carbon budgets: Undergraduate students' perceptions of mass balance. *Journal of Geoscience Education* 62 (3): 460–468.

CHAPTER 15

C39. The disciplinary core ideas provide context for investigation of smaller-scale phenomena.

The focus on explaining phenomena represents an important shift in the goals of instruction. Rather than teaching ideas in the abstract or in isolation, the new aim is to engage students in using these ideas to explain interesting phenomena. For example, instead of having students describe the water cycle and its components, students should be explaining cloud formation or precipitation patterns by using understandings about the water cycle and thermal-energy transfer to describe how weather events come about.

Duncan, R. G., and V. L. Cavera. 2015. DCIs, SEPs, and CCs, oh my! Understanding the three dimensions of the *NGSS*. *The Science Teacher* 82 (7): 67–71.

C40. Different parts of the enterprise of science serve different purposes and need integration.

The unit design in the case studies [from NGSS *Appendix D] promotes three-dimensional learning through the following important shifts: (1) the phenomenon anchors and frames the unit, (2) the engaging driving question guides the direction for learning, and (3) the place-based context offers relevance ... When we involve students in phenomena and task them to explain and predict those phenomena, they discover the usefulness of science ideas as a means to make sense of the world ... [The driving question] allows the teacher to place students in the role of sense makers.*

Miller, E., R. Januszyk, and O. Lee. 2015. Using the case studies to inform unit design. In NGSS *for All Students*, ed. O. Lee, E. Miller, and R. Januszyk. Arlington, VA: NSTA Press, p. 171.

C41. Curricular narrative connects small ideas to bigger ideas.

The storyline should show how the DCIs, science and engineering practices, and crosscutting concepts develop overtime. It should also show how learners build sophisticated ideas from prior ideas, using evidence that builds to the understanding described in the PEs as students engage in the practices to explain phenomena.

Krajcik, J., S. Codere, C. Dahsah, R. Bayer, and K. Mun. 2014. Planning instruction to meet the intent of the *Next Generation Science Standards. Journal of Science Teacher Education* 25 (2):157–175. Also available online at *http://link.springer.com/article/10.1007%2Fs10972-014-9383-2/fulltext.html.*

C42. Leaving the lecture exposes teachers to the messiness of learning.

More challenging for me was the regular confrontation with the messiness of learning. Lecture content may be causing great confusion, but students are so good at faking attention and not asking questions that we discover there's a problem when we grade their exams. If students are engaged in group work and executing the assigned task poorly, that feedback is in your face. And recognizing the mess is only the first step. You've got to do something about it.

Weimer, M. 2013. *Learner-centered teaching: Five key changes to practice.* San Francisco: John Wiley & Sons, p. 71.

C43. *NGSS* performance expectations are not suggestions for stand-alone curricular units.

We do not recommend developing lessons that focus only on one performance expectation. Focusing on a bundle helps students see connections among the elements of DCIs and the various scientific and engineering practices that would not be seen by focusing on one performance expectation at a time. Focusing on one performance expectation could contribute to learners developing compartmentalized understanding that the Framework for K–12 Science Education was trying to avoid.

Krajcik, J., S. Codere, C. Dahsah, R. Bayer, and K. Mun. 2014. Planning instruction to meet the intent of the *Next Generation Science Standards. Journal of Science Teacher Education* 25 (2):157–175. Also available online at *http://link.springer.com/article/10.1007%2Fs10972-014-9383-2/fulltext.html.*

CHAPTER 16

C44. The importance of earth science at your location

Earth science is an inherently local subject. No two places share exactly the same sequence of events that led to the way they are today. In this sense, Earth science is a subject to be explored in one's own neighborhood, examining the detailed sequence of rocks for the history that has gone on under our feet. What is not possible from only one location is making sense of why this particular sequence of events took place when and where it did, particularly relative to sequences in other places around it.

Paleontological Research Institution. n.d. *The teacher friendly guides to the earth science of the United States. http://geology.teacherfriendlyguide.org.*

> *Context for science is essential to providing access to diverse student groups. The phenomenon that we engage in should be interesting, familiar, and place based (i.e. located in students' local contexts). Place-based science means that science is situated in the local ecology and environment, which encourages students to see themselves as citizens who have a responsibility toward their local community.*

Miller, E., R. Januszyk, and O. Lee. 2015. Using the case studies to inform unit design. In NGSS *for All Students,* ed. O. Lee, E. Miller, and R. Januszyk. Arlington, VA: NSTA Press, p. 175.

CHAPTER 17

C45. Teacher: Facilitator, supporter, partner

> *But how do we get from transmission of information to construction of meaning? Such a change can entail a considerable shift in roles for the professor [or teacher], who must move away from being the one who has all the answers and does most of the talking toward being a facilitator who orchestrates the context, provides sources, and poses questions to stimulate students to think up their own answers.*

King, A. 1993. From sage of the stage to guide on the side. *College Teaching* 41 (1): 30–35.

> *Successful strategies for science learning engage students in scientific tasks that explore ideas and problems that are meaningful to them with carefully structured support from teachers.*

National Research Council (NRC) and Committee on Science Learning, Kindergarten Through Eighth Grade. 2007. *Taking science to school: Learning and teaching science in grades K–8.* R. A. Duschl, H. A. Schweingruber, and A. W. Shouse, eds. Washington, DC: National Academies Press, p. 333.

> *Engaging the vision of the NGSS is like climbing a mountain: daunting yet transformational, resulting in capacity building, changing perspectives, and understanding built from connecting knowledge to experience. As with mountaineering, chances of success increase with a guide and a team ... We found joy in the partnership ... We welcomed the challenge because we were experiencing the best way to learn: a focus on figuring out together.*

Shelton, T. 2015. Climbing the *NGSS* mountain. *The Science Teacher* 82 (9): 65.

C46. Even facilitating discussion requires command of the discipline.

> *There is also evidence from case studies of science teachers that teacher knowledge influences instructional practice and, in particular, that classroom discourse—an integral component of science learning environments—is sensitive to teachers' knowledge of science.… For example, Sanders and colleagues (1993) conducted an in-depth analysis of three secondary science teachers teaching inside and outside their areas of certification. They reported that when teachers had limited knowledge of the content, they often struggled to sustain discussions with students and found themselves trying to field student questions that they could not address.*

National Research Council (NRC) and Committee on Science Learning, Kindergarten Through Eighth Grade. 2007. *Taking science to school: Learning and teaching science in grades K–8.* R. A. Duschl, H. A. Schweingruber, and A. W. Shouse, eds. Washington, DC: National Academies Press, p. 298.

Sanders, L. R., H. Borko, and J. D. Lockard. 1993. Secondary science teachers' knowledge base when teaching science courses in and out of their area of certification. *Journal of Research in Science Teaching* 30 (7): 723–736.

AFTERWORD

C47. Assessment in *How Students Learn*

Learning to Read the Earth and Sky does not deal with assessment, but …

> *The word "assessment" rarely appears in [Chapters 10–12 of How Students Learn], but in fact the chapters are rich in assessment opportunities. Students are helped to assess the quality of their hypotheses and models, the adequacy of their methods and conclusions, and the effectiveness of their efforts as learners and collaborators. These assessments are extremely important for students, but also help teachers see the degree to which students are making progress toward the course goals and use this information in deciding what to do next. It is noteworthy that these are formative assessments, complete with opportunities for students (and teachers) to use feedback to revise their thinking; they are not merely summative assessments that give students a grade on one task (e.g., a presentation about an experiment) and then go on to the next task.*

National Research Council (NRC). 2005. *How students learn: History, mathematics, and science in the classroom.* Washington, DC: National Academies Press, p. 415.

C48. Standardized assessments need to go beyond testing of information.

> *A decade of emphasis on coverage of relatively low-level standards in preparation for right answer standardized tests has done little to commend creation of rich, robust curriculum and flexible instruction, and less to grow teachers who have the competence and confidence to create them.*

Tomlinson, C. A. 2015. Teaching for excellence in academically diverse classrooms. *Society* 52 (3): 203–209.

INDEX

Page numbers in **boldface** type refer to figures or tables.

INDEX

INDEX

INDEX

INDEX